Linux虚拟化数据中心实战

何坤源 编著

人民邮电出版社
北京

图书在版编目（CIP）数据

Linux虚拟化数据中心实战 / 何坤源编著. -- 北京 : 人民邮电出版社, 2021.3
ISBN 978-7-115-55521-2

Ⅰ. ①L… Ⅱ. ①何… Ⅲ. ①Linux操作系统－虚拟处理机 Ⅳ. ①TP338

中国版本图书馆CIP数据核字(2020)第245700号

内 容 提 要

本书共 8 章，采用循序渐进的方式，帮助读者掌握 Linux 虚拟化架构的部署和使用，包括开源虚拟化平台介绍，实验环境搭建，部署使用 KVM 虚拟化，部署使用 oVirt 平台、OpenStack、Docker 和 Hadoop，以及认识 SDN 架构等内容。

本书以实战操作为主，理论讲解为辅，通过讲解搭建各种物理环境的方法，详细介绍在生产环境中如何使用 Linux 部署虚拟化数据中心，可以迅速提高读者的实际动手能力和操作水平。

本书语言通俗易懂，具有很强的可操作性，不仅适合 Linux 虚拟化架构的管理人员阅读，还可供其他虚拟化平台的管理人员阅读参考。

◆ 编　　著　何坤源
责任编辑　王峰松
责任印制　王　郁　焦志炜

◆ 人民邮电出版社出版发行　　北京市丰台区成寿寺路 11 号
邮编　100164　　电子邮件　315@ptpress.com.cn
网址　https://www.ptpress.com.cn
山东百润本色印刷有限公司印刷

◆ 开本：787×1092　1/16
印张：18
字数：423 千字　　2021 年 3 月第 1 版
印数：1 – 2 000 册　　2021 年 3 月山东第 1 次印刷

定价：79.00 元

读者服务热线：(010)81055410　印装质量热线：(010)81055316
反盗版热线：(010)81055315
广告经营许可证：京东市监广登字 20170147 号

作者简介

何坤源，知名讲师，黑色数据网络实验室创始人，持有CCIE（RS/DC/SEC）、VCP-DCV（4/5/6）、H3CSE、ITIL 等证书，目前担任多家企业、学校的 IT 咨询顾问，主讲 VMware、oVirt 等虚拟化课程。

早在 2006 年，作者就将工作重心转向虚拟化、数据中心以及灾难备份中心的建设，2008 年创建 Cisco 路由交换远程实验室，2009 年创建虚拟化远程实验室，2015 年创建云计算远程实验室。到目前为止，作者已经参与了多个企业虚拟化建设和改造项目，在虚拟化的设计、设备选型、运营维护等方面积累了丰富的经验。

工作之余，作者注重经验的总结和分享，近几年来编写了《VMware vSphere 5.0 虚拟化架构实战指南》《Linux KVM 虚拟化架构实战指南》《VMware vSphere 6.0 虚拟化架构实战指南》等图书，并有多种图书被各地的高校选作教材。

前　　言

作为云计算、大数据等技术的底层应用，服务器虚拟化有着不可替代的作用。而在企业级虚拟化方面，以开源 Linux 为代表的虚拟化解决方案占有不少的市场份额。

客观地说，商业软件与开源软件各具特色。商业软件的购置成本是一笔很大的开销，而开源软件基本上不需要花费购置成本，但运维成本是必须考虑的。总的来说，两者的结合使用是发展趋势。从目前的态势来看，不存在谁取代谁的问题，两者包容并存会持续很长一段时间。

本书一共 8 章，采用循序渐进的方式，帮助读者掌握 Linux 虚拟化架构的部署。希望这本书能够让虚拟化架构的管理人员在虚拟化的部署中得到一定的参考和指引。

需要指出的是，本书在介绍一些网络技术和网络术语时，参考了官方文档以及百度百科和 CSDN 的一些文章，在此一并表示感谢。

由于本书涉及的知识较多，作者水平有限，书中难免有不妥和错漏之处，欢迎大家批评指正。有关本书的任何问题、意见和建议，可以发邮件到 heky@vip.sina.com 与作者联系，也可与本书编辑（wangfengsong@ptpress.com.cn）联系。

以下是作者的技术交流方式。

技术交流 QQ：44222798。

技术交流 QQ 群：240222381。

技术交流微信：bdnetlab。

资源与支持

本书由异步社区出品，社区（https://www.epubit.com/）为您提供相关资源和后续服务。

配套资源

提供书中彩图文件。

要获得该配套资源，请在异步社区本书页面中单击 配套资源 ，跳转到下载界面，按提示进行操作即可。注意：为保证购书读者的权益，该操作会给出相关提示，要求输入提取码进行验证。

提交勘误

作者和编辑尽最大努力来确保书中内容的准确性，但难免会存在疏漏。欢迎您将发现的问题反馈给我们，帮助我们提升图书的质量。

当您发现错误时，请登录异步社区，按书名搜索，进入本书页面，单击“提交勘误”，输入勘误信息，单击“提交”按钮即可，如下图所示。本书的作者和编辑会对您提交的勘误进行审核，确认并接受后，您将获赠异步社区的 100 积分。积分可用于在异步社区兑换优惠券、样书或奖品。

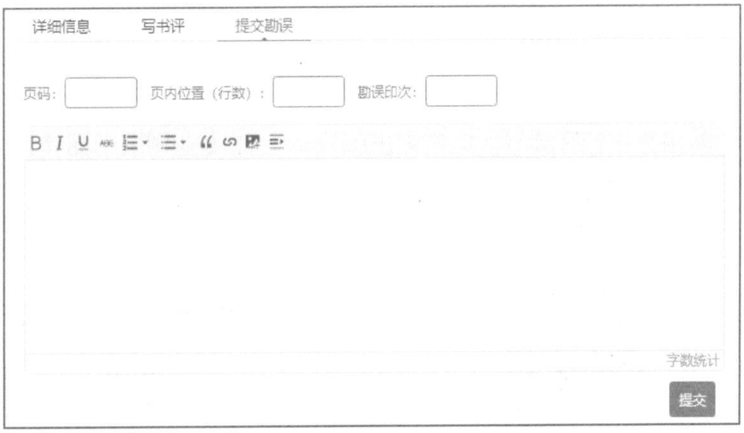

扫码关注本书

扫描下方二维码，您将会在异步社区微信服务号中看到本书信息及相关的服务提示。

与我们联系

我们的联系邮箱是 contact@epubit.com.cn。

如果您对本书有任何疑问或建议，请您发邮件给我们，并请在邮件标题中注明本书书名，以便我们更高效地做出反馈。

如果您有兴趣出版图书、录制教学视频，或者参与图书翻译、技术审校等工作，可以发邮件给我们；有意出版图书的作者也可以到异步社区在线投稿（直接访问 www.epubit.com/selfpublish/submission 即可）。

如果您是学校、培训机构或企业用户，想批量购买本书或异步社区出版的其他图书，也可以发邮件给我们。

如果您在网上发现有针对异步社区出品图书的各种形式的盗版行为，包括对图书全部或部分内容的非授权传播，请您将怀疑有侵权行为的链接发邮件给我们。您的这一举动是对作者权益的保护，也是我们持续为您提供有价值的内容的动力之源。

关于异步社区和异步图书

“异步社区”是人民邮电出版社旗下 IT 专业图书社区，致力于出版精品 IT 图书和相关学习产品，为作译者提供优质出版服务。异步社区创办于 2015 年 8 月，几年来提供了大量精品 IT 图书和电子书，以及高品质技术文章和视频课程。更多详情请访问异步社区官网 https://www.epubit.com。

“异步图书”是由异步社区编辑团队策划出版的精品 IT 专业图书的品牌，依托于人民邮电出版社 30 余年的计算机图书出版积累和专业编辑团队，相关图书在封面上印有异步图书的 LOGO。异步图书的出版领域包括软件开发、大数据、AI、测试、前端、网络技术等。

异步社区

微信服务号

目　录

第 1 章 开源虚拟化平台介绍

从 2006 年 Amazon 公司第一次把云计算进行商用开始，虚拟化平台已经发展了 10 多个年头。特别是最近几年，作为云计算核心的虚拟化平台开始在生产环境中被大量使用。到目前为止，虚拟化平台主要可以分为两大类：一类是 VMware（早期版本使用 Linux 内核）、Microsoft 等厂商提供的商业平台，另一类是主要基于 Linux 的 Xen、KVM 等开源平台。本章对开源虚拟化平台进行介绍。

本章要点

- Xen 虚拟化介绍。
- KVM 虚拟化介绍。
- oVirt 虚拟化介绍。
- OpenStack 平台介绍。

1.1 Xen 虚拟化介绍

学习开源虚拟化平台前，需要了解什么是虚拟化以及为什么要使用虚拟化。以一台物理服务器为例，如果不使用虚拟化技术，这台物理服务器只能安装一个 Windows 或 Linux 操作系统（不讨论双操作系统）；如果使用虚拟化技术，这台物理服务器可以安装多个操作系统并且同时运行，每个操作系统独立运行且互相不受影响，这就是虚拟化技术。

Xen 虚拟化技术是英国剑桥大学计算机实验室开发的一个虚拟化开源项目，通过它可以在一套物理硬件上安全地运行多个虚拟机。Xen 和操作系统平台结合得极为密切，占用的资源较少。

Xen 以高性能、占用资源少著称，赢得了 IBM、AMD、HP、Red Hat 以及 Novell 等众多软、硬件厂商的高度认可和大力支持，已被国内外众多企、事业用户用来搭建高性能的虚拟化平台。

1.1.1 Xen 虚拟化简介

Xen 采用独立计算体系结构（Intelligent Console Architecture，ICA）协议，通过一种叫作准虚拟化的技术获得高性能，甚至在某些对传统虚拟化技术“极度不友好”的架构（如 x86）上，Xen 也有上佳的表现。与传统通过软件模拟实现硬件的虚拟机不同，在 Intel VT-x 的支持下，3.0 版本之前的 Xen 需要系统的来宾权限和 Xen API 进行连接。到目前为止，这种技术已经可以运用在 NetBSD、GNU、Linux、FreeBSD 以及 Plan 9 等操作系统上。

Xen 虚拟机可以在不停止工作的情况下在多个物理主机之间实时迁移。在操作过程中，虚拟机在没有停止工作的情况下内存被反复地复制到目标机器上。虚拟机在最终目的开始执行之前，会有一次 60～300ms 的暂停以执行最终的同步，给人“无缝迁移”的感觉。类似的技术被用来暂停一台正在运行的虚拟机，并切换到另一台虚拟机，第一台虚拟机在切换后可以恢复工作。

Xen 是一种基于 x86 架构、发展较快、性能较稳定、占用资源较少的开源虚拟化技术。Xen 可以在一套物理硬件上安全地运行多个虚拟机，与 Linux 形成一个完美的开源组合。Novell SUSE Linux Enterprise Server 最先采用了 Xen 虚拟化技术。Xen 特别适用于服务器应用整合，可有效节省运营成本，提高设备利用率，最大化利用数据中心的 IT 基础架构。

1.1.2 Xen 虚拟化类型

在介绍 Xen 虚拟化类型之前，需要了解一下 x86 平台指令集的模式。x86 平台指令集使用 Ring 0、Ring 1、Ring 2、Ring 3 共 4 种级别来管理和使用物理服务器硬件，如图 1-1-1 所示。其中操作系统内核使用 Ring 0 级别，驱动程序使用 Ring 1、Ring 2 级别，应用程序使用 Ring 3 级别。对于不使用虚拟化技术的操作系统来说，这样的机制没有任何问题。但如果使用虚拟化技术，如何让虚拟机越级使用 x86 平台指令集是需要解决的问题。

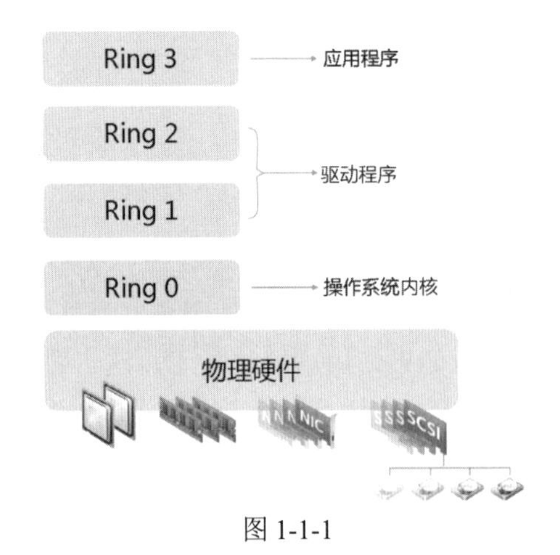

图 1-1-1

为解决虚拟机越级使用 x86 平台指令集的问题，Xen 虚拟化使用两种技术：半虚拟化（如图 1-1-2 所示）和全虚拟化（如图 1-1-3 所示）。

1）半虚拟化（Para Virtualization）：半虚拟化也可以称为超虚拟化。使用这种虚拟化技术，虚拟机操作系统认为自己运行在 Hypervisor 上而不是运行在物理服务器上，Hypervisor 对虚拟机操作的 Ring 0 级别的指令进行转换，模拟 CPU 给虚拟机使用（实际使用的是 Ring 1 级别），虚拟机不直接使用真实的 CPU。在半虚拟化环境下，虚拟机操作系统感应到自己是虚拟机，因此需要安装半虚拟化驱动程序，数据直接发送给半虚拟化设备，经过特殊处理再发给物理硬件。

2）全虚拟化（Full Virtualization）：全虚拟化也可以称为硬件虚拟化。需要注意，全虚拟化技术需要 Intel VT 和 AMD-V 的支持，相当于 Intel VT 和 AMD-V 创建了一个新的 Ring 1 级别单独给 Hypervisor 使用，但虚拟机操作系统认为自己直接运行在 Ring 0 级别上。在全虚拟化的环境下，虚拟机操作系统不知道自己是虚拟机，其数据的传输方式与在物理服务器上一致，但数据会被 Hypervisor 拦截再转发给物理硬件。

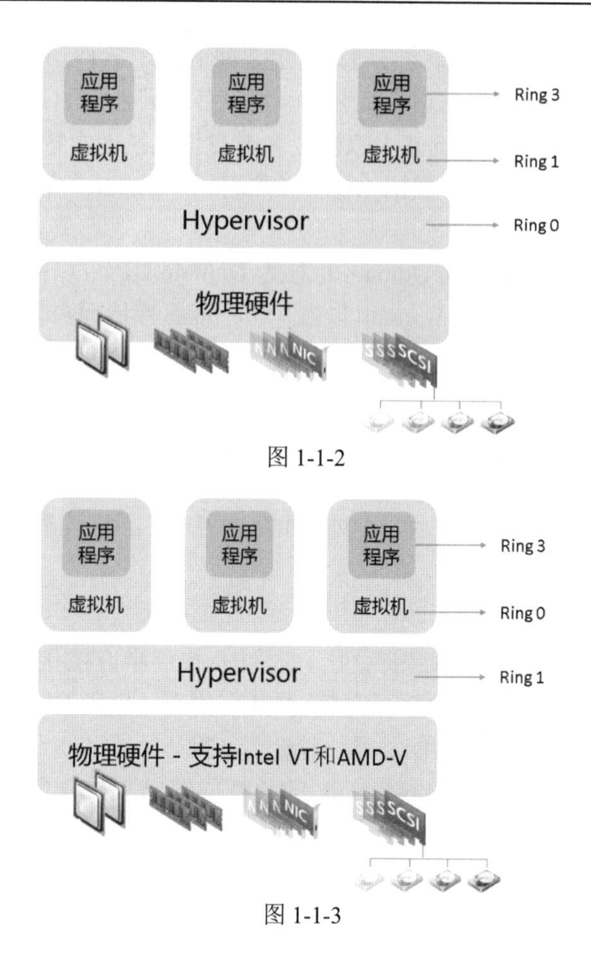

图 1-1-2

图 1-1-3

1.1.3 Xen 虚拟化组件

Xen 虚拟化组件主要包括图 1-1-4 所示的几个部分。

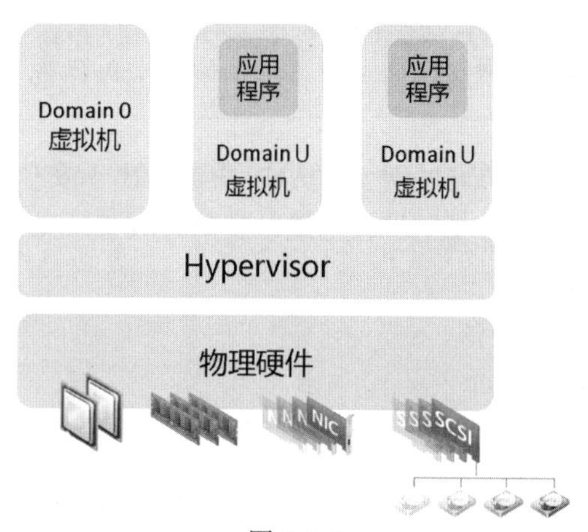

图 1-1-4

1）物理硬件：物理硬件层是最底层，包括物理服务器配置的 CPU、内存、硬盘以及网卡等硬件资源。

2）Hypervisor：运行在物理硬件与虚拟机层之间的基本软件层，其本身也是一种特殊的操作系统，负责为运行在上层的虚拟机分配、调度各种硬件资源。

3）Domain 0 虚拟机：Domain 0 虚拟机是 Xen 虚拟化技术中特殊的虚拟机，具有访问物理资源的特权，简单来说，Xen 虚拟化环境必须运行 Domain 0 虚拟机后，才能够安装运行其他虚拟机。

4）Domain U 虚拟机：无特权 Domain 也称为 Domain U，可以把除 Domain 0 虚拟机外的虚拟机称为 Domain U 虚拟机。Domain U 虚拟机不能直接访问物理硬件，每个 Domain U 虚拟机拥有独立的虚拟硬件资源并独立存在，一个 Domain U 虚拟机出现问题不影响其他 Domain U 虚拟机。

1.1.4 Xen 虚拟化的优缺点

Xen 作为一种企业级虚拟化技术，其功能相对完善。在了解其基本原理后，再了解一下它的优缺点。

1. Xen 虚拟化技术的优点

Xen 构建于开源的虚拟机管理程序之上，是结合使用半虚拟化和硬件协助的虚拟化。Xen 提供了复杂的工作负载平衡功能，可捕获 CPU、内存、磁盘 I/O 以及网络 I/O 数据，它提供了两种优化模式：一种针对性能，另一种针对密度。

Xen 拥有一种名为 Citrix Storage Link 的独特的存储集成功能。使用 Citrix Storage Link，系统管理员可直接利用来自 HP、Dell Equal Logic、NetApp、EMC 等公司的存储产品。

Xen 包含多核处理器支持、实时迁移、物理服务器到虚拟机转换、虚拟机到虚拟机转换工具，具有集中化的多服务器管理、实时性能监控功能。

2. Xen 虚拟化技术的缺点

Xen 会占用相对较大的空间，且依赖于 Domain 0 虚拟机中的 Linux。

Xen 依靠第三方解决方案来管理硬件设备驱动程序、存储、备份、恢复，以及容错。

任何具有高 I/O 速率的操作或任何会“吞噬”资源的操作都会使 Xen 陷入困境，使其他虚拟机缺乏资源。

Xen 缺少 IEEE 802.1Q 虚拟局域网（VLAN）中继，出于安全考虑，它没有提供目录服务集成、基于角色的访问控制、安全日志记录以及审计或管理操作。

1.2 KVM 虚拟化介绍

基于内核的虚拟机（Kernel-based Virtual Machine，KVM），最初是由一个以色列的创业公司 Qumranet 开发的，作为他们的 VDI 产品的虚拟机，从 Linux 2.6.20 内核之后集成在 Linux 的各个主要发行版本中。它使用 Linux 自身的调度器进行管理，相对于 Xen 其核心源代码很少。KVM 目前在开源系统中被大规模使用。

1.2.1 KVM 虚拟化简介

为简化开发，KVM 的开发人员并没有选择从底层开始新写一个 Hypervisor，而是选择基于 Linux Kernel，通过加载新的模块使 Linux Kernel 本身变成一个 Hypervisor。

2006 年 10 月，在完成基本功能、动态迁移以及主要的性能优化之后，Qumranet 公司正式对外宣布了 KVM 的诞生。同年 10 月，KVM 模块的源代码被正式接纳入 Linux Kernel，成为 Linux 内核源代码的一部分。作为一个当时在功能和成熟度上都逊于 Xen 的项目，在这么短的时间内被 Linux 内核社区接纳，主要原因在于：当时虚拟化方兴未艾，Linux 内核社区急于将虚拟化的支持包含在内，但是 Xen 取代内核由自身管理系统资源的架构引起了 Linux 内核开发人员的不满和抵触。

2008 年 9 月 4 日，Linux 发行版提供商 Red Hat 公司出人意料地出资上亿美元，收购了 Qumranet 公司，成为 KVM 开源项目的新东家。由于此次收购，Red Hat 公司有了自己的虚拟机解决方案，于是开始在自己的产品中用 KVM 替换 Xen。

2010 年 11 月，Red Hat 公司推出了 Red Hat Enterprise Linux 6，在这个发行版中集成了最新的 KVM，而去掉了在 Red Hat Enterprise Linux 5 中集成的 Xen。

1.2.2 KVM 虚拟化架构

与 Xen 虚拟化架构相比较，KVM 虚拟化架构非常简单，图 1-2-1 为 KVM 虚拟化架构示意。KVM 直接通过加载相关模块将 Linux Kernel 转换为 Hypervisor，KVM 在安装完成后就可以通过 QEMU 将模拟硬件提供给虚拟机使用。

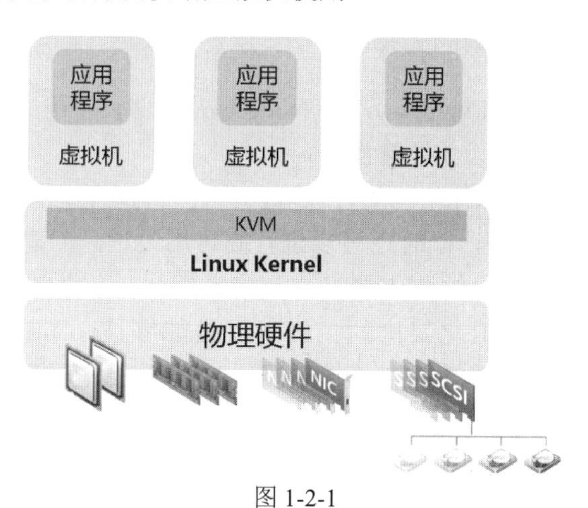

图 1-2-1

1.2.3 KVM 虚拟化的优缺点

KVM 作为一种企业级虚拟化技术，功能相对完善。其基本的运行原理是在 Linux 中加载 KVM 模块。需要注意的是 KVM 虚拟化需要 Intel VT 和 AMD-V 的支持，KVM 本身包含为处理器提供底层虚拟化的模块 kvm-intel.ko、kvm-amd.ko，当在 Linux 上安装 KVM 后可以创建并运行虚拟机，一台虚拟机可以理解为一个 Linux 单一进程，通过管理工具对这个进程进行管理就相当于对虚拟机进行管理。在了解其基本原理后，再了解一下它的优缺点。

1. KVM 虚拟化技术的优点

利用 Linux 的功能，与 Linux 操作系统紧密结合，构建在稳定的企业级平台之上，直接使用 Linux 进行进程调度、内存管理，广泛的硬件支持等。

KVM 是开源项目，很多成熟的解决方案也是免费的。对于中小企业或者是小微企业来说，成熟并且免费的解决方案是企业的首选。

2. KVM 虚拟化技术的缺点

KVM 虚拟化需要硬件支持，只能在具有虚拟化功能的 CPU 上才能运行。

1.3 oVirt 虚拟化介绍

1.3.1 oVirt 虚拟化简介

针对企业级虚拟化需求，Red Hat 公司发布的 Red Hat Enterprise Virtualization（简称 RHEV）是一套完整的服务器虚拟化管理产品。但这套产品不能直接使用，需要订阅 Red Hat RHN 服务才能下载使用。

oVirt 是开源的分布式虚拟化解决方案，也可以理解为 Red Hat Enterprise Virtualization 的社区版本。oVirt 虚拟化平台由 libvirt、Gluster、PatternFly 以及 Ansible 多个社区进行维护并发布新的版本，从使用上看，oVirt 是适用于企业的免费开源虚拟化解决方案。

1.3.2 oVirt 虚拟化架构

oVirt 虚拟化架构由 oVirt Engine、oVirt Node 以及存储设备组成，能够统一对 KVM 虚拟机进行创建、删除、迁移以及快照等操作。

oVert Engine 是 oVirt 虚拟化架构的管理端，负责控制、管理虚拟化平台，能够管理虚拟机、硬盘、网络等硬件资源，同时还可以进行高可用设置，创建虚拟机模板、快照等，这些都可通过 oVirt Engine 提供的 Web 界面来完成。

oVrit Node 是 oVirt 虚拟化架构的主机端，是能够被 oVirt Engine 连接并管理的 Hypervisor，在 oVirt 虚拟化架构中提供运算功能。oVirt Node 有两种实现方式：一种方式是直接安装包含 Hypervisor 代码的微型操作系统，这是一个专为 oVirt Node 设计的微型操作系统，能够更加充分地使用物理服务器资源，这也是 oVirt 社区推荐的做法；另一种方式是在 CentOS 主机上安装 Hypervisor 软件，将已有的 CentOS 主机配置成 oVirt Node。

1.3.3 oVirt 虚拟化的优缺点

oVirt 虚拟化作为开源企业级免费虚拟化平台，其优缺点非常明显。

1. oVirt 虚拟化技术的优点

作为 Red Hat Enterprise Virtualization 的社区版本，其内核架构的稳定性非常不错，基于 Web 方式的管理容易操作，同时版本更新更快。

2. KVM 虚拟化技术的缺点

oVirt 虚拟化是社区版本，意味着没有官方提供技术支持，只能通过社区获取有限的技术支持。oVirt 虚拟化在企业中部署使用时，建议配备专业、成熟的运维团队。

1.4 OpenStack 平台介绍

1.4.1 OpenStack 简介

OpenStack 是一个开源的云计算管理平台项目，是一系列软件开源项目的组合，由美国国家航空航天局和 Rackspace 合作研发并发起，是拥有 Apache 许可证的开源代码项目。

OpenStack 为私有云和公有云提供可扩展的、弹性的云计算服务。项目目标是提供实施简单、可大规模扩展、丰富、标准统一的云计算管理平台。

1.4.2 OpenStack 的主要组件

OpenStack 覆盖了网络、虚拟化、操作系统、服务器等各个方面。它是一个正在开发中的云计算平台项目，根据成熟和重要程度的不同，被分解成多个组件。

1. Nova

计算组件，用于为单个用户或使用群组管理虚拟机实例的整个生命周期，根据用户需求来提供虚拟服务，负责虚拟机的创建、开机、关机、挂起、暂停、调整、迁移、重启、销毁等操作，配置 CPU、内存等信息规格。

2. Swift

对象存储组件，用于大规模可扩展系统，通过内置冗余及高容错机制实现对象存储的系统，允许进行存储或者检索文件。可为 Glance 提供镜像存储，为 Cinder 提供卷备份服务。

3. Glance

镜像组件，用于虚拟机镜像查找和检索系统，支持多种虚拟机镜像格式（AKI、AMI、ARI、ISO、QCOW2、Raw、VDI、VHD、VMDK 等），有创建镜像、上传镜像、删除镜像以及编辑镜像基本信息等功能。

4. Keystone

身份认证组件，为 OpenStack 其他服务提供身份验证、服务规则以及服务令牌的功能，用于管理 Domains、Projects、Users、Groups、Roles 等。

5. Neutron

网络组件，提供云计算的网络虚拟化技术，为 OpenStack 其他服务提供网络连接服务。为用户提供接口，可以定义 Network、Subnet、Router，配置 DHCP、DNS、负载均衡、L3 服务，网络支持 GRE、VLAN。插件架构支持许多厂商的产品和技术，如 OpenvSwitch。

6. Cinder

块存储组件，为运行实例提供稳定的数据块存储服务，它的插件驱动架构有利于块设备的创建和管理，如创建卷、删除卷，在实例上挂载和卸载卷等。

7. Horizon

UI组件，OpenStack中各种服务的Web管理门户，用于简化用户对服务的操作，如启动实例、分配IP地址、配置访问控制等。

8. Ceilometer

测量组件，能把OpenStack内部发生的几乎所有的事件都收集起来，然后为计费、监控以及其他服务提供数据支撑。

9. Heat

部署编排组件，提供了一种通过模板定义的协同部署方式，实现云基础设施软件运行环境（计算、存储以及网络资源）的自动化部署。

10. Trove

数据库组件，为用户在OpenStack的环境下提供可扩展、可靠的关系和非关系数据库引擎服务。

1.5 本章小结

本章对基于开源Linux操作系统的虚拟化平台进行了介绍，包括Xen、KVM、oVirt以及OpenStack，同时分析了各自的优缺点，用户可以根据实际情况进行选择。

Xen虚拟化目前使用范围相对较小，KVM目前被作为主流虚拟化技术在企业中大量使用。作者推荐使用KVM和oVirt搭建开源虚拟化平台。另外，作者想提醒，相对于使用商业软件，虽然使用开源虚拟化平台前期不需要购买授权，但是需要专业的运维团队进行搭建和后期的运维，而建立专业的运维团队也是一笔不小的开支。

第 2 章　实验环境搭建

实验环境对于开源 Linux 虚拟化数据中心的学习非常重要。为最大程度地模拟企业生产环境架构的情况，作者使用全物理服务器、存储设备构建完整的实验环境。

本章要点

- 实验环境介绍。
- 物理服务器安装 Linux 操作系统。
- 常见 Linux 服务器搭建。

2.1　实验环境介绍

为了最大程度地模拟企业生产环境，作者使用了全物理设备构建实验环境，其中包括服务器、网络交换机、存储等设备。

2.1.1　实验物理设备简介

实验环境一共使用多台物理服务器安装多种 Linux 操作系统，使用 Dell PowerVault MD3620f 构建 FC SAN 存储，使用 OPEN-E 操作系统构建了 IP SAN 存储（iSCSI 存储），使用 Cisco 公司的 Nexus 系列交换机作为网络交换，部分设备的配置如表 2-1-1 所示。

表 2-1-1　实验环境部分硬件配置

设备名称	CPU 型号	内存大小	硬盘容量	操作系统
HOST 7 服务器	Intel Xeon L5620（2 个）	64GB	1TB	CentOS 7
HOST 8 服务器	Intel Xeon L5620（2 个）	64GB	1TB	Ubuntu 18.04
HOST 9 服务器	Intel Xeon L5620（2 个）	64GB	1TB	oVirt 4.2 系统
HOST 10 服务器	Intel Xeon L5620（2 个）	64GB	1TB	oVirt 4.2 系统
iSCSI 存储	Intel Xeon E3-1220 v3	12GB	3TB SAS（8 个）	OPEN-E

2.1.2　实验物理拓扑简介

实验使用作者数据中心实验室的部分设备，整体拓扑如图 2-1-1 所示。

拓扑说明：数据中心实验室是作者自建的实验室，托管于互联网数据中心（Internet Data Center，IDC）机房，所有实验均通过远程进行操作，包括安装服务器操作系统等。

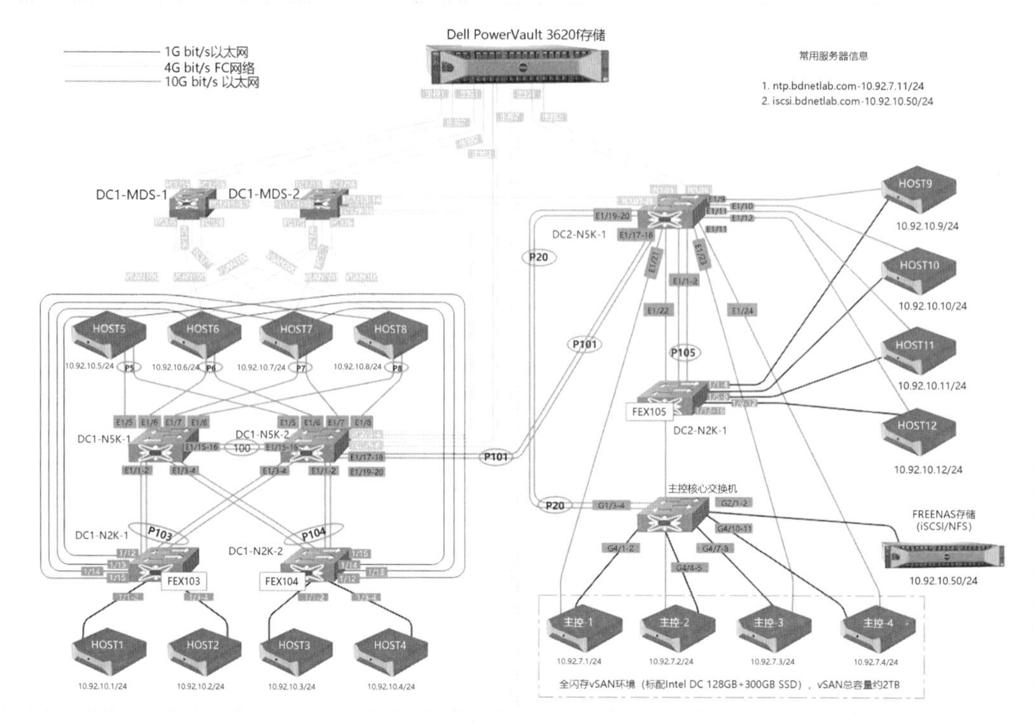

图 2-1-1

2.1.3 自建学习环境建议

作者推荐使用物理设备学习搭建企业级 Linux 虚拟化数据中心，有两方面原因：一方面 Linux 操作系统底层技术模拟环境可能会出现很多未知的问题；另一方面很多生产环境的设备是无法模拟的，如网络交换机、存储设备等。模拟环境与真实环境的差异很大，也并不是每个读者都可以有真实环境，如果没有真实环境也可以使用模拟环境进行一些基础的学习。

我们介绍一下两种环境的搭建，读者可以参考。

1. 真实环境搭建

采用真实环境学习搭建企业级 Linux 虚拟化数据中心，推荐至少 2 台物理服务器用于安装 Linux 操作系统，1 台服务器用于存储，同时配置 1 台物理交换机用于连接物理服务器。物理服务器硬件推荐配置如表 2-1-2 所示。

表 2-1-2　物理服务器硬件推荐配置

CPU	内存	硬盘	网卡	说明
Intel Xeon L56xx Intel Xeon E5-2xxx	64GB 或以上	300GB	4 个 8Gbit/s 以太网接口	虚拟化计算 资源服务器
Intel Xeon L56xx	8GB 或以上	1TB 或以上	2 个 8Gbit/s 以太网接口	存储服务器

实验用物理服务器需要安装部署操作系统和服务，具体的版本可以参考表 2-1-3。

表 2-1-3 物理服务器操作系统和需要安装的服务

服务器	操作系统	安装服务	说明
服务器 1	CentOS 7 Ubuntu 18.4	KVM VDSM oVirt Engine	将普通 Linux 主机升级为节点主机运行虚拟机并加入管理平台
服务器 2	CentOS 7 Ubuntu 18.4		
服务器 3	CentOS 7 Ubuntu 18.4	NFS iSCSI	用于提供存储服务

2. 模拟环境搭建

采用模拟环境学习搭建企业级 Linux 虚拟化数据中心，推荐使用高配置计算机，当然也可以购置 1 台物理服务器安装 Linux 操作系统，再运行虚拟机，但这样属于多层嵌套模拟，部分实验可能无法操作。模拟环境硬件推荐配置如表 2-1-4 所示。

表 2-1-4 模拟环境硬件配置

CPU	内存	硬盘	网卡	其他
Intel Core I5-65xx 或以上 Intel Core I7-67xx 或以上	32GB 或以上	1TB 或以上	1 个 8Gbit/s 以太网接口	

模拟环境也需要安装部署操作系统和服务，具体的版本可以参考表 2-1-5。

表 2-1-5 模拟环境操作系统以及需要安装的服务

服务器	操作系统	安装服务	说明
物理服务器	Windows 7/10	Workstation VirtualBox	操作设备
虚拟服务器 1	CentOS 7 Ubuntu 18.4	KVM VDSM oVirt Engine	将普通 Linux 主机升级为节点主机运行虚拟机并加入管理平台
虚拟服务器 2	CentOS 7 Ubuntu 18.4		
虚拟服务器 3	CentOS 7 Ubuntu 18.4		
虚拟服务器 4	CentOS 7 Ubuntu 18.4	NFS iSCSI	用于提供存储服务

以上配置可以作为参考，读者可以根据自己的情况进行搭建，低于该推荐配置可能会影响实验效果。

2.2 物理服务器安装 Linux

2.2.1 IPMI 简介

物理服务器远程管理工具也可以称为智能平台管理接口（Intelligent Platform Management Interface，IPMI）。IPMI 标准是 1998 年由 Intel、Dell、HP 及 NEC 等公司共同提出的一种通过网络远程控制温度和电压等的工业标准，后续版本不断升级，增加了通过网络远程控制服务器等的标准。服务器管理人员可以依据 IPMI 标准监视服务器的物理健康特征，如温度、电压、风扇工作状态、电源状态等。

IPMI 标准从 1998 年创建以来，已经得到了 170 多家厂商的支持，这使得其逐渐成为一个完整的包括服务器和其他系统（如存储设备、网络以及通信设备）的硬件管理规范。目前该标准最新版本为 2.0，该版本在原有基础上有不少改进，包括通过串口、Modem 以及网络等远程环境管理服务器系统（包括远程开关机），在安全、VLAN 以及“刀片支持”等方面的标准的升级。

IPMI 针对大量监控和自动回复服务器的作业，提供智能的管理方式，适用于不同拓扑的服务器，以及 Windows、Linux、Solaris、macOS 或是混合型的操作系统。此外，IPMI 可在不同的属性值下运作，即使服务器本身的运作不正常或是由于任何原因而无法提供服务，IPMI 仍可正常运作。

IPMI 的核心是一个专用芯片/控制器（Baseboard Management Controller，BMC），它并不依赖于服务器的处理器、BIOS 或操作系统来工作，可谓非常独立，是一个在系统内单独运行的无代理管理子系统，只要有 BMC 与 IPMI 固件其便可开始工作。而 BMC 通常是一个安装在服务器主板上的独立板卡，现在也有服务器主板提供对 IPMI 固件的支持。IPMI 良好的自治特性克服了以往基于操作系统的管理方式所受的限制，如在操作系统不响应或未加载的情况下其仍然可以进行开关机、信息提取等操作。

在工作时， IPMI 的所有功能都是通过向 BMC 发送命令来完成的，命令使用 IPMI 标准中规定的指令，BMC 接收并在系统事件日志中记录事件消息，维护描述系统中传感器情况的传感器数据记录。在需要远程访问系统时，IPMI 新的 LAN 上串行（Serial Over LAN，SOL）特性很有用。SOL 改变 IPMI 会话过程中本地串口的传送方向，从而提供对紧急管理服务、Windows 专用管理控制台或 Linux 串行控制台的远程访问。BMC 通过在 LAN 上改变传送给串行端口的信息方向来做到这一点，提供了一种与厂商无关的远程查看启动、操作系统加载器或紧急管理控制台来诊断和维修故障的标准方式。

2.2.2 常用的服务器远程管理工具

IPMI 标准是一个开放的免费标准，各大服务器厂商对其进行了大量的二次开发，使得服务器远程管理工具更好地适用于自己的服务器。下面介绍 HP、Dell、Cisco 以及其他厂商的服务器远程管理工具。需要注意的是，不少厂商的服务器远程管理工具需要 Java 控件支持，Java 版本的不同也可能影响到服务器远程管理工具的使用。

1. HP 服务器远程管理工具

HP 服务器远程管理工具是 HP Integrated Lights-Out 2（iLO 2），配置 IP 地址后可以远程控制服务器，进行开启/关闭服务器电源、挂载/安装镜像、收集日志等操作。图 2-2-1 为 HP 服务器远程管理工具登录界面，图 2-2-2 为远程启动服务器进入 BIOS 自检界面。

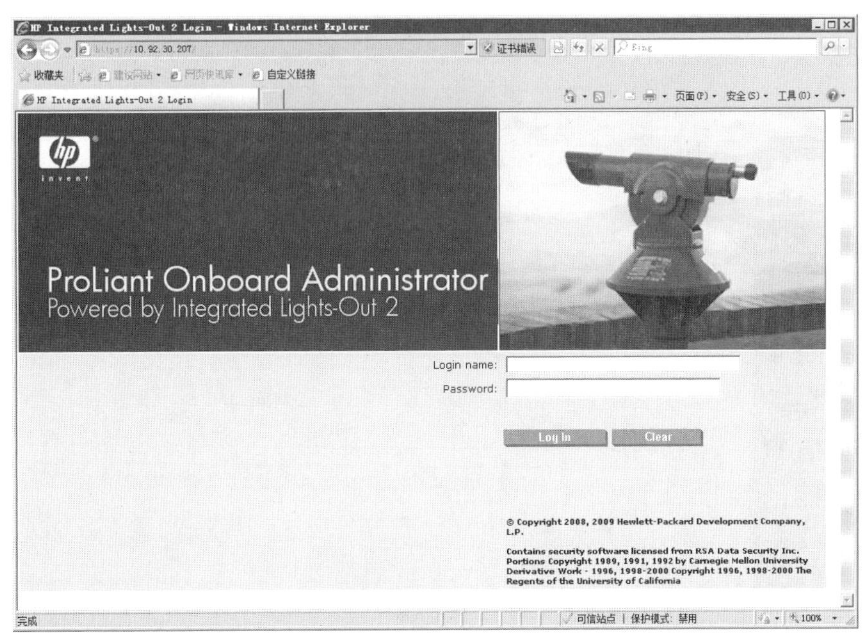

图 2-2-1

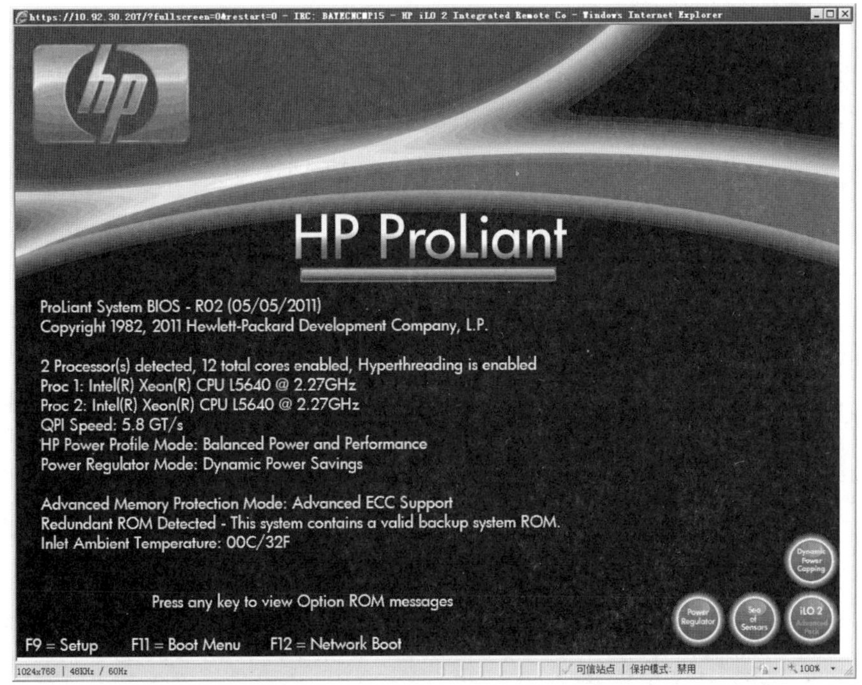

图 2-2-2

2. Dell 服务器远程管理工具

Dell 服务器远程管理工具是 Integrated Dell Remote Access Controller（iDRAC），配置 IP 地址后可以远程控制服务器，进行开启/关闭服务器电源、挂载/安装镜像、收集日志等操作。图 2-2-3 为 Dell 服务器远程管理工具登录界面，图 2-2-4 为远程启动服务器进入 BIOS 自检界面。

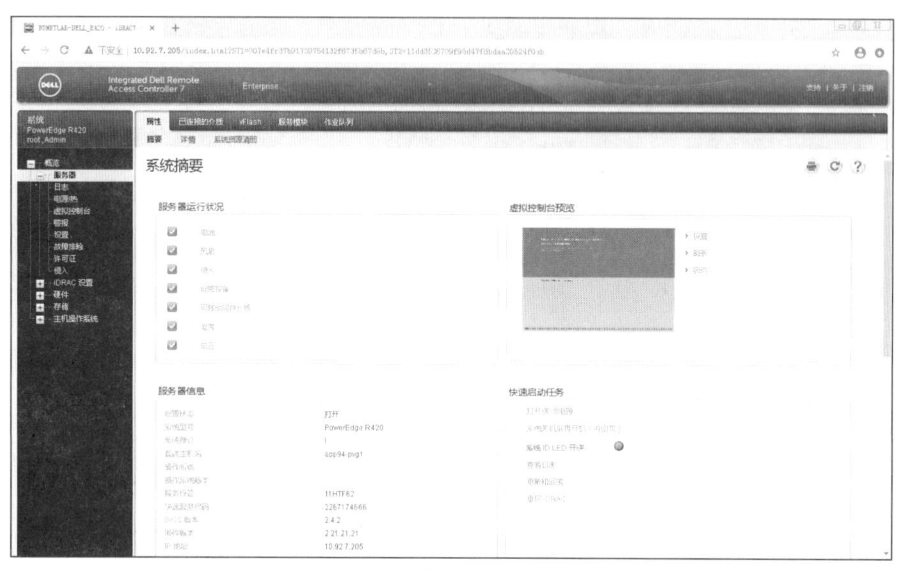

图 2-2-3

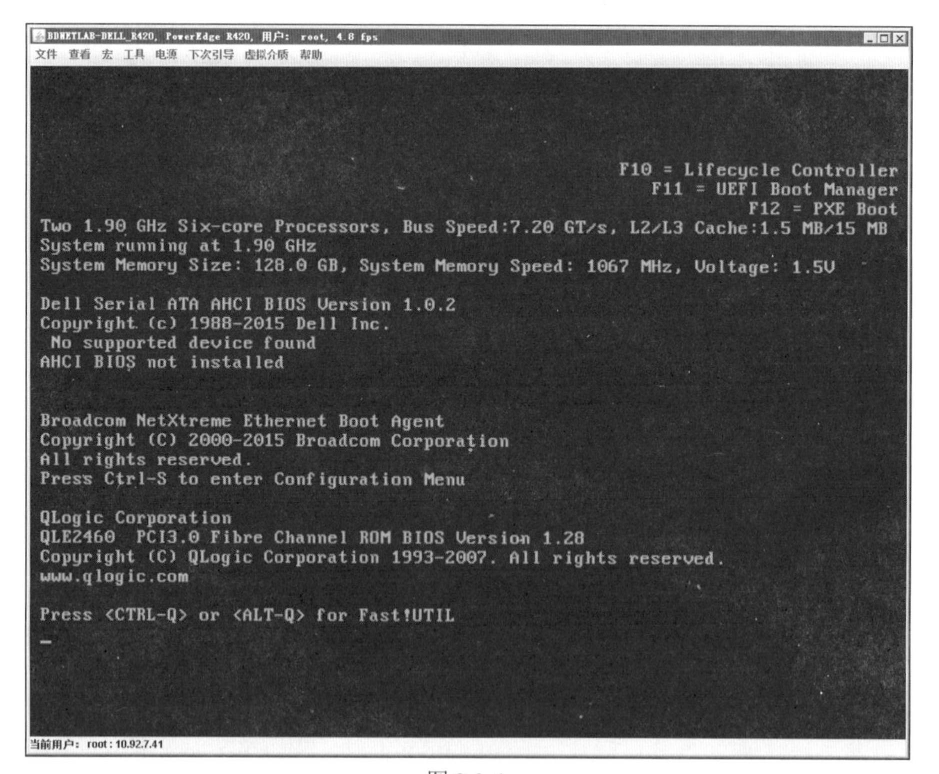

图 2-2-4

3. Cisco 服务器远程管理工具

Cisco 服务器远程管理工具是 Cisco Integrated Management Controller（IMC），配置 IP 地址后可以远程控制服务器，进行开启/关闭服务器电源、挂载/安装镜像、收集日志等操作。图 2-2-5 为 Cisco 服务器远程管理工具登录界面，图 2-2-6 为 Cisco 服务器远程管理工具操作界面。

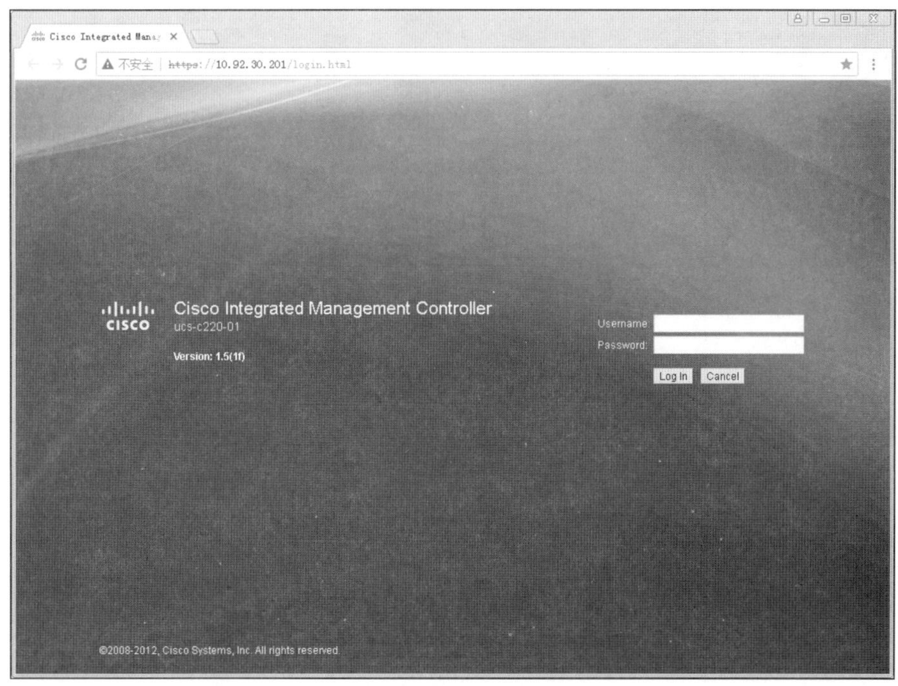

图 2-2-5

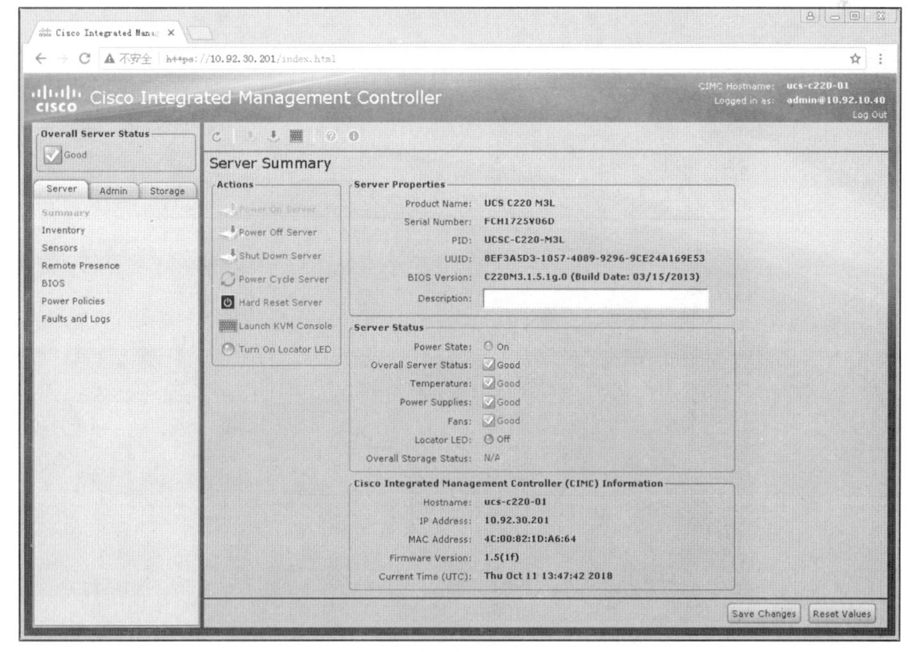

图 2-2-6

4. 其他厂商服务器远程管理工具

其他厂商服务器远程管理工具大多依赖于 IPMI 开发，配置 IP 地址后可以远程控制服务器，进行开启关闭/服务器电源、挂载/安装镜像、收集日志等操作。图 2-2-7 为广达服务器远程管理工具登录界面，图 2-2-8 为广达服务器运行 CentOS 操作系统界面。

图 2-2-7

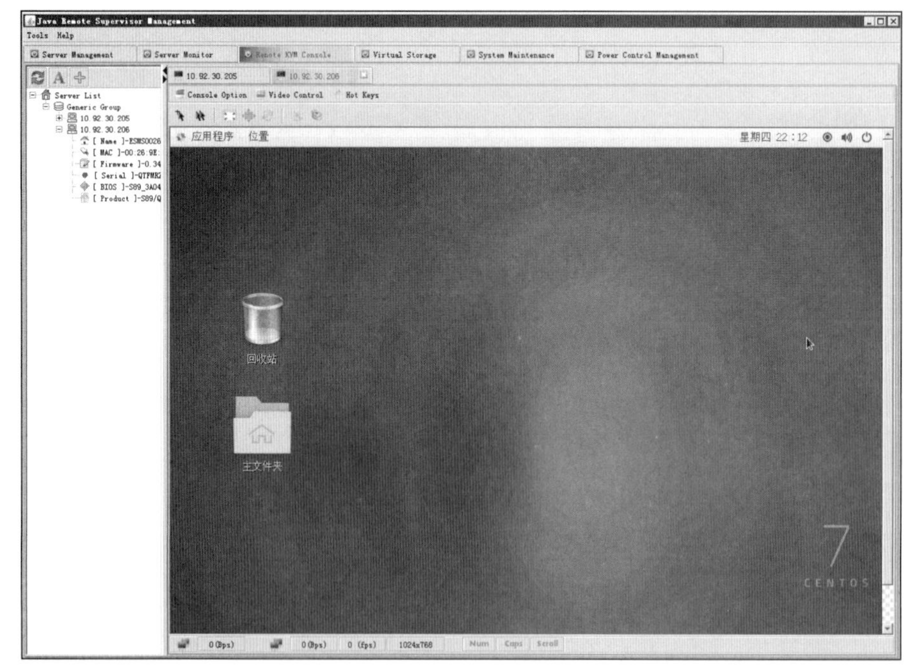

图 2-2-8

本节介绍的 HP 和 Dell 服务器远程管理工具版本相对较老，但对服务器电源的管理、远程安装操作系统等基本功能是具备的，新的版本功能更加丰富强大。需要注意的是，一些服务器厂商的远程管理工具需要购买许可后才能使用。

2.2.3 安装 CentOS 操作系统

准备好实验设备后需要为服务器安装操作系统，本书使用的开源 Linux 操作系统，所以需要下载常用的 Linux 操作系统。企业常用的 Linux 操作系统包括 Red Hat Enterprise Linux（RHEL）、Centos 以及 Ubuntu 等，国内阿里云、网易等网站也提供相应 ISO 文件的下载。

本节实战操作使用物理服务器通过远程管理工具安装 CentOS 7 操作系统，其安装文件为 CentOS-7-x86_64-DVD-1804.ISO。

第 1 步，使用 HP iLo 2 远程管理工具挂载下载好的 CentOS 7 操作系统安装 ISO 文件，在“Image”中选择“Mount”，如图 2-2-9 所示。

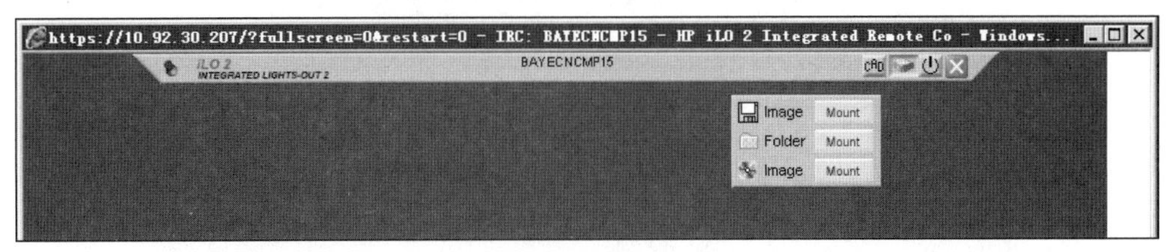

图 2-2-9

第 2 步，选择下载好的 CentOS-7-x86_64-DVD-1804.ISO 文件，如图 2-2-10 所示，单击“打开”按钮。

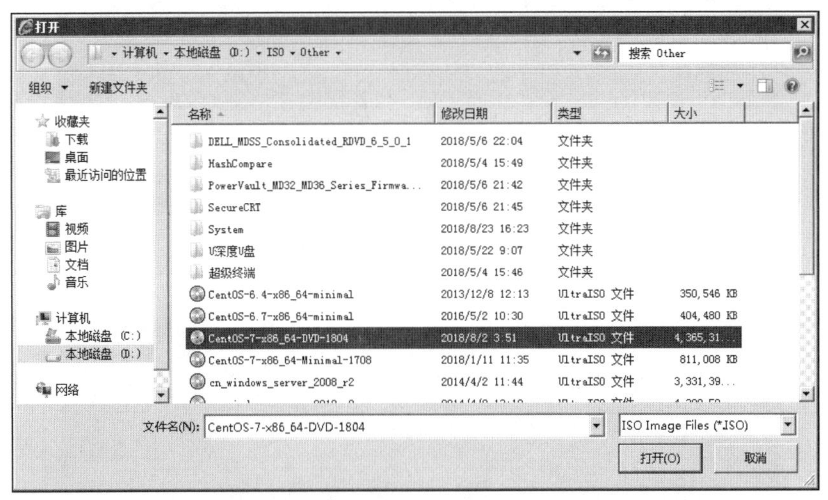

图 2-2-10

第 3 步，使用 HP iLO 2 远程管理工具打开服务器电源，按 F11 键手动选择引导，如图 2-2-11 所示。

第 4 步，按数字键 1 选择从 CD-ROM 引导，如图 2-2-12 所示。

第 5 步，进入 CentOS 7 操作系统安装界面，选择“Install CentOS 7”，如图 2-2-13 所示，按 Enter 键继续。

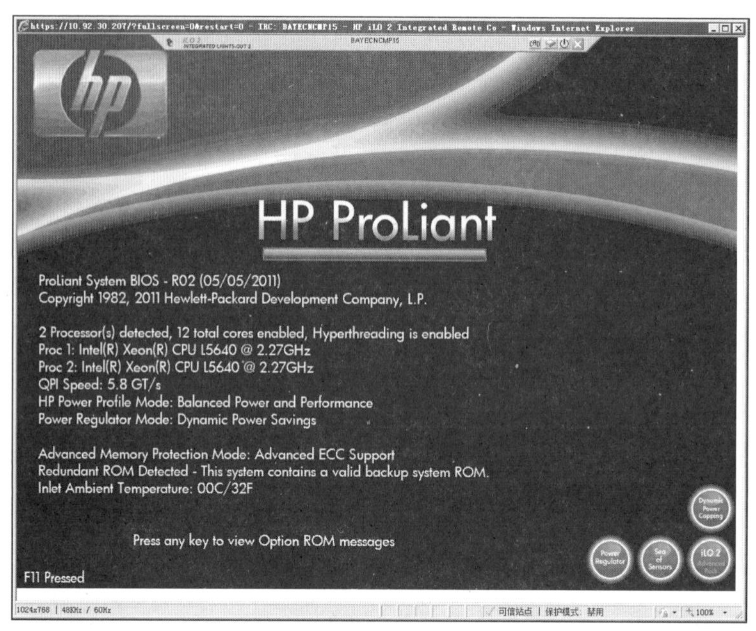

图 2-2-11

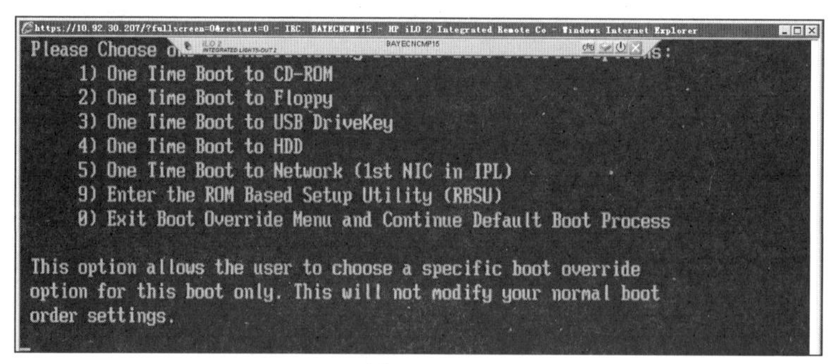

图 2-2-12

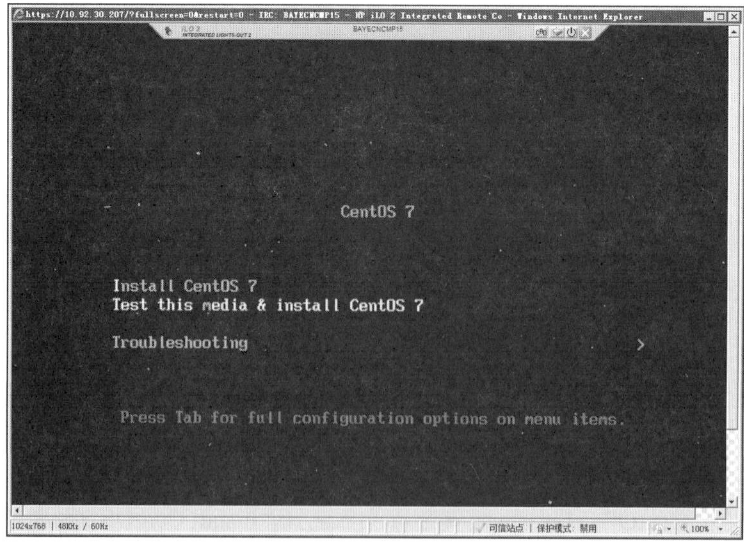

图 2-2-13

第 6 步，进入 CentOS 7 安装核心配置界面，“本地化”组根据情况选择即可，特别注意“软件”组中的“安装源”、“软件选择”以及“系统”组中的“安装位置”的参数配置，如图 2-2-14 所示，下面对这几个参数进行介绍。

图 2-2-14

第 7 步，单击“安装源”进入配置界面，CentOS 7 支持多种安装源，根据实际情况进行选择，如图 2-2-15 所示，选择“自动检测到的安装介质”，单击“完成”按钮。

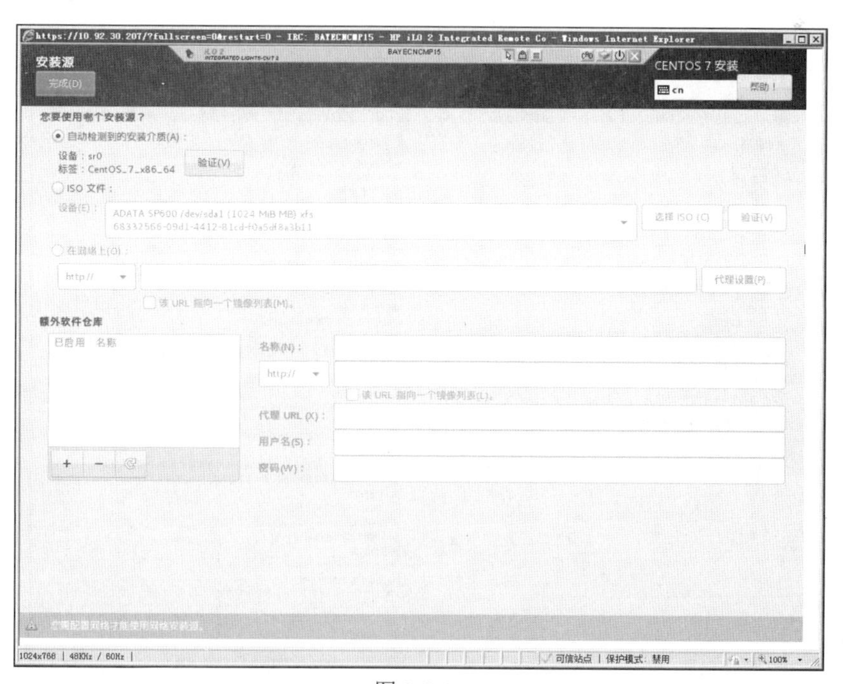

图 2-2-15

第8步，单击“软件选择”进入配置界面。实验挂载的是DVD安装ISO文件，其自身带有的软件包较多。如果使用mini安装ISO文件，基本环境默认采用“最小安装”。在初始安装的时候可以根据实际情况选择，如图2-2-16所示。

图2-2-16

第9步，本节实战操作后续会使用图形用户界面（Graphical User Interface，GUI）安装配置虚拟机，因此选择“带GUI的服务器”，如图2-2-17所示，单击“完成”按钮。

图2-2-17

第 10 步，单击“安装位置”进入配置界面，CentOS 7 支持多种安装模式，特别需要注意的是，如果本地硬盘有其他操作系统，强烈建议清空原分区再安装，以避免后续出现问题，如图 2-2-18 所示，单击“完成”按钮。

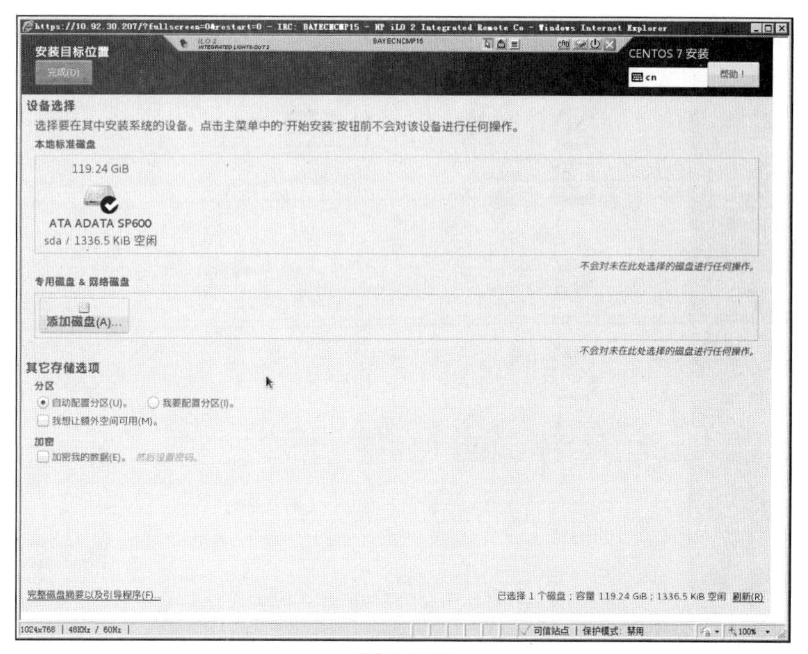

图 2-2-18

第 11 步，单击“网络和主机名”进入配置界面，安装过程中的网络配置为图形化界面，本节不配置，待安装完成后通过命令行进行配置，如图 2-2-19 所示，单击“完成”按钮。

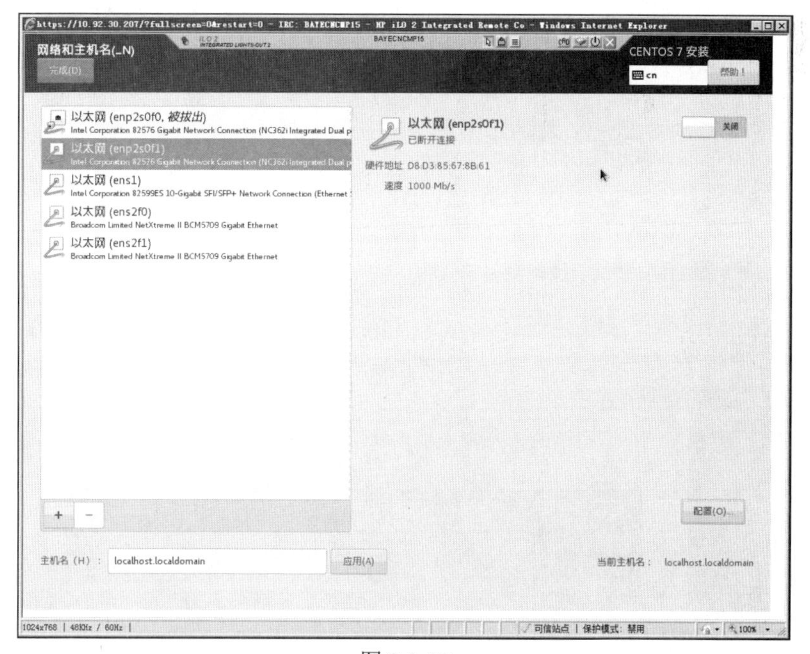

图 2-2-19

第 12 步，完成 CentOS 7 操作系统基本的信息配置后，如图 2-2-20 所示，单击“开始安装”按钮。

图 2-2-20

第 13 步，进入“用户设置”界面，需要配置 root 用户密码和创建非 root 用户，如图 2-2-21 所示，单击“ROOT 密码”进入配置。

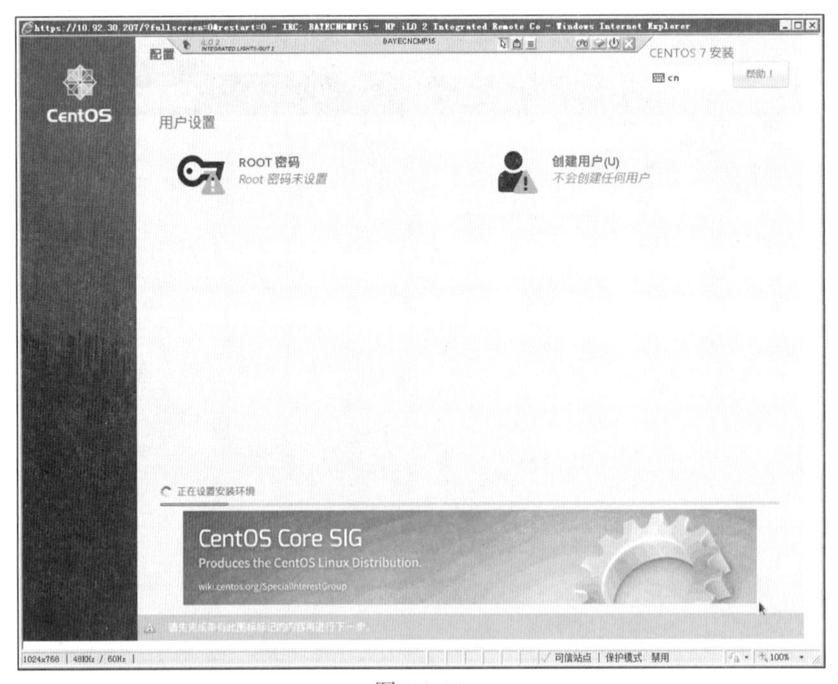

图 2-2-21

第 14 步，配置 root 用户密码，如图 2-2-22 所示，单击“完成”按钮。

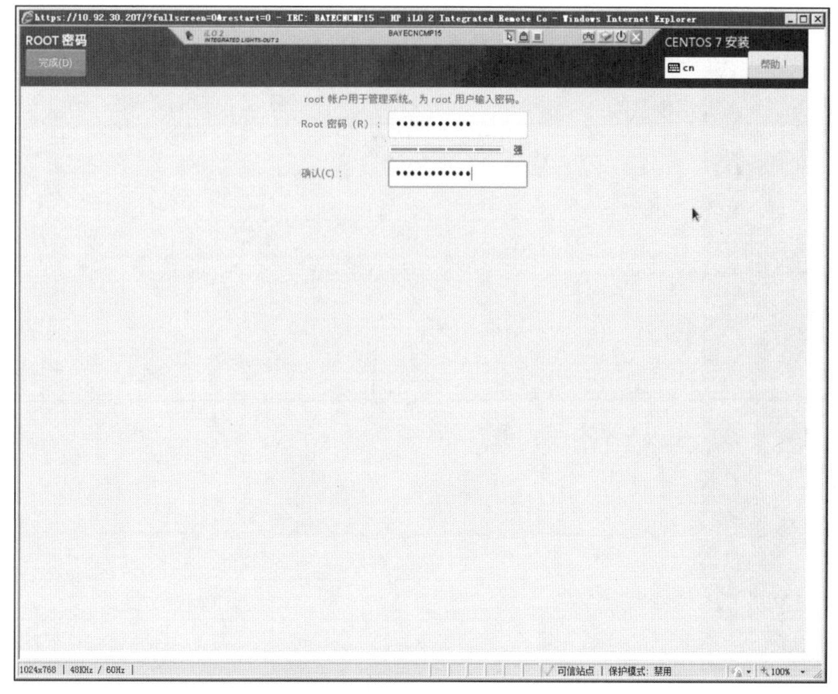

图 2-2-22

第 15 步，非 root 用户可以根据实际情况选择是否创建，本节创建非 root 用户 admin，如图 2-2-23 所示。

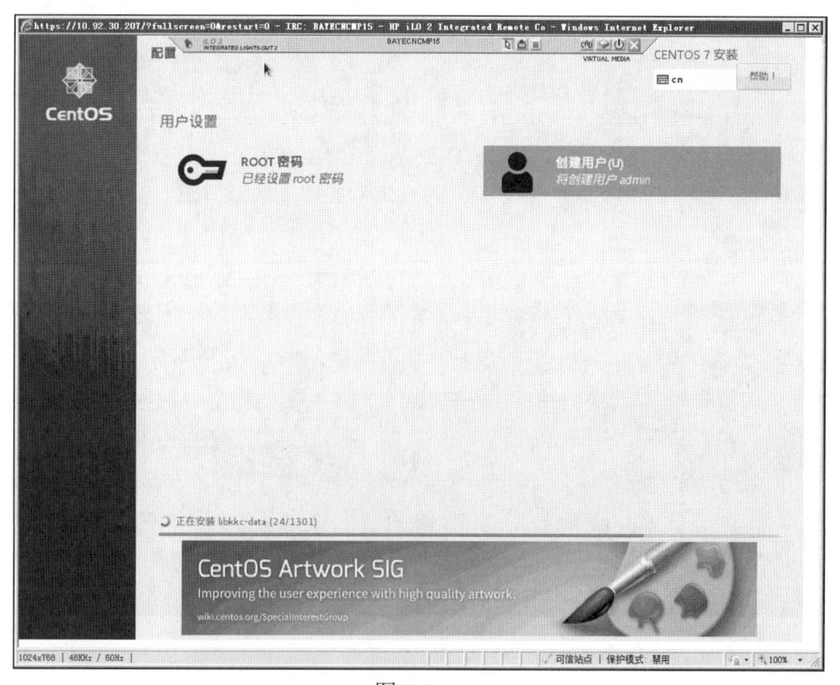

图 2-2-23

第 16 步，结束 CentOS 7 操作系统的安装后需要重启，如图 2-2-24 所示，单击“重启”按钮。

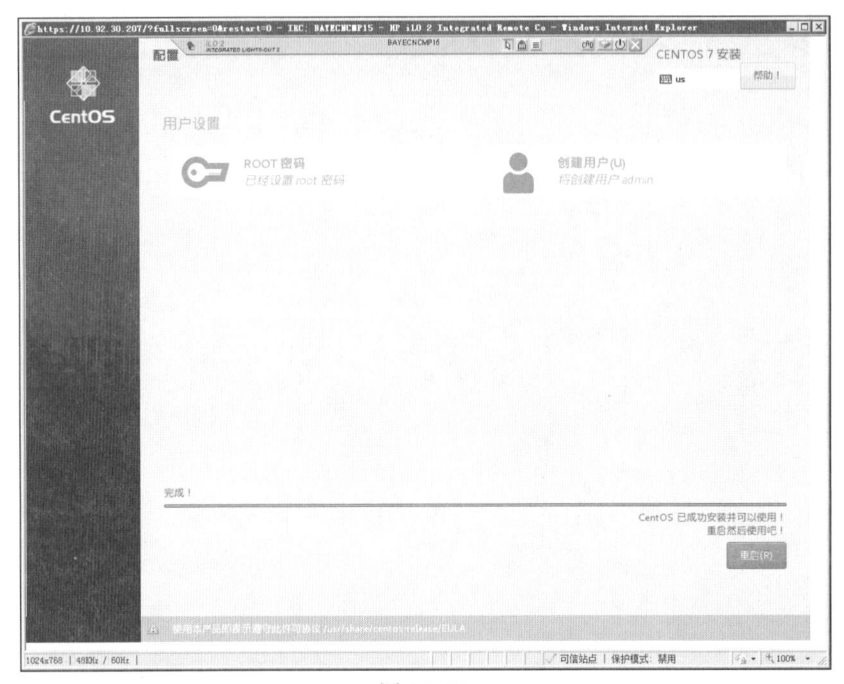

图 2-2-24

第 17 步，物理服务器重启过程如图 2-2-25 所示。

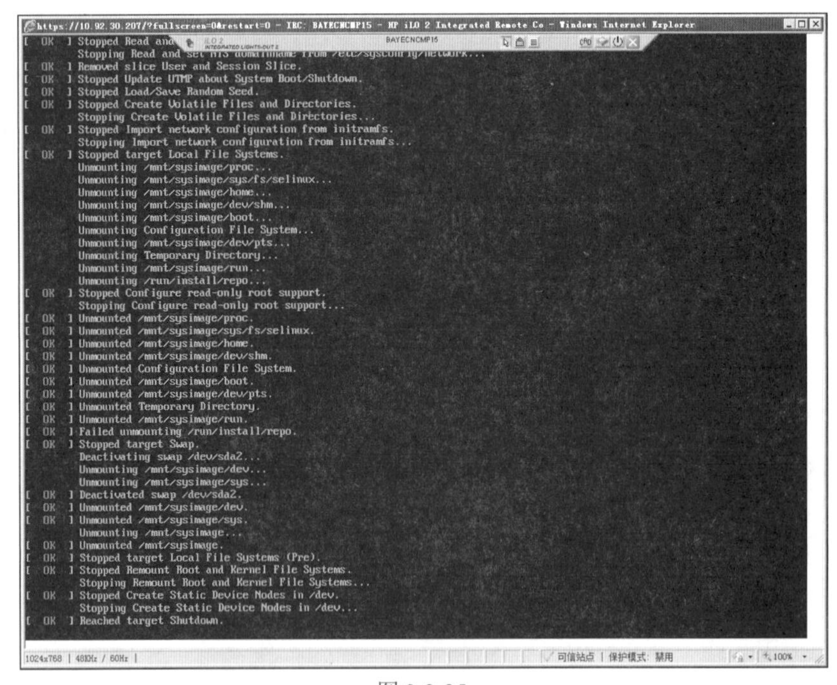

图 2-2-25

第 18 步，物理服务器重启后经过一些简单的对话配置即可进入操作系统，使用创建的用户 admin 登录 CentOS 7 操作系统的 GUI，如图 2-2-26 所示。

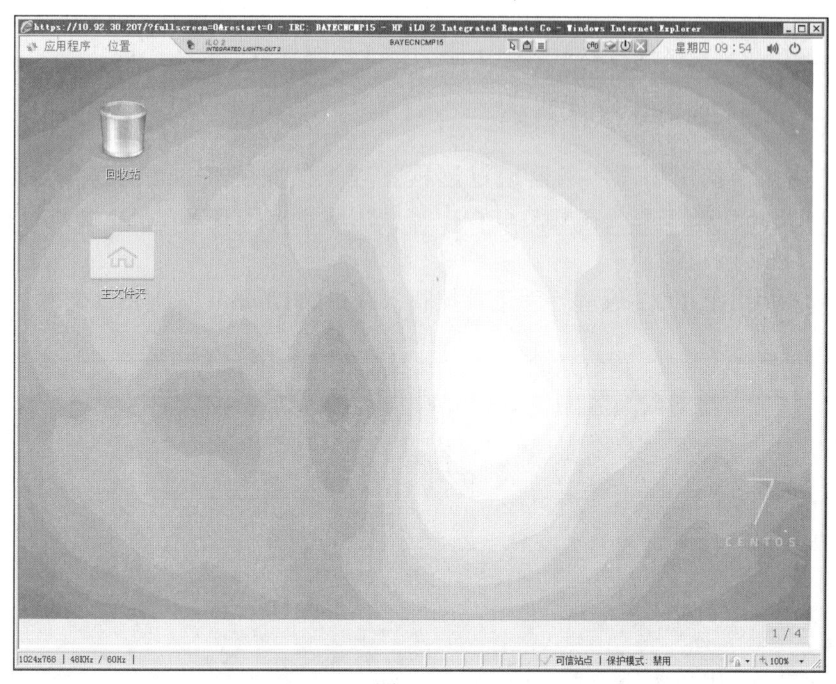

图 2-2-26

第 19 步，在 GUI 打开终端命令窗口，命令可正常执行，如图 2-2-27 所示。

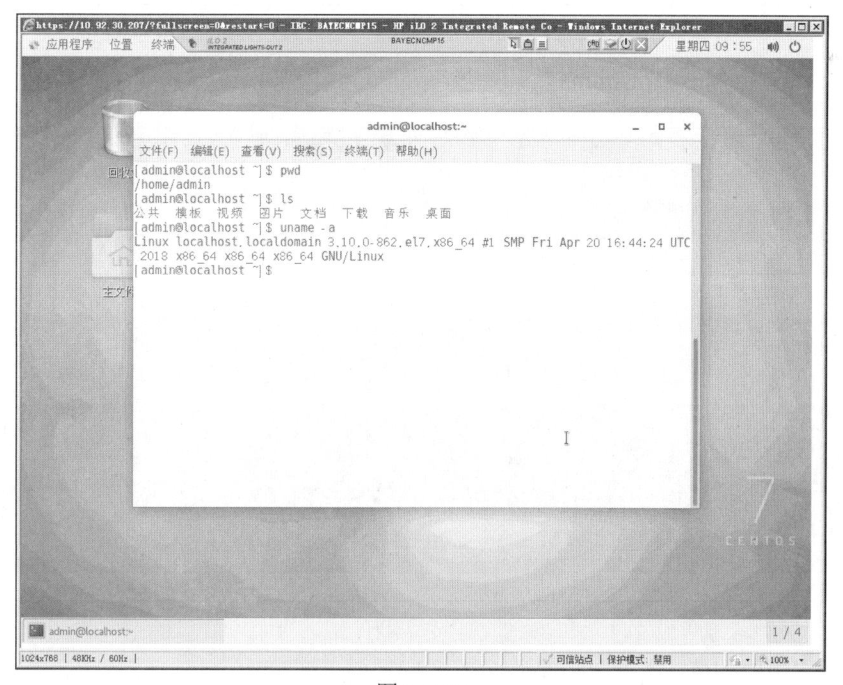

图 2-2-27

至此，通过远程管理工具安装 CentOS 7 操作系统完成。目前新的操作系统对硬件的支持或兼容性更好，基本上不存在过多的驱动程序问题。一些厂商也提供自行研发的引导光盘，引导光盘加载了厂商服务器硬件驱动程序，如果物理服务器有引导光盘，推荐使用引导光盘安装操作系统，这样可以避免由于驱动程序问题导致的安装不成功。

2.2.4 安装 Ubuntu 操作系统

本节实战操作使用物理服务器通过远程管理工具安装 Ubuntu 操作系统，版本为 18.04。注意 Ubuntu 操作系统在安装过程中需使用键盘方向键和 Enter 键进行操作，不能使用鼠标。

第 1 步，使用 HP iLO 2 远程管理工具挂载下载好的 Ubuntu 安装 ISO 文件，如图 2-2-28 所示。

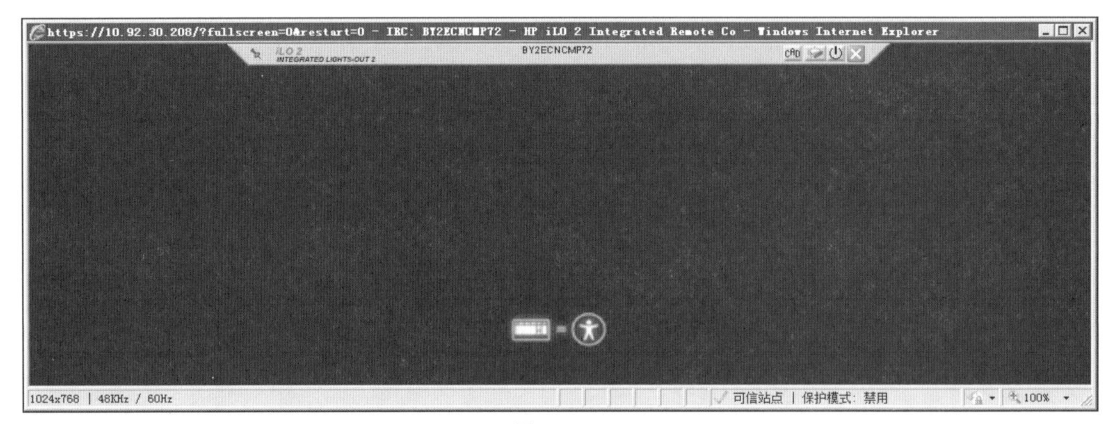

图 2-2-28

第 2 步，选择“Install Ubuntu”开始安装 Ubuntu 操作系统，如图 2-2-29 所示。

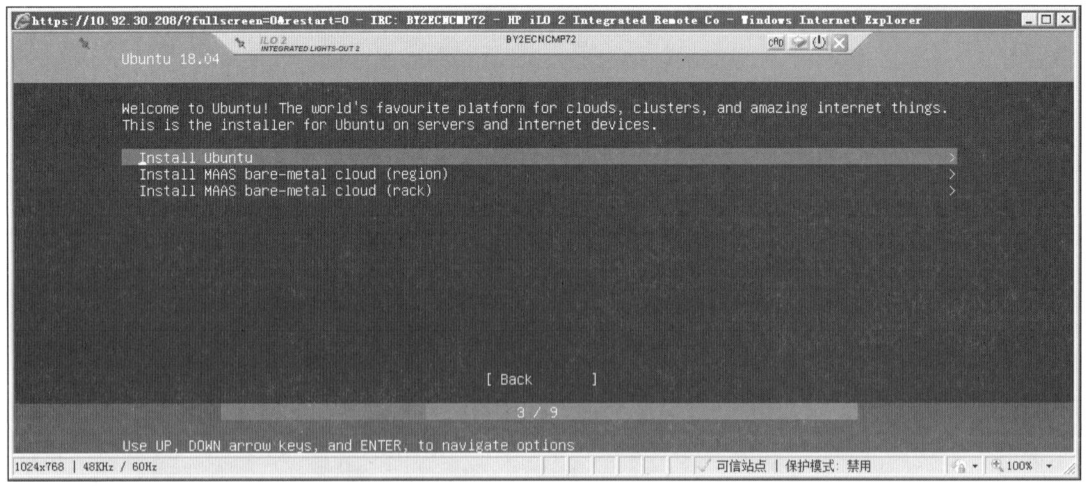

图 2-2-29

第 3 步，配置网络，如图 2-2-30 所示，选择“enp2s0f0”并按 Enter 键进行配置，默认使用 DHCP 服务器获取地址。注意安装 Ubuntu 与 CentOS 操作系统存在区别，如果 Ubuntu

服务器未连接网络，无法进行下一步安装操作。

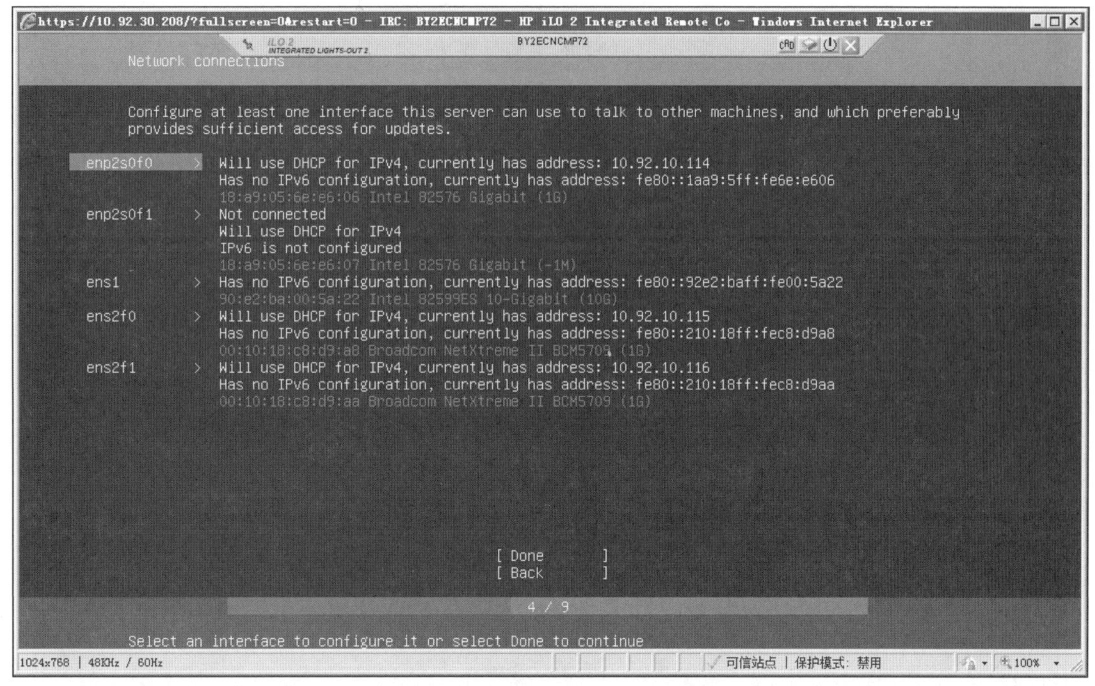

图 2-2-30

第 4 步，如图 2-2-31 所示，选择“Use a static IPv4 configuration”并按 Enter 键手动配置静态 IP 地址。

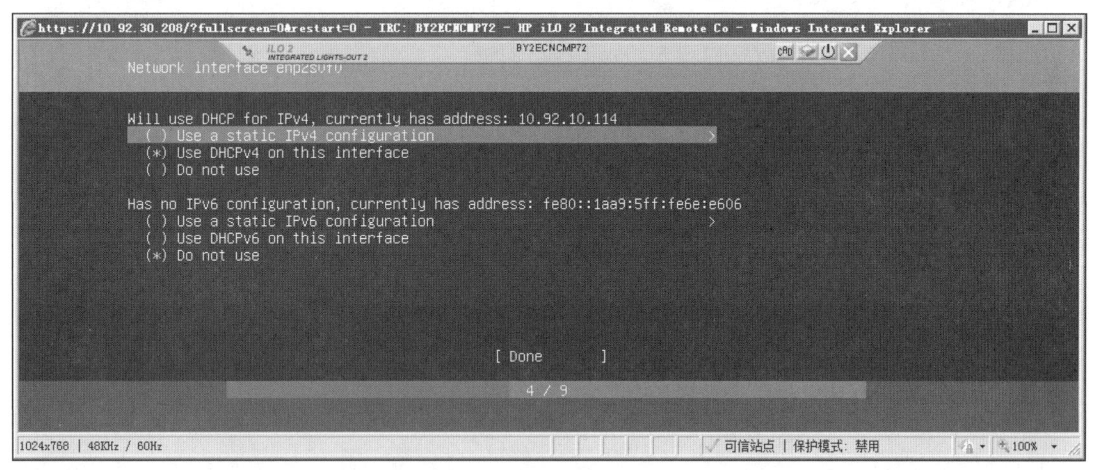

图 2-2-31

第 5 步，根据需求配置 IP 地址等，如图 2-2-32 所示，配置完成后选择“Save”保存配置。

第 6 步，完成配置后的界面如图 2-2-33 所示，选择“Done”继续安装。

第 7 步，进入代理地址配置界面，该配置为可选项，如果环境中没有代理服务器无须配置，如图 2-2-34 所示，选择“Done”继续安装。

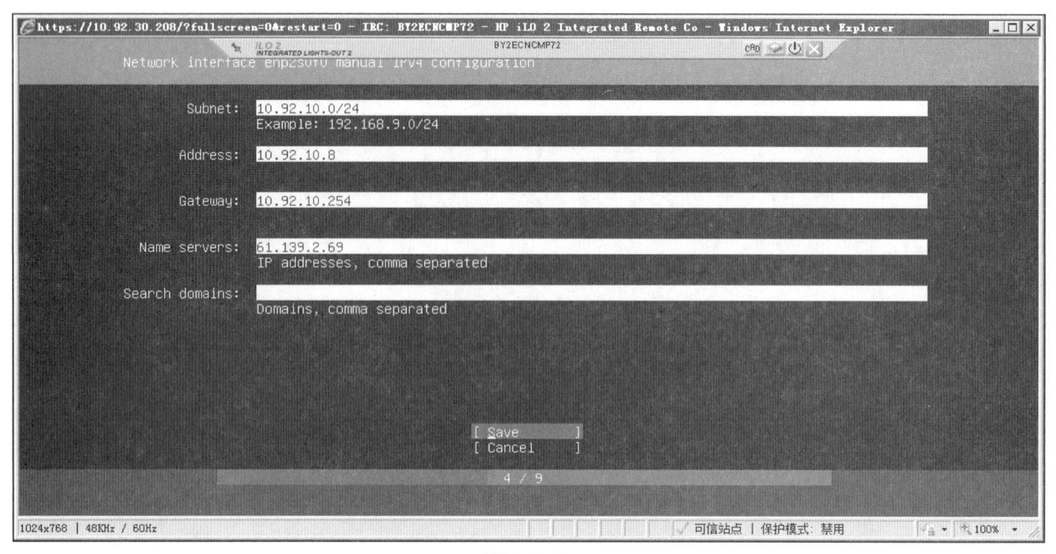

图 2-2-32

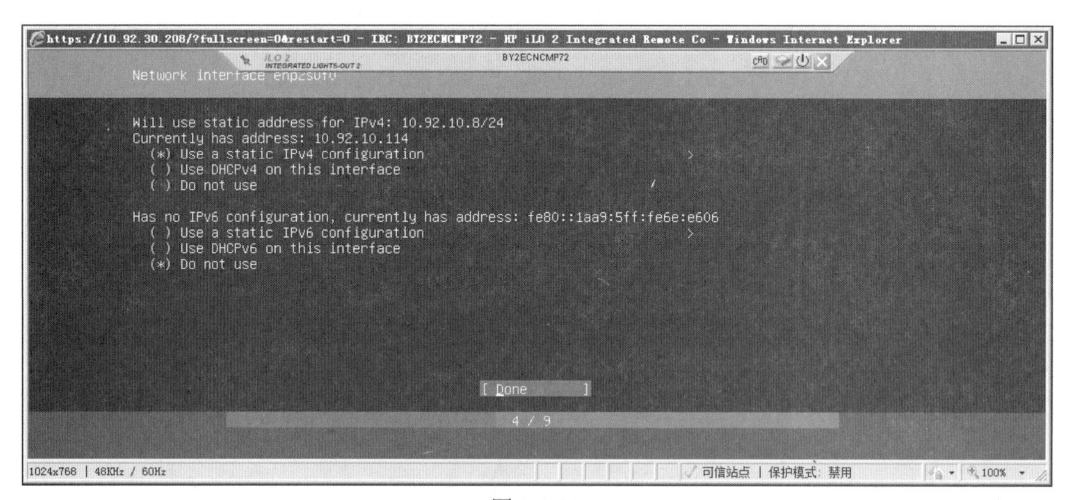

图 2-2-33

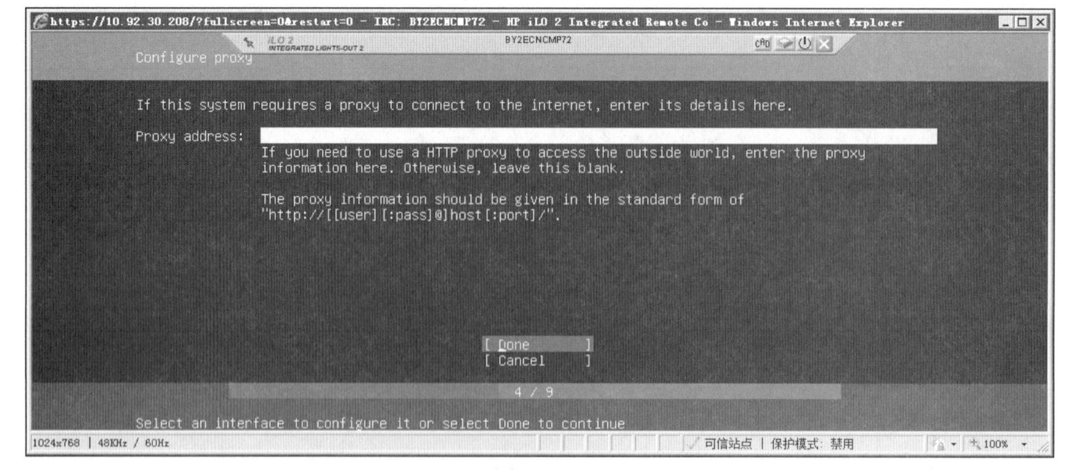

图 2-2-34

第 8 步，对操作系统安装位置进行选择，如图 2-2-35 所示，选择“Use An Entire Disk”可以看到物理服务器硬盘信息。

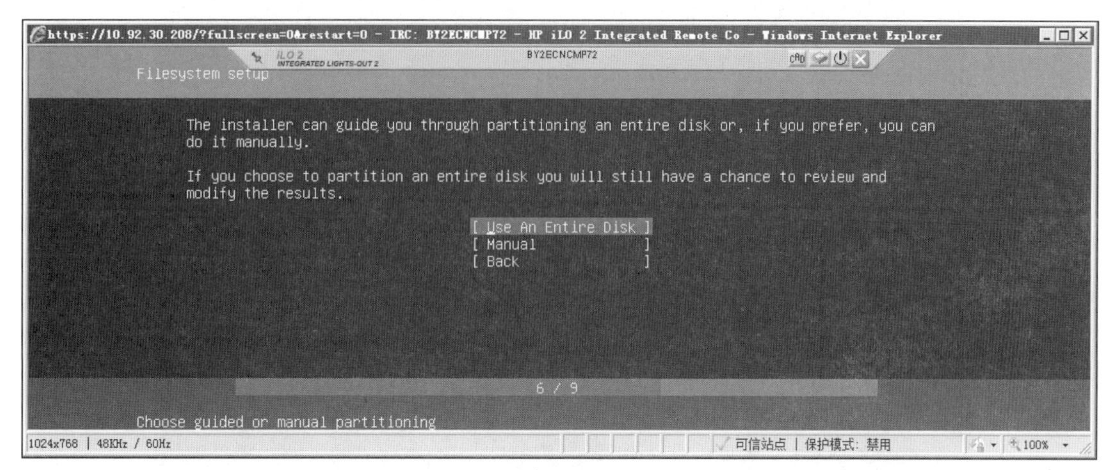

图 2-2-35

第 9 步，在本地硬盘列表中选择安装操作系统的硬盘，如图 2-2-36 所示。

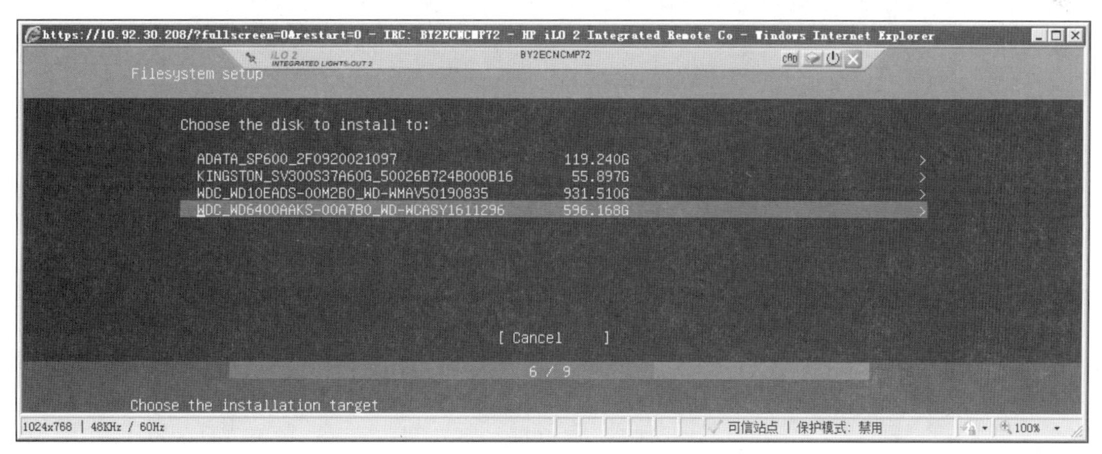

图 2-2-36

第 10 步，如果所选硬盘中有其他操作系统或分区，系统会提示是否继续，如图 2-2-37 所示，选择“Continue”继续安装。

第 11 步，配置登录操作系统的用户名和服务器名等信息，如图 2-2-38 所示，配置完成后选择“Done”继续安装。

第 12 步，开始安装 Ubuntu 操作系统，如图 2-2-39 所示，可以选择“view full log”查看日志信息。

第 13 步，Ubuntu 操作系统安装完成后需要重启，如图 2-2-40 所示，选择“Reboot Now”重启服务器。

第 14 步，物理服务器重启后，使用创建的 admin 用户登录，如图 2-2-41 所示，注意不是 root 用户。

第 15 步，使用命令“sudo passwd root”设置 root 用户密码，如图 2-2-42 所示。

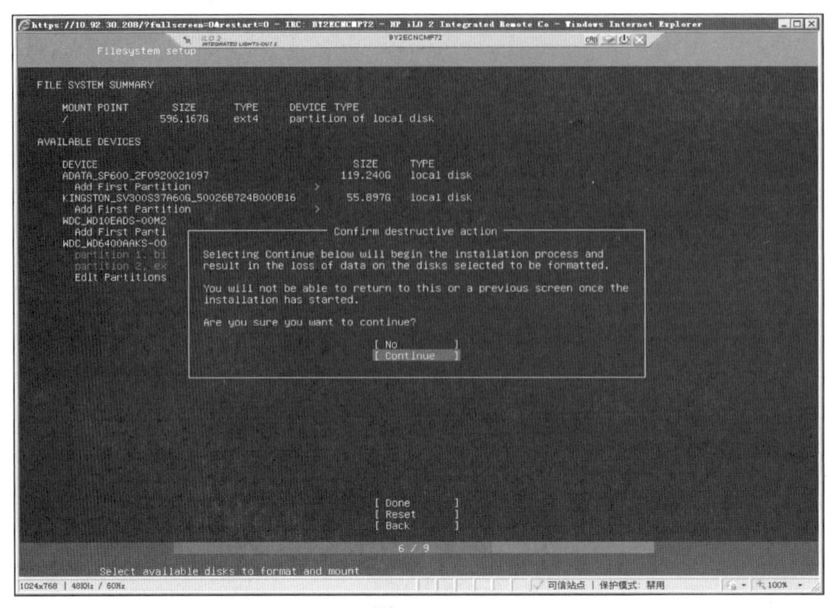

图 2-2-37

```
Profile setup

Enter the username and password (or ssh identity) you will use to log in to the system.

              Your name: admin
     Your server's name: kvm-server
                         The name it uses when it talks to other computers.
        Pick a username: admin
      Choose a password: ************
  Confirm your password: ************
    Import SSH identity: (+) No
                         You can import your SSH keys from Github or Launchpad.
        Import Username:

[ Done ]
6 / 9
Install in progress: removing previous storage devices
```

图 2-2-38

```
Installing system

curtin command install
  preparing for installation
  configuring storage
    running 'curtin block-meta custom'
      curtin command block-meta
        removing previous storage devices
        configuring disk: disk-0
        configuring partition: part-2
        configuring partition: part-3
        configuring format: fs-1 /

[ View full log ]
8 / 9
Thank you for using Ubuntu!
```

图 2-2-39

Installation complete!
Finished install!
curtin command install
preparing for installation
configuring storage
running 'curtin block-meta custom'
curtin command block-meta
removing previous storage devices
configuring disk: disk-0
configuring partition: part-2
configuring partition: part-3
configuring format: fs-1
configuring mount: mount-1
configuring network
running 'curtin net-meta auto'
curtin command net-meta
writing install sources to disk
running 'curtin extract'
curtin command extract
acquiring and extracting image from cp:///media/filesystem
configuring installed system
running 'curtin curthooks'
curtin command curthooks
configuring apt configuring apt
installing missing packages
installing kernel
setting up swap
apply networking config
writing etc/fstab
configuring multipath
updating packages on target system
configuring pollinate user-agent on target system
finalizing installation
running 'curtin hook'
curtin command hook
executing late commands
[View full log]
[Reboot Now]
9 / 9
Thank you for using Ubuntu!

图 2-2-40

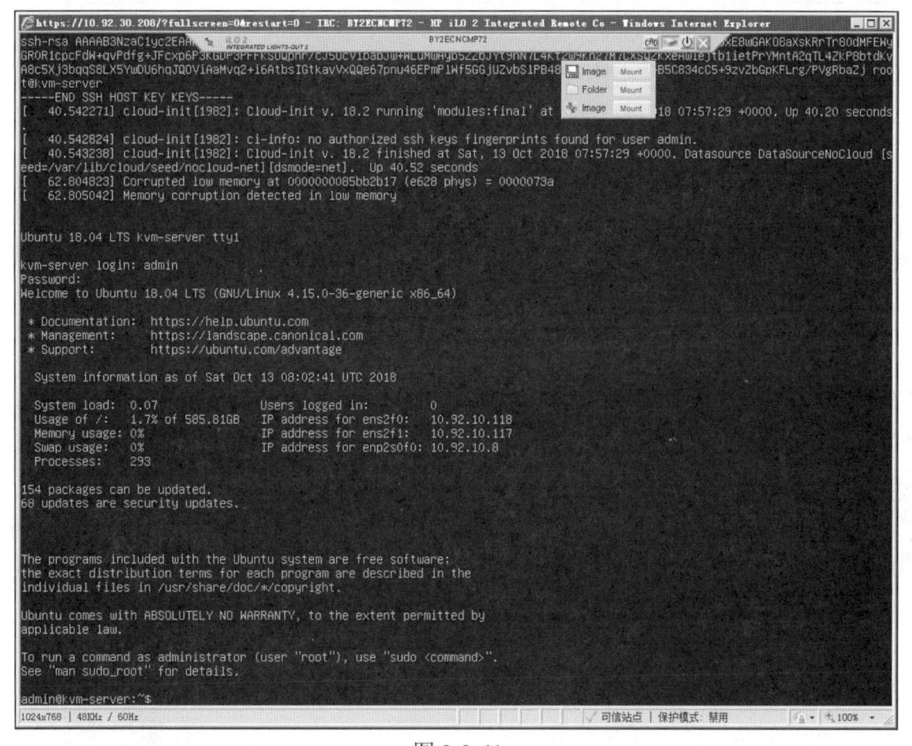
-----END SSH HOST KEY KEYS-----
[40.542824] cloud-init[1982]: ci-info: no authorized ssh keys fingerprints found for user admin.
[62.805042] Memory corruption detected in low memory
Ubuntu 18.04 LTS kvm-server tty1
kvm-server login: admin
Password:
Welcome to Ubuntu 18.04 LTS (GNU/Linux 4.15.0-36-generic x86_64)
* Documentation: https://help.ubuntu.com
* Management: https://landscape.canonical.com
* Support: https://ubuntu.com/advantage
System information as of Sat Oct 13 08:02:41 UTC 2018
System load: 0.07 Users logged in: 0
Usage of /: 1.7% of 585.81GB IP address for ens2f0: 10.92.10.118
Memory usage: 0% IP address for ens2f1: 10.92.10.117
Swap usage: 0% IP address for enp2s0f0: 10.92.10.8
Processes: 293
154 packages can be updated.
68 updates are security updates.
The programs included with the Ubuntu system are free software;
the exact distribution terms for each program are described in the
individual files in /usr/share/doc/*/copyright.
Ubuntu comes with ABSOLUTELY NO WARRANTY, to the extent permitted by
applicable law.
To run a command as administrator (user "root"), use "sudo <command>".
See "man sudo_root" for details.
admin@kvm-server:~$

图 2-2-41

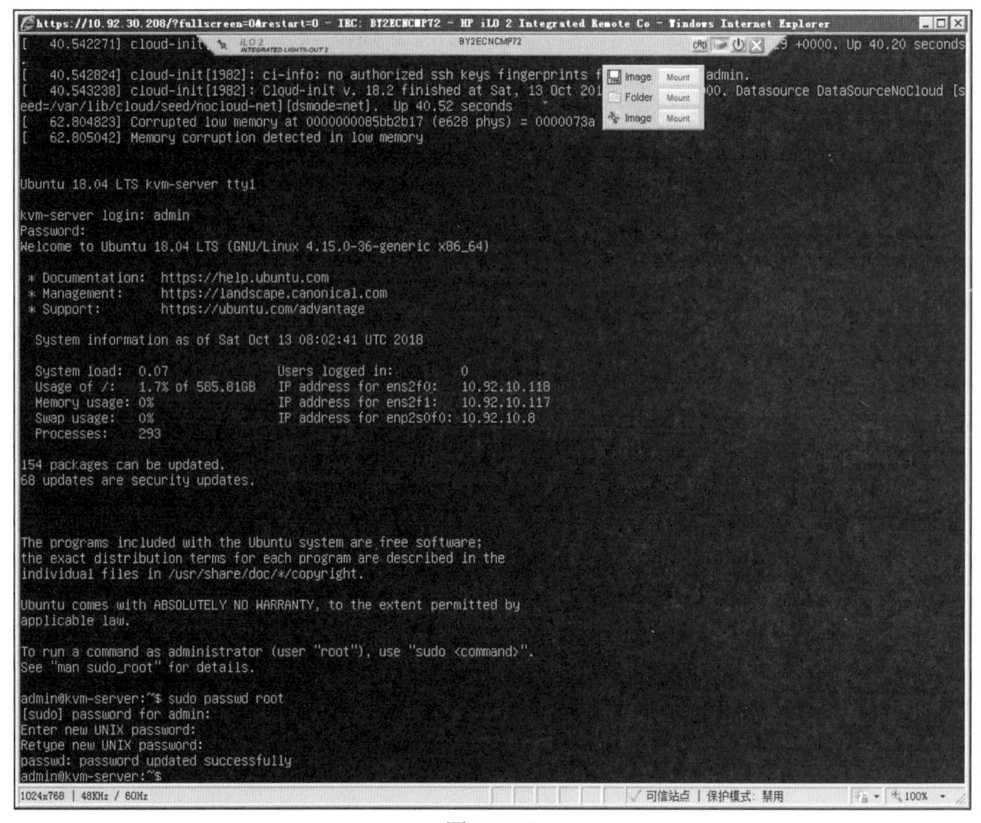

图 2-2-42

第 16 步，使用命令“su”切换到 root 用户，通过图 2-2-43 可以看到 Ubuntu 操作系统的版本信息和网络访问情况。

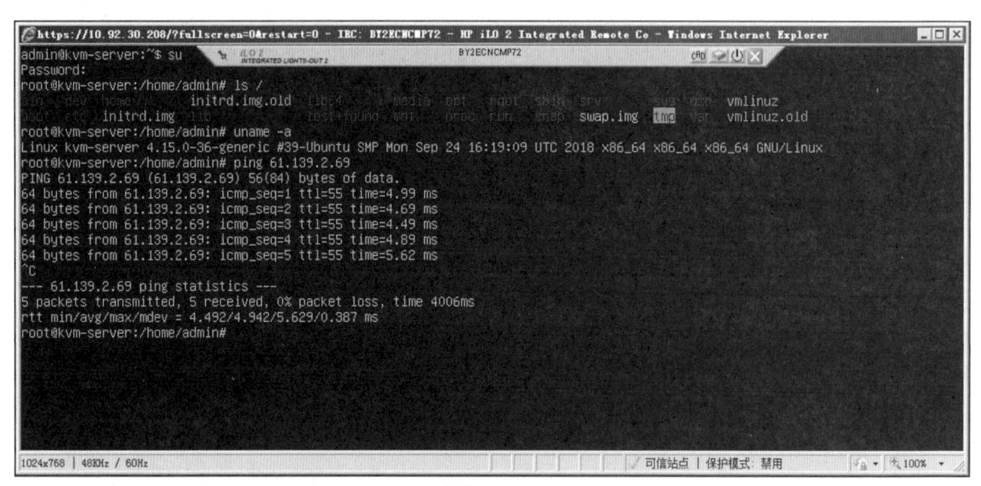

图 2-2-43

至此，Ubuntu 操作系统安装完成，虽然都是 Linux 操作系统，但从安装过程来看，Ubuntu 操作系统与 CentOS 操作系统还是存在一些区别，读者多进行安装操作即可熟练掌握操作系统的安装。

2.2.5 基本网络配置

操作系统安装完成后，需要进行一些基础的配置才能开始使用，前文介绍安装 CentOS 7 操作系统时没有配置网络，意味着该操作系统无法通过网络进行管理。本节介绍 CentOS 操作系统和 Ubuntu 操作系统的网络配置。

1. CentOS 操作系统网络配置

第 1 步，使用命令“ip addr”查看当前网络情况，如图 2-2-44 所示。注意 CentOS 6.x 使用命令“ifconfig”查看。

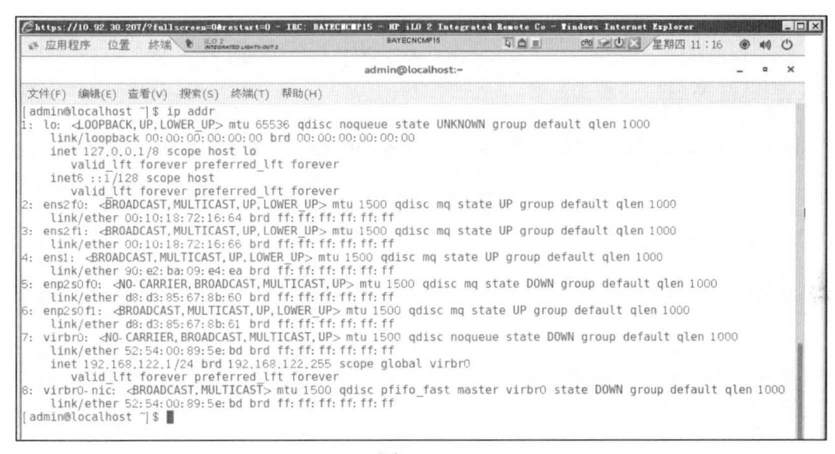

图 2-2-44

第 2 步，Linux 操作系统的网络都以文件形式配置，配置网络的实质就是修改配置文件，使用命令“ls /etc/sysconfig/network-scripts/”查看配置文件位置，如图 2-2-45 所示。

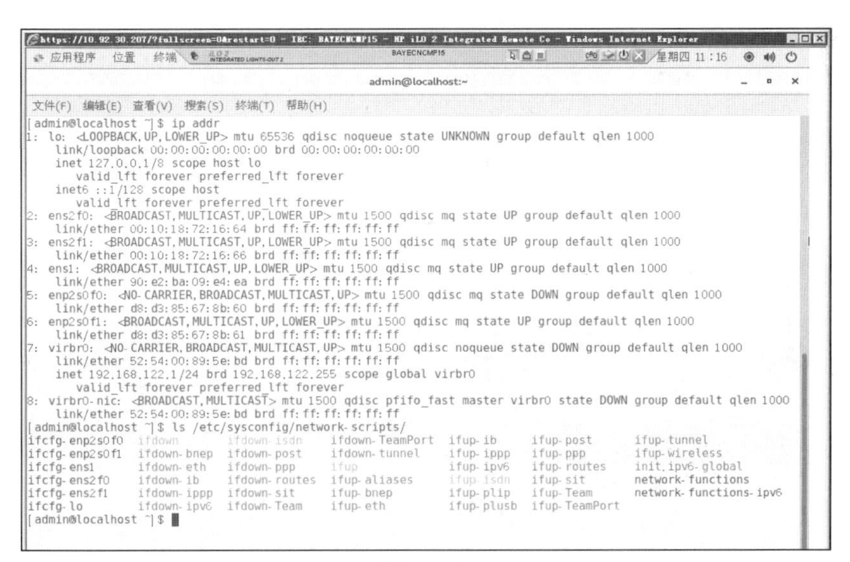

图 2-2-45

第 3 步，使用 Vi 编辑器编辑配置文件“vi /etc/sysconfig/network-scripts/ifcfg-ens2f0”，如图 2-2-46 所示。

图 2-2-46

第 4 步，配置 IP 地址相关信息，注意修改 ONBOOT 的值为“yes”，如图 2-2-47 所示，修改完成后保存配置文件。

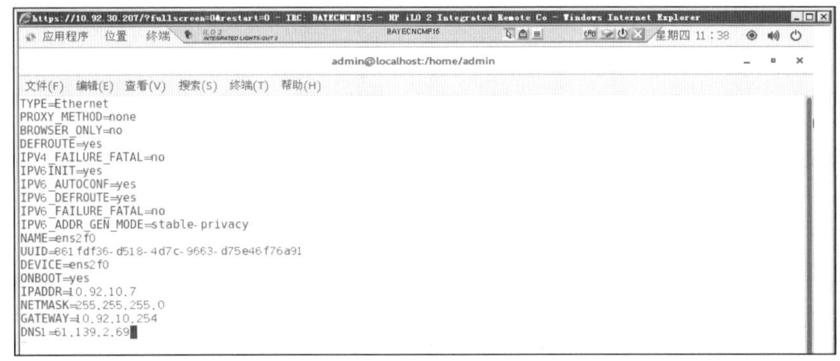

图 2-2-47

第 5 步，重启网络或操作系统后，使用“ping”命令检测网络的连通性，如图 2-2-48 所示，访问外部网络正常，如果无法连接外部网络请检查配置文件。

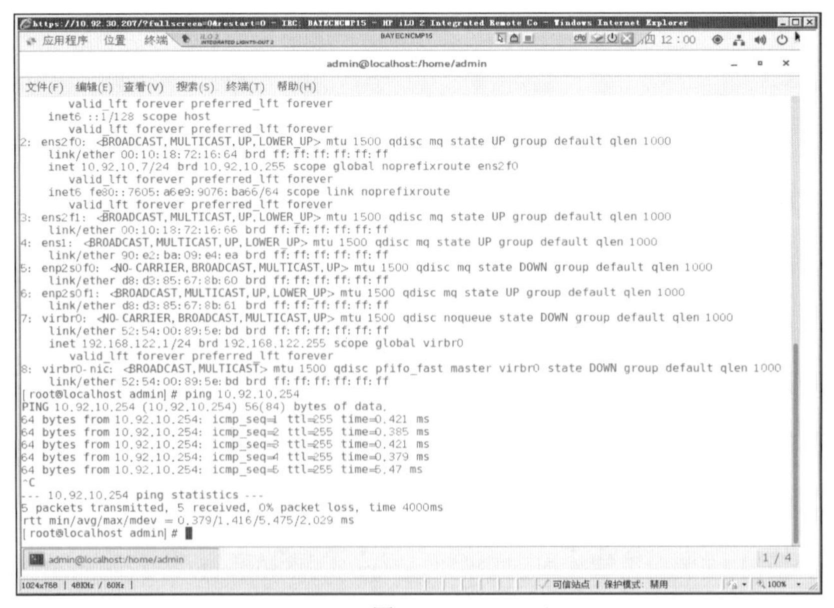

图 2-2-48

2. Ubuntu 操作系统网络配置

Ubuntu 18.04 使用了新的网络工具 netplan，网络配置文件是“/etc/netplan/50-cloud-init.yaml”，需要特别注意。

第 1 步，使用命令“su”切换到 root 用户，使用命令“ip addr”查看网络信息，在安装物理服务器操作系统时已经配置了 IP 地址，因此可以看到已配置的 IP 地址，如图 2-2-49 所示。

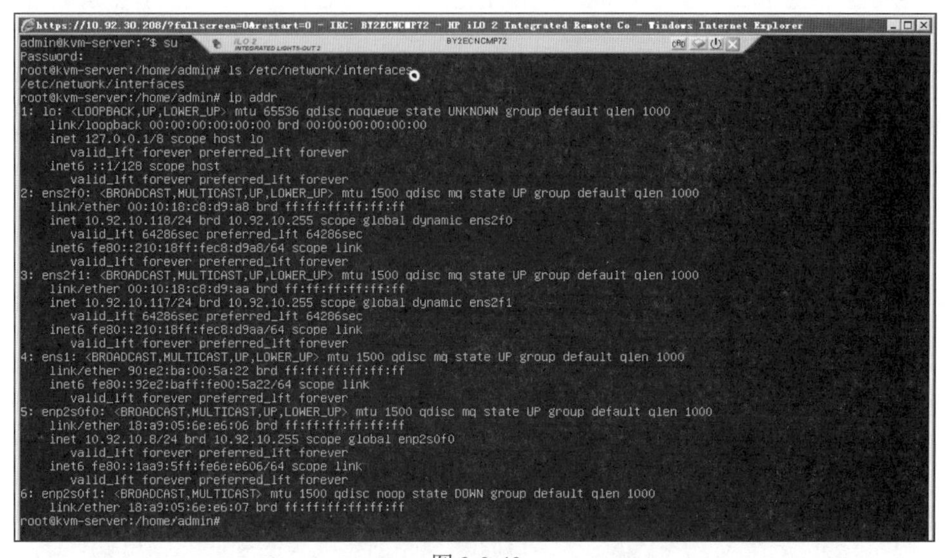

图 2-2-49

第 2 步，使用 Vi 编辑器编辑配置文件“vi /etc/netplan/50-cloud-init.yaml”，调整网络配置，如图 2-2-50 所示。

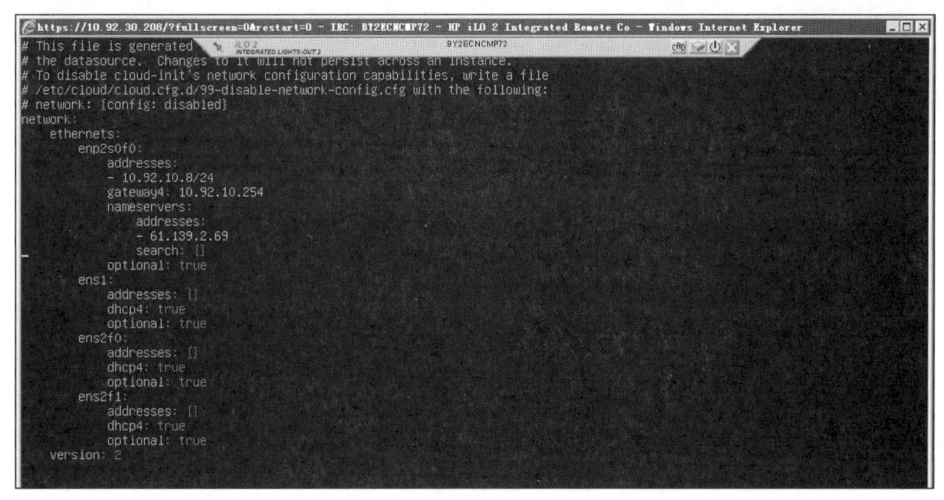

图 2-2-50

第 3 步，使用 Vi 编辑器编辑 DNS 配置文件“vi /etc/resolv.conf”，调整 DNS 服务器地址，如图 2-2-51 所示。

```
# This file is managed by
#
# This is a dynamic resolv.conf file for connecting local clients to the
# internal DNS stub resolver of systemd-resolved. This file lists all
# configured search domains.
#
# Run "systemd-resolve --status" to see details about the uplink DNS servers
# currently in use.
#
# Third party programs must not access this file directly, but only through the
# symlink at /etc/resolv.conf. To manage man:resolv.conf(5) in a different way,
# replace this symlink by a static file or a different symlink.
#
# See man:systemd-resolved.service(8) for details about the supported modes of
# operation for /etc/resolv.conf.

nameserver 61.139.2.69
```

图 2-2-51

第 4 步，完成网络配置，使用“ping”命令检测网络的连通性，如图 2-2-52 所示，如果无法连接外部网络请检查配置文件。

图 2-2-52

2.2.6　修改 Linux 系统 YUM 源

除 RHEL 操作系统本身的 YUM 源需订阅使用外，其他 Linux 操作系统厂商都提供自己的 YUM 源。通过 YUM 源可以方便、快捷地安装软件包，同时可以解决软件包的依赖问题，但由于多数 Linux 操作系统厂商都在国外，使用国外的 YUM 源更新比较缓慢且容易失败，因此推荐将其 YUM 源修改为网易或阿里云 YUM 源，这样效率更高。同时，后续安装使用 KVM 也需要外部的 YUM 源，所以学习修改 YUM 源非常重要。

1. 修改 CentOS 操作系统 YUM 源

第 1 步，使用命令“ls /etc/yum.repos.d”查看系统自带的 YUM 源文件。

```
[root@localhost ~]# ls /etc/yum.repos.d/
CentOS-Base.repo  CentOS-Debuginfo.repo  CentOS-Media.repo   CentOS-Vault.repo
CentOS-CR.repo    CentOS-fasttrack.repo  CentOS-Sources.repo
```

第 2 步，使用命令“cat /etc/yum.repos.d/CentOS-Base.repo”查看系统自带的 YUM 源文件信息，可以看到访问地址为 mirrorlist.centos.org，这是 CentOS 官方的服务器，国内的访问速度较慢。

```
[root@localhost ~]# cat /etc/yum.repos.d/CentOS-Base.repo
# CentOS-Base.repo
#
[base]
name=CentOS-$releasever - Base
mirrorlist=http://mirrorlist.centos.org/?release=$releasever&arch=$basearch&repo=
os&infra=$infra
#baseurl=http://mirror.centos.org/centos/$releasever/os/$basearch/
gpgcheck=1
gpgkey=file:///etc/pki/rpm-gpg/RPM-GPG-KEY-CentOS-7

#released updates
……（省略）
```

第 3 步，修改 YUM 源地址为阿里云 YUM 源地址前，请确保物理服务器可以访问阿里云，如果不能访问将不能使用阿里云 YUM 源。

```
[root@localhost ~]# ping www.aliyun.com
PING v6wagbridge.aliyun.com.gds.alibabadns.com (106.11.93.21) 56(84) bytes of data.
64 bytes from 106.11.93.21 (106.11.93.21): icmp_seq=1 ttl=37 time=47.2 ms
64 bytes from 106.11.93.21 (106.11.93.21): icmp_seq=2 ttl=37 time=42.3 ms
……
--- v6wagbridge.aliyun.com.gds.alibabadns.com ping statistics ---
3 packets transmitted, 2 received, +1 errors, 33% packet loss, time 6103ms
rtt min/avg/max/mdev = 42.379/44.800/47.222/2.430 ms
```

第 4 步，使用命令“mv /etc/yum.repos.d/CentOS-*.repo /tmp”将系统自带的 YUM 源文件移动到 tmp 目录。

```
[root@localhost ~]# mv /etc/yum.repos.d/CentOS-*.repo /tmp
```

第 5 步，使用命令“wget http://mirrors.aliyun.com/repo/Centos-7.repo”下载 CentOS 操作系统 YUM 源文件。

```
[root@localhost yum.repos.d]# wget http://mirrors.aliyun.com/repo/Centos-7.repo
--2018-10-13 22:41:13--  http://mirrors.aliyun.com/repo/Centos-7.repo
正在解析主机 mirrors.aliyun.com (mirrors.aliyun.com)... 180.95.171.106, 180.95.171.108,
113.200.111.221, ...
正在连接 mirrors.aliyun.com (mirrors.aliyun.com)|180.95.171.106|:80... 已连接。
已发出 HTTP 请求，正在等待回应... 200 OK
长度：2523 (2.5K) [application/octet-stream]
正在保存至："Centos-7.repo"
100%[=========================================>] 2,523 --.-K/s 用时 0.05s
2018-10-13 22:41:19 (51.6 KB/s) - 已保存 "Centos-7.repo" [2523/2523])
```

第 6 步，使用命令“ls”查看下载的 YUM 源文件。

```
[root@localhost yum.repos.d]# ls
Centos-7.repo
```

第 7 步，使用命令“cat Centos-7.repo”查看 YUM 源文件的详细信息，可以看到访问地址为 mirrors.aliyun.com，也就是使用阿里云 YUM 源。

```
[root@localhost yum.repos.d]# cat Centos-7.repo
# CentOS-Base.repo
#
[base]
name=CentOS-$releasever - Base - mirrors.aliyun.com
failovermethod=priority
baseurl=http://mirrors.aliyun.com/centos/$releasever/os/$basearch/
       http://mirrors.aliyuncs.com/centos/$releasever/os/$basearch/
       http://mirrors.cloud.aliyuncs.com/centos/$releasever/os/$basearch/
gpgcheck=1
gpgkey=http://mirrors.aliyun.com/centos/RPM-GPG-KEY-CentOS-7
……（省略）
```

第8步，使用命令“yum clean all”清理YUM源。

```
[root@localhost yum.repos.d]# yum clean all
已加载插件：fastestmirror, langpacks
正在清理软件源： base extras updates
Cleaning up everything
Cleaning up list of fastest mirrors
```

第9步，使用命令“yum makecache”重建YUM源缓存。

```
[root@localhost yum.repos.d]# yum makecache
已加载插件：fastestmirror, langpacks
Determining fastest mirrors
 * base: mirrors.aliyun.com
 * extras: mirrors.aliyun.com
 * updates: mirrors.aliyun.com
Base                                          | 3.6 kB      00:00:00
Extras                                        | 3.4 kB      00:00:00
Updates                                       | 3.4 kB      00:00:00
(1/12): base/7/x86_64/group_gz                | 166 kB      00:00:00
(2/12): base/7/x86_64/filelists_db            | 6.9 MB      00:00:20
……（省略）
元数据缓存已建立
```

2. 修改Ubuntu操作系统YUM源

第1步，使用命令“ls /etc/apt”查看系统自带的YUM源文件。

```
root@kvm-server:/home/admin# ls /etc/apt/
apt.conf.d  preferences.d  sources.list  sources.list.d  trusted.gpg.d
```

第2步，使用命令“mv /etc/apt/sources.list /tmp”将系统自带的YUM源文件移到tmp目录。

```
root@kvm-server:/home/admin# mv /etc/apt/sources.list /tmp
```

第3步，使用Vi编辑器新建YUM源文件，可以看到访问地址为mirrors.163.com，也就是使用网易YUM源。

```
root@kvm-server:/home/admin# vi /etc/apt/sources.list
deb-src http://mirrors.163.com/Ubuntu/ bionic main restricted universe multiverse
deb-src http://mirrors.163.com/Ubuntu/ bionic-security main restricted universe multiverse
deb-src http://mirrors.163.com/Ubuntu/ bionic-updates main restricted universe multiverse
deb-src http://mirrors.163.com/Ubuntu/ bionic-proposed main restricted universe multiverse
```

第4步，使用命令“apt-get update”更新YUM源。

```
vroot@kvm-server:/home/admin# apt-get update
Get:1 http://mirrors.163.com/Ubuntu bionic InRelease [242 kB]
……
```

```
Get:47 http://mirrors.163.com/Ubuntu bionic-backports/universe Translation-en [1,200 B]
Fetched 28.3 MB in 1min 8s (415 kB/s)
Reading package lists... Done
```

第 5 步，使用命令“apt-get upgrade”更新组件。

```
root@kvm-server:/home/admin# apt-get upgrade
Reading package lists... Done
Building dependency tree
Reading state information... Done
Calculating upgrade... Done
The following packages have been kept back:
  initramfs-tools initramfs-tools-bin initramfs-tools-core linux-generic
  linux-headers-generic linux-image-generic lxd lxd-client netplan.io
The following packages will be upgraded:
  apparmor apport apt apt-utils base-files bind9-host bsdutils cloud-init
  cloud-initramfs-copymods cloud-initramfs-dyn-netconf command-not-found
  command-not-found-data console-setup console-setup-linux cryptsetup cryptsetup-bin curl
……（省略）
152 upgraded, 0 newly installed, 0 to remove and 9 not upgraded.
Need to get 130 MB of archives.
After this operation, 15.5 MB of additional disk space will be used.
Do you want to continue? [Y/n] y
Get:1 http://mirrors.163.com/Ubuntu bionic-updates/main amd64 base-files amd64 10.1Ubuntu2.3 [60.4 kB]
……（省略）
Get:5 http://mirrors.163.com/Ubuntu bionic-updates/main amd64 perl amd64 5.26.1-6Ubuntu0.2 [201 kB]
```

至此，CentOS 操作系统和 Ubuntu 操作系统 YUM 源修改完成，分别使用阿里云 YUM 源和网易 YUM 源。

2.3　常见 Linux 服务器搭建

在企业生产环境中，比较常见的就是 NTP 服务器和 DNS 服务器的搭建，其中 NTP 服务器为生产环境的设备提供同步时间服务，DNS 服务器提供域名解析服务。

2.3.1　搭建 NTP 服务器

网络时间协议（Network Time Protocol，NTP）服务器，也就是日常所说的 NTP 服务器，用来提供同步时间服务。在生产环境中，很多人都会忽略时间问题，实际上服务器、网络设备等，特别是 Linux 操作系统和虚拟化平台的时间不同步会导致很多问题。那么搭建一台 NTP 服务器就非常重要，生产环境中的设备可以直接与 NTP 服务器进行时间同步，NTP 服务器本身也可以访问互连的 NTP 服务器进行同步。NTP 服务器可以是物理服务器，也可以是虚拟机。

生产环境中推荐使用 Linux 操作系统搭建 NTP 服务器，因为基于 Windows 的 NTP 服务器在一些设备上同步可能会出现问题。整体来说，NTP 服务器配置简单，本节介绍如何在 Linux 操作系统下搭建 NTP 服务器。

第 1 步，准备好 Linux 服务器（物理服务器和虚拟机），使用命令“apt-get install ntp”安装 NTP 组件。

```
root@kvm-server:/home/admin# apt-get install ntp
Reading package lists... Done
Building dependency tree
```

```
Reading state information... Done
The following additional packages will be installed:
  libopts25 sntp
Suggested packages:
  ntp-doc
The following NEW packages will be installed:
  libopts25 ntp sntp
0 upgraded, 3 newly installed, 0 to remove and 90 not upgraded.
Need to get 785 kB of archives.
After this operation, 2,393 kB of additional disk space will be used.
Do you want to continue? [Y/n] y
……(省略)
ntp-systemd-netif.service is a disabled or a static unit, not starting it.
Processing triggers for libc-bin (2.27-3ubuntu1) ...
Processing triggers for systemd (237-3ubuntu10.6) ...
Processing triggers for ureadahead (0.100.0-20) ...
……(省略)
```

第2步，推荐使用阿里云NTP服务器，可以同时使用多个NTP服务器地址，使用之前需要确保能够访问外部的NTP服务器。

```
root@kvm-server:/home/admin# ping ntp.aliyun.com
PING ntp.aliyun.com (203.107.6.88) 56(84) bytes of data.
64 bytes from 203.107.6.88 (203.107.6.88): icmp_seq=1 ttl=53 time=49.2 ms
64 bytes from 203.107.6.88 (203.107.6.88): icmp_seq=2 ttl=53 time=47.6 ms

root@kvm-server:/home/admin# ping ntp1.aliyun.com
PING ntp1.aliyun.com (120.25.115.20) 56(84) bytes of data.
64 bytes from 120.25.115.20: icmp_seq=1 ttl=50 time=52.5 ms
64 bytes from 120.25.115.20: icmp_seq=2 ttl=50 time=50.1 ms
```

第3步，使用命令“vi /etc/ntp.conf”修改NTP配置文件，将原配置文件的NTP服务器地址取消，添加阿里云NTP服务器地址。

```
root@kvm-server:/home/admin# vi /etc/ntp.conf
……(省略)
#pool 0.ubuntu.pool.ntp.org iburst    #取消原NTP服务器
#pool 1.ubuntu.pool.ntp.org iburst    #取消原NTP服务器
#pool 2.ubuntu.pool.ntp.org iburst    #取消原NTP服务器
#pool 3.ubuntu.pool.ntp.org iburst    #取消原NTP服务器
# Use Ubuntu's ntp server as a fallback.
#pool ntp.ubuntu.com                  #取消原NTP服务器
pool ntp.aliyun.com                   #添加阿里云NTP服务器
……(省略)
```

第4步，使用命令“service ntp start”启动NTP服务器。

```
root@kvm-server:/home/admin# service ntp start
```

第5步，使用命令“ntpq -p”查看2个NTP源，203.107.6.88是阿里云NTP服务器的IP地址。

```
root@kvm-server:/home/admin# ntpq -p
    remote        refid          st  t  when  poll  reach  delay   offset  jitter
==============================================================================
ntp.aliyun.com  .POOL.          16  p   -     64     0    0.000   0.000   0.000
*203.107.6.88   10.137.55.181  2   u  40     64    37    48.040  -1.717   8.569
```

参数解释如下。

1）remote：表示NTP服务器的IP地址或域名。

2）refid：表示上一层NTP服务器的地址，和DNS服务器的结构类似，一层一层递归，当 remote 处已经是根 NTP 服务器的 IP 地址或域名的时候，此处就不会显示 IP 地址或域名。

3）st：远程服务器的级别。当此值为 1 时表示为根 NTP 服务器。

4）t：服务器类型。u 代表使用单播。

5）when：距上次同步的时间，单位为 s。

6）poll：同步间隔时长，单位为 s，默认为 128。

7）reach：已同步次数。

8）delay：延迟时长，单位为 ms。

9）offset：时间偏差，显示为负值表示负偏差，单位为 ms。

10）jitter：与远程 NTP 服务器的平均时间偏差，单位为 ms。

第 6 步，选择一台 Linux 主机作为客户端，使用命令“date”查看时间。

```
[root@host11 ~]# date
2018年 12月 20日 星期四 19:23:32 CST
```

第 7 步，使用命令“ntpdate –d 10.92.10.8”进行同步。

```
[root@host11 ~]# ntpdate -d 10.92.10.8
20 Dec 19:23:36 ntpdate[9376]: ntpdate 4.2.6p5@1.2349-o Fri Apr 13 12:52:28 UTC 2018 (1)
Looking for host 10.92.10.8 and service ntp
host found : 10.92.10.8
transmit    (10.92.10.8)
receive     (10.92.10.8)
transmit    (10.92.10.8)
receive     (10.92.10.8)
transmit    (10.92.10.8)
receive     (10.92.10.8)
transmit    (10.92.10.8)
receive     (10.92.10.8)
server       10.92.10.8, port 123
stratum 3, precision -23, leap 00, trust 000
refid [10.92.10.8], delay 0.02589, dispersion 0.00005
transmitted 4, in filter 4
reference time:        dfc5fa1d.8e0690da  Thu, Dec 20 2018 19:23:09.554
originate timestamp:   dfc5fa3e.a1aebb41  Thu, Dec 20 2018 19:23:42.631
transmit timestamp:    dfc5fa3e.9734d90b  Thu, Dec 20 2018 19:23:42.590
filter delay:          0.02591   0.02589   0.02589   0.02589
                       0.00000   0.00000   0.00000   0.00000
filter offset:         0.040594  0.040614  0.040665  0.040701
                       0.000000  0.000000  0.000000  0.000000
delay 0.02589, dispersion 0.00005
offset 0.040614
20 Dec 19:23:42 ntpdate[9376]: adjust time server 10.92.10.8 offset 0.040614 sec
```

第 8 步，使用命令“hwclock -w”将系统时间同步到硬件，防止系统重启后时间不准确。

```
[root@host11 ~]# hwclock -w
[root@host11 ~]# date
2018年 12月 20日 星期四 19:28:12 CST
```

第 9 步，使用 Linux 操作系统搭建的 NTP 服务器也支持 Windows 操作系统同步，选择一台 Windows 主机，Windows 时间服务器使用 time.windows.com 且服务未运行，如图 2-3-1 所示，单击“更改设置”按钮。

第 10 步，修改 NTP 服务器地址为自建 NTP 服务器地址，如图 2-3-2 所示，单击“立即更新”按钮。

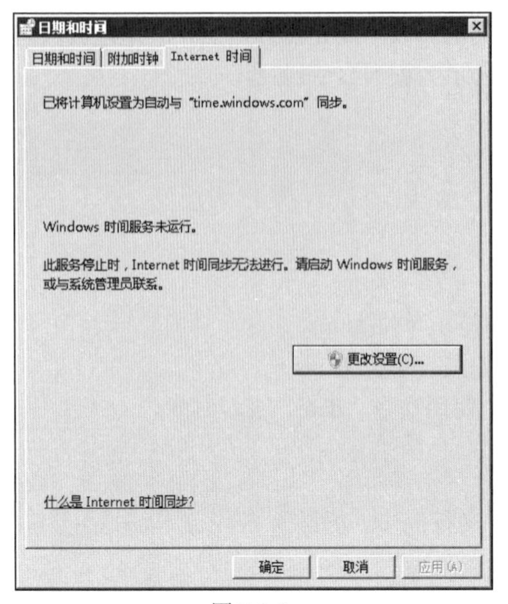

图 2-3-1

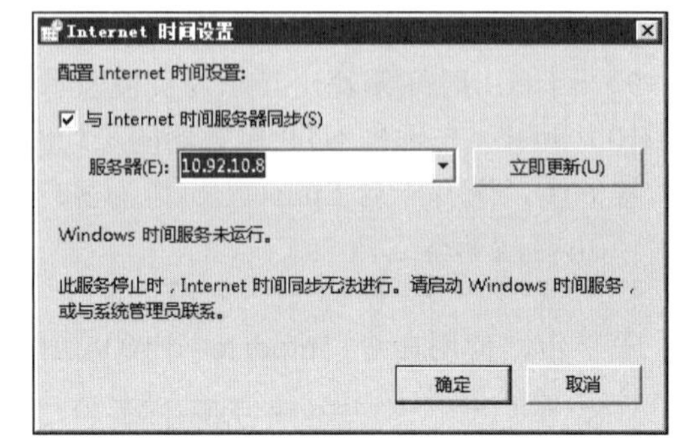

图 2-3-2

第 11 步，Windows 主机时间与自建 NTP 服务器同步成功，如图 2-3-3 所示，单击“确定”按钮。

第 12 步，Windows 主机时间服务器配置完成且服务成功启动，如图 2-3-4 所示。

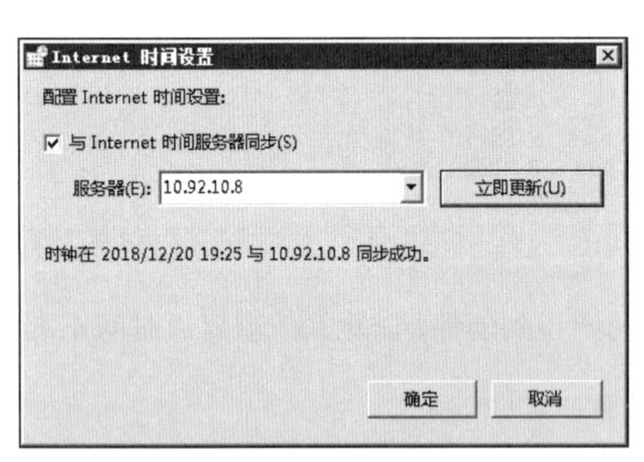

图 2-3-3

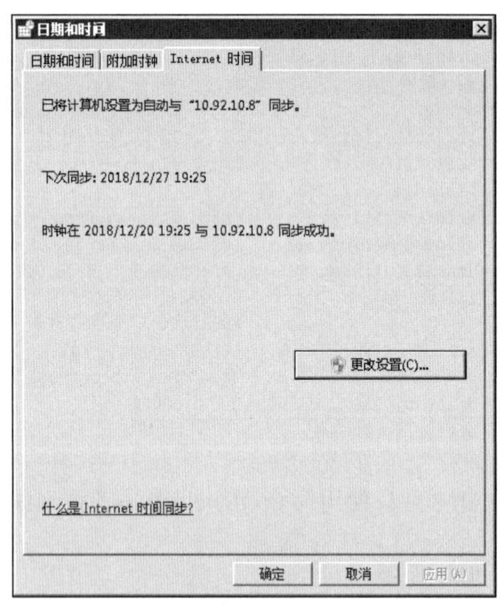

图 2-3-4

至此，NTP 服务器搭建完成，需要注意的是，当 NTP 服务器完成同步后，客户端要添加相应的 NTP 服务器地址进行同步。

2.3.2 搭建 DNS 服务器

在生产环境中，内部 DNS 服务器也是基础服务器之一。特别是在服务器较多的环境中，由于技术人员不可能记住每台服务器的 IP 地址，因此，在生产环境部署 DNS 服务器也是技术人员常见的工作之一。本节介绍使用 Linux 操作系统搭建 DNS 服务器。

第 1 步，使用命令“yum install bind”安装 DNS 组件。

```
[root@localhost ~]# yum install bind
Loaded plugins: fastestmirror
Base                            | 3.6 kB  00:00:00
centos-sclo-rh-release          | 3.0 kB  00:00:00};
……（省略）
Dependencies Resolved
==============================================================================
Package              Arch            Version             Repository      Size
==============================================================================
Installing:
 bind                x86_64          32:9.9.4-72.el7     base            1.8 M
Installing for dependencies:
 audit-libs-python   x86_64          2.8.4-4.el7         base            76 k
Transaction Summary
==============================================================================
Install  1 Package       (+ 9 Dependent packages)
Upgrade                  ( 10 Dependent packages)

Total download size: 7.6 M
Is this ok [y/d/N]: y
Downloading packages:
Delta RPMs disabled because /usr/bin/applydeltarpm not installed.
(1/20): audit-libs-python-2.8.4-4.el7.x86_64.rpm       |  76 kB    00:00:00
(2/20): audit-libs-2.8.4-4.el7.x86_64.rpm              |  100 kB   00:00:02
……（省略）
Complete!
```

第 2 步，使用命令“vi /etc/named.conf”编辑 DNS 服务器文件。

```
[root@localhost ~]# vi /etc/named.conf
options {
       #listen-on port 53 { 127.0.0.1; };   #注释掉仅允许本机使用端口
       #listen-on-v6 port 53 { ::1; };      #注释掉仅允许本机使用端口
       Directory              "/var/named";
       dump-file              "/var/named/data/cache_dump.db";
       statistics-file        "/var/named/data/named_stats.txt";
       memstatistics-file     "/var/named/data/named_mem_stats.txt";
       recursing-file         "/var/named/data/named.recursing";
       secroots-file          "/var/named/data/named.secroots";
       #allow-query     { localhost; };   #注释掉仅允许本机查询
```

第 3 步，使用命令“vi /etc/named.rfc1912.zones”编辑 zones 文件，新增 bdnetlab.cn 本地域。

```
[root@localhost ~]# vi /etc/named.rfc1912.zones
zone "bdnetlab.cn" IN {
       type master;
       file "bdnetlab.cn.zone";
};
```

第 4 步，使用命令“vi /var/named/bdnetlab.cn.zone”创建 bdnetlab.cn.zone 文件，配置域名与 IP 地址对应。

```
[root@localhost ~]# vi /var/named/bdnetlab.cn.zone
$TTL 3600
@      IN SOA ns1.bdnetlab.cn.  dnsadmin.bdnetlab.cn. (
```

```
                        2046     ; serial
                        1D       ; refresh
                        1H       ; retry
                        1W       ; expire
                        3H )     ; minimum

        NS     ns1
ns1     A      10.92.10.171
www     A      10.92.10.11
edu     A      10.92.10.8
```

第 5 步，使用命令“named-checkzone bdnetlab.cn /var/named/bdnetlab.cn.zone”校验配置文件是否存在问题，注意如果出现错误提示，请检查配置文件。

```
[root@localhost ~]# named-checkzone bdnetlab.cn /var/named/bdnetlab.cn.zone
zone bdnetlab.cn/IN: loaded serial 2046 OK
```

第 6 步，使用命令“vi /etc/sysconfig/network-scripts/ifcfg-eth0”修改本地 DNS 服务器地址。

```
[root@localhost ~]# vi /etc/sysconfig/network-scripts/ifcfg-eth0
TYPE=Ethernet
PROXY_METHOD=none
BROWSER_ONLY=no
DEFROUTE=yes
IPV4_FAILURE_FATAL=no
NAME=eth0
UUID=09aedc75-3451-46c6-b477-50bcd7e7e3f7
DEVICE=eth0
ONBOOT=yes
IPV6_PRIVACY=no
IPADDR=10.92.10.171
NETMASK=255.255.255.0
GATEWAY=10.92.10.254
DNS1=10.92.10.171    #使用内部 DNS 服务器地址
```

第 7 步，使用命令“systemctl start named.service”启动 DNS 服务器。

```
[root@localhost ~]# systemctl start named.service
```

第 8 步，使用命令“systemctl status named”查看 DNS 服务器状态。

```
[root@localhost ~]# systemctl status named
 named.service - Berkeley Internet Name Domain (DNS)
   Loaded: loaded (/usr/lib/systemd/system/named.service; enabled; vendor preset: disabled)
   Active: active (running) since Sun 2018-12-23 15:20:39 CST; 8min ago
  Process: 1006 ExecStart=/usr/sbin/named -u named -c ${NAMEDCONF} $OPTIONS (code=exited, status=0/SUCCESS)
  Process: 875 ExecStartPre=/bin/bash -c if [ ! "$DISABLE_ZONE_CHECKING" == "yes" ]; then /usr/sbin/named-checkconf -z "$NAMEDCONF"; else echo "Checking of zone files is disabled"; fi (code=exited, status=0/SUCCESS)
 Main PID: 1025 (named)
   CGroup: /system.slice/named.service
           └─1025 /usr/sbin/named -u named -c /etc/named.conf
……（省略）
```

第 9 步，使用“ping”命令解析 www.bdnetlab.cn 和 edu.bdnetlab.cn 两个域名，查看是否解析到内部网络。

```
[root@host11 ~]# ping www.bdnetlab.cn
PING www.bdnetlab.cn (10.92.10.11) 56(84) bytes of data.
64 bytes from host11 (10.92.10.11): icmp_seq=1 ttl=64 time=0.042 ms
```

```
64 bytes from host11 (10.92.10.11): icmp_seq=2 ttl=64 time=0.035 ms
[root@host11 ~]# ping edu.bdnetlab.cn
PING edu.bdnetlab.cn (10.92.10.8) 56(84) bytes of data.
64 bytes from 10.92.10.8 (10.92.10.8): icmp_seq=1 ttl=64 time=0.241 ms
64 bytes from 10.92.10.8 (10.92.10.8): icmp_seq=2 ttl=64 time=0.223 ms
```

至此，基本的 DNS 服务器配置完成。本章介绍的是基础配置，如果需要对 DNS 服务器进行复杂配置，可参考相关文档。

2.3.3 搭建 HTTP 服务器

在生产环境中，HTTP 服务器也是基本服务器之一，比较常见的是基于 Apache 和 Nginx 进行搭建，生产环境一般不会单独使用。常见的 HTTP 服务架构有 LAMP（Linux+Apache+MySQL+PHP）和 LNMP（Linux+Nginx+MySQL+PHP），架构的搭建不在本书的范围，本节介绍基于 Apache 的 HTTP 服务器的搭建。

第 1 步，使用命令"yum install httpd"安装 Apache 服务。

```
[root@localhost ~]# yum install httpd
Loaded plugins: fastestmirror
Loading mirror speeds from cached hostfile
 * base: mirrors.cn99.com
 * extras: mirrors.163.com
……（省略）
Dependencies Resolved
==============================================================================
 Package    Arch       Version                  Repository          Size
==============================================================================
Installing:
 httpd      x86_64     2.4.6-88.el7.centos      base                2.7 M
……（省略）
Transaction Summary
==============================================================================
Install  1 Package (+4 Dependent packages)

Total download size: 3.0 M
Installed size: 10 M
Is this ok [y/d/N]: y
Downloading packages:
(1/5): apr-util-1.5.2-6.el7.x86_64.rpm        | 92 kB  00:00:00
(2/5): mailcap-2.1.41-2.el7.noarch.rpm        | 31 kB  00:00:00
……（省略）
Installed:
  httpd.x86_64 0:2.4.6-88.el7.centos

Dependency Installed:
  apr.x86_64     0:1.4.8-3.el7_4.1        apr-util.x86_64     0:1.5.2-6.el7     httpd-tools.x86_64 
0:2.4.6-88.el7.centos
  mailcap.noarch 0:2.1.41-2.el7
Complete!
```

第 2 步，使用命令"systemctl start httpd"启动 Apache 服务。

```
[root@localhost ~]# systemctl start httpd
```

第 3 步，使用浏览器测试 Apache 服务是否启动成功，如果出现"Testing 123..."界面说明服务器启动成功，如图 2-3-5 所示。

第 4 步，使用命令"vi /var/www/html/index.html"编辑 HTTP 服务器首页。

```
[root@localhost ~]# vi /var/www/html/index.html
Welcome to www.bdnetlab.cn
```

第 5 步，刷新浏览器查看首页是否正常，如图 2-3-6 所示。

第 6 步，修改内部 DNS 服务器解析，使用域名访问 HTTP 服务器，如图 2-3-7 所示。

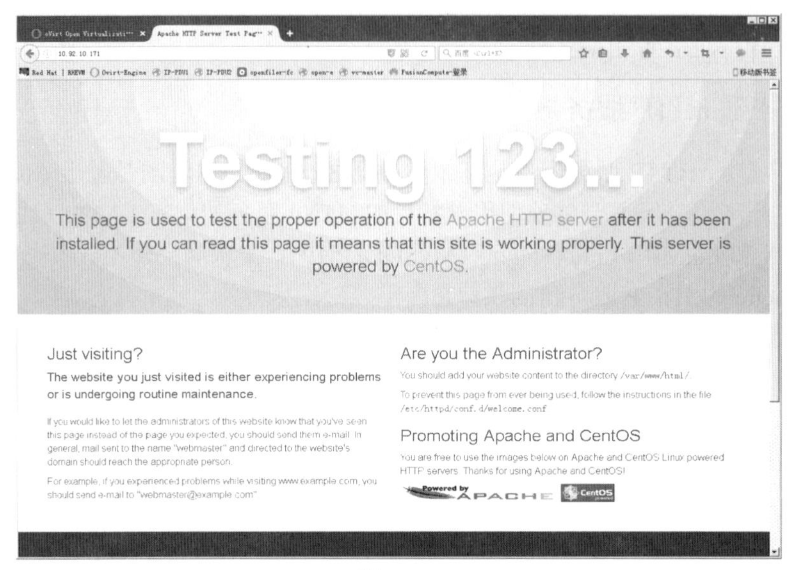

图 2-3-5

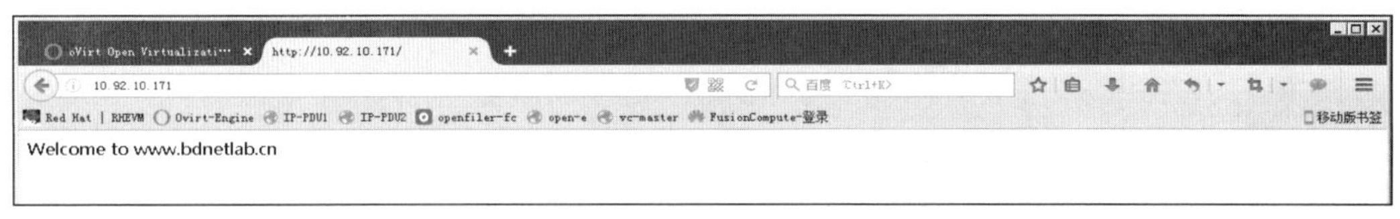

图 2-3-6

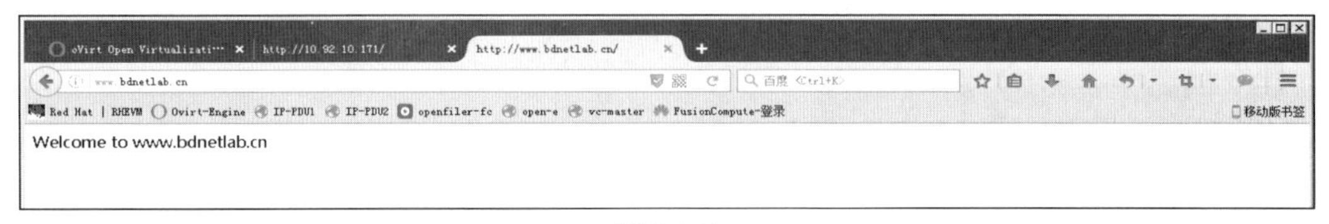

图 2-3-7

至此，基于 Apache 的 HTTP 服务器搭建完成。与 DNS 服务器配置的介绍一样，本章介绍的是最基本的配置，读者如果需要进行复杂的架构配置，可参考相关文档。

2.4　本章小结

本章对实验环境进行了详细的介绍，一个好的实验环境对于学习非常重要。同时介绍了如何通过服务器远程管理工具远程安装 Linux 操作系统和常见 Linux 服务器的搭建。特别需要注意的是，如果读者没有 Linux 基础知识，推荐读者学习一些 Linux 基础知识以满足后续学习的需要。

第 3 章　部署使用 KVM 虚拟化

通过安装 Linux 操作系统和对常用命令的学习，相信读者对 Linux 操作系统有了一些基本的认识。当然基础知识不仅仅是这些，读者还需要持续学习，理解并掌握基础知识，这对今后的学习和运维会有非常大的帮助。看到本章标题相信不少读者会有些害怕，其实可以负责任地告诉大家，使用 YUM 部署 KVM 并不复杂，因为在使用命令后系统会自动查找所依赖的软件包进行安装。难点在于后续虚拟机的创建、管理等其他方面。本章介绍如何在 CentOS 和 Ubuntu 操作系统上部署 KVM，以及虚拟机的创建和使用。

本章要点

- 在 Linux 操作系统上部署 KVM。
- 使用命令行部署 Linux 虚拟机。
- 使用命令行部署 Windows 虚拟机。
- 使用 GUI 部署虚拟机。
- 使用模板部署虚拟机。
- 虚拟机硬盘格式。
- 虚拟机网络架构。
- 虚拟机日常操作。

3.1　在 Linux 操作系统上部署 KVM

由于 KVM 的特殊性，不建议使用模拟环境进行部署，推荐使用物理服务器部署 Linux 系统，而后再部署 KVM。

3.1.1　在 CentOS 操作系统上部署 KVM

第 1 步，正式部署前需要检查 CPU 是否开启了硬件虚拟化支持。使用命令“cat /proc/cpuinfo”，若在 flags 输出信息中有 vmx 和 smx 等，说明 CPU 开启了硬件虚拟化支持。需要注意的是，某些服务器的 BIOS 默认未开启硬件虚拟化支持，需要手动开启。

```
[root@localhost yum.repos.d]# cat /proc/cpuinfo
Processor       : 0
vendor_id       : GenuineIntel
cpu family      : 6
model           : 44
model nam       : Intel(R) Xeon(R) CPU   L5640  @ 2.27GHz
stepping        : 2
microcode       : 0x14
cpu MHz         : 1600.000
cache size      : 12288 KB
physical id     : 1
siblings        : 12
```

```
core id         : 0
cpu cores       : 6
apicid          : 32
initial apicid  : 32
fpu             : yes
fpu_exception   : yes
cpuid level     : 11
wp              : yes
flags           : fpu vme de pse tsc msr pae mce cx8 apic sep mtrr pge mca cmov pat pse36 clflush dts
acpi mmx fxsr sse sse2 ss ht tm pbe syscall nx pdpe1gb rdtscp lm constant_tsc arch_perfmon pebs bts rep_good
nopl xtopology nonstop_tsc aperfmperf pni pclmulqdq dtes64 monitor ds_cpl vmx smx est tm2 ssse3 cx16 xtpr
pdcm pcid dca sse4_1 sse4_2 popcnt aes lahf_lm epb tpr_shadow vnmi flexpriority ept vpid dtherm ida arat
bogomips         : 4533.58
clflush size     : 64
cache_alignment  : 64
address sizes    : 40 bits physical, 48 bits virtual
```

第2步，使用命令“yum list qemu-kvm”检查qemu-kvm软件包信息，系统通过YUM源找到相应的软件包。如果YUM源存在问题，在线安装将无法进行。

```
[root@localhost yum.repos.d]# yum list qemu-kvm
已加载插件：fastestmirror, langpacks
Loading mirror speeds from cached hostfile
 * base: mirrors.aliyun.com
 * extras: mirrors.aliyun.com
 * updates: mirrors.aliyun.com
已安装的软件包
qemu-kvm.x86_64       10:1.5.3-156.el7       @anaconda
可安装的软件包
qemu-kvm.x86_64       10:1.5.3-156.el7_5.5    updates
```

第3步，使用命令“yum list libvirt”检查管理工具libvirt软件包信息，系统通过YUM源找到相应的软件包。如果YUM源存在问题，在线安装将无法进行。

```
[root@localhost yum.repos.d]# yum list libvirt
已加载插件：fastestmirror, langpacks
Loading mirror speeds from cached hostfile
 * base: mirrors.aliyun.com
 * extras: mirrors.aliyun.com
 * updates: mirrors.aliyun.com
可安装的软件包
libvirt.x86_64          3.9.0-14.el7_5.8           updates
```

第4步，使用命令“yum install qemu-kvm”安装qemu-kvm组件，注意需要安装的组件数量较大，下载安装过程中不能出现任何错误提示。

```
[root@localhost yum.repos.d]# yum install qemu-kvm
已加载插件：fastestmirror, langpacks
Loading mirror speeds from cached hostfile
 * base: mirrors.aliyun.com
 * extras: mirrors.aliyun.com
 * updates: mirrors.aliyun.com
正在解决依赖关系
--> 正在检查事务
---> 软件包 qemu-kvm.x86_64.10.1.5.3-156.el7             将被升级
---> 软件包 qemu-kvm.x86_64.10.1.5.3-156.el7_5.5         将被更新
--> 正在处理依赖关系 qemu-kvm-common = 10:1.5.3-156.el7_5.5，它被软件包 10:qemu-kvm-1.5.3-156.
el7_5.5.x86_64         需要
--> 正在处理依赖关系 qemu-img = 10:1.5.3-156.el7_5.5，它被软件包 10:qemu-kvm-1.5.3-156.el7_5.5.
x86_64                 需要
--> 正在检查事务
```

```
---> 软件包 qemu-img.x86_64.10.1.5.3-156.el7                将被升级
---> 软件包 qemu-img.x86_64.10.1.5.3-156.el7_5.5            将被更新
---> 软件包 qemu-kvm-common.x86_64.10.1.5.3-156.el7         将被升级
---> 软件包 qemu-kvm-common.x86_64.10.1.5.3-156.el7_5.5     将被更新
--> 解决依赖关系完成
依赖关系解决
================================================================================
 Package              架构         版本                 源            大小
================================================================================
正在更新:
 qemu-kvm             x86_64    10:1.5.3-156.el7_5.5    updates       1.9 M
为依赖而更新:
 qemu-img             x86_64    10:1.5.3-156.el7_5.5    updates       693 k
 qemu-kvm-common      x86_64    10:1.5.3-156.el7_5.5    updates       430 k
事务概要
================================================================================
升级  1 软件包 (+2 依赖软件包)

总下载量: 3.0 M
Is this ok [y/d/N]: y
Downloading packages:
Delta RPMs reduced 430 k of updates to 277 k (35% saved)
(1/3): qemu-kvm-common-1.5.3-156.el7_1.5.3-156.el7_5.5.x86_64.drpm   | 277 kB    00:00:02
警告: /var/cache/yum/x86_64/7/updates/packages/qemu-img-1.5.3-156.el7_5.5.x86_64.rpm: 头 V3
RSA/SHA256 Signature, 密钥 ID f4a80eb5: NOKEY
qemu-img-1.5.3-156.el7_5.5.x86_64.rpm 的公钥尚未安装
(2/3): qemu-img-1.5.3-156.el7_5.5.x86_64.rpm                          | 693 kB    00:00:06
(3/3): qemu-kvm-1.5.3-156.el7_5.5.x86_64.rpm                          | 1.9 MB    00:00:06
----------------------------------------------------------------------------------------------
总计                                                     348 kB/s     | 2.9 MB    00:00:08
从 http://mirrors.aliyun.com/centos/RPM-GPG-KEY-CentOS-7 检索密钥
导入 GPG key 0xF4A80EB5:
 用户 ID     : "CentOS-7 Key (CentOS 7 Official Signing Key) <security@centos.org>"
 指纹        : 6341 ab27 53d7 8a78 a7c2 7bb1 24c6 a8a7 f4a8 0eb5
 来自        : http://mirrors.aliyun.com/centos/RPM-GPG-KEY-CentOS-7
是否继续? [y/N]: y
Running transaction check
Running transaction test
Transaction test succeeded
Running transaction
  正在更新    : 10:qemu-img-1.5.3-156.el7_5.5.x86_64                1/6
  正在更新    : 10:qemu-kvm-common-1.5.3-156.el7_5.5.x86_64         2/6
  正在更新    : 10:qemu-kvm-1.5.3-156.el7_5.5.x86_64                3/6
  清理        : 10:qemu-kvm-1.5.3-156.el7.x86_64                    4/6
  清理        : 10:qemu-img-1.5.3-156.el7.x86_64                    5/6
  清理        : 10:qemu-kvm-common-1.5.3-156.el7.x86_64             6/6
  验证中      : 10:qemu-kvm-1.5.3-156.el7_5.5.x86_64                1/6
  验证中      : 10:qemu-kvm-common-1.5.3-156.el7_5.5.x86_64         2/6
  验证中      : 10:qemu-img-1.5.3-156.el7_5.5.x86_64                3/6
  验证中      : 10:qemu-kvm-1.5.3-156.el7.x86_64                    4/6
  验证中      : 10:qemu-kvm-common-1.5.3-156.el7.x86_64             5/6
  验证中      : 10:qemu-img-1.5.3-156.el7.x86_64                    6/6
更新完毕:
  qemu-kvm.x86_64 10:1.5.3-156.el7_5.5
作为依赖被升级:
  qemu-img.x86_64 10:1.5.3-156.el7_5.5      qemu-kvm-common.x86_64 10:1.5.3-156.el7_5.5
完毕!
```

第 5 步，使用命令“yum install libvirt virtinst virt-manager”安装管理工具。

```
[root@localhost yum.repos.d]# yum install libvirt virtinst virt-manager
已加载插件: fastestmirror, langpacks
```

```
Loading mirror speeds from cached hostfile
 * base: mirrors.aliyun.com
 * extras: mirrors.aliyun.com
 * updates: mirrors.aliyun.com
正在解决依赖关系
--> 正在检查事务
---> 软件包 libvirt.x86_64.0.3.9.0-14.el7_5.8       将被安装
……（省略）
--> 解决依赖关系完成
依赖关系解决
=============================================================================
 Package            架构          版本                   源              大小
=============================================================================
正在安装:
 Libvirt            86_64         3.9.0-14.el7_5.8       updates         174 k
 virt-manager       noarch        1.4.3-3.el7            base            652 k
为依赖而安装:
 autogen-libopts    x86_64        5.18-5.el7             base            66 k
 gnutls-dane        x86_64        3.3.26-9.el7           base            34 k
……（省略）
为依赖而更新:
 libvirt-daemon     x86_64        3.9.0-14.el7_5.8       updates         852 k
……（省略）
事务概要
=============================================================================
安装  2 软件包 (+11 依赖软件包)
升级          ( 19 依赖软件包)
总下载量: 13 M
Is this ok [y/d/N]: y
Downloading packages:
Delta RPMs reduced 5.1 M of updates to 908 k (82% saved)
(1/32): libvirt-daemon-driver-nwfilter-3.9.0-14.el7_3.9.0-14.el7_5.8.x8    | 178 kB  00:00:02
 ……（省略）
 (32/32): virt-manager-common-1.4.3-3.el7.noarch.rpm                       | 1.2 MB  00:00:03
-----------------------------------------------------------------------------------------------
总计                                                              514 kB/s  | 9.3 MB  00:00:18
Running transaction check
Running transaction test
Transaction test succeeded
Running transaction
  正在更新    : libvirt-libs-3.9.0-14.el7_5.8.x86_64                1/51
……（省略）
已安装:
  libvirt.x86_64 0:3.9.0-14.el7_5.8            virt-manager.noarch 0:1.4.3-3.el7
……（省略）
完毕!
```

第6步，使用命令“qemu-img”查看安装是否正确，出现参数提示说明安装正常。

```
[root@localhost /]# qemu-img
qemu-img version 1.5.3, Copyright (c) 2004-2008 Fabrice Bellard
usage: qemu-img command [command options]
QEMU disk image utility

Command syntax:
  check       [-q] [-f fmt] [--output=ofmt] [-r [leaks | all]] [-T src_cache] filename
  create      [-q] [-f fmt] [-o options] filename [size]
  commit      [-q] [-f fmt] [-t cache] filename
  compare     [-f fmt] [-F fmt] [-T src_cache] [-p] [-q] [-s] filename1 filename2
  convert     [-c] [-p] [-q] [-n] [-f fmt] [-t cache] [-T src_cache] [-O output_fmt] [-o options] [-s
snapshot_name] [-S sparse_size] filename [filename2 [...]] output_filename
```

```
    info        [-f fmt] [--output=ofmt] [--backing-chain] filename
    map         [-f fmt] [--output=ofmt] filename
    snapshot    [-q] [-l | -a snapshot | -c snapshot | -d snapshot] filename
    rebase      [-q] [-f fmt] [-t cache] [-T src_cache] [-p] [-u] -b backing_file [-F backing_fmt] filename
    resize      [-q] filename [+ | -]size
    amend       [-q] [-f fmt] [-t cache] -o options filename
```

3.1.2 在 Ubuntu 操作系统上部署 KVM

第 1 步，与 CentOS 操作系统一样，Ubuntu 操作系统也需要检查 CPU 是否开启了硬件虚拟化支持。

```
root@kvm-server:/home/admin# cat /proc/cpuinfo
processor           : 0
vendor_id           : GenuineIntel
cpu family          : 6
model               : 44
model name          : Intel(R) Xeon(R) CPU   L5640  @ 2.27GHz
stepping            : 2
microcode           : 0x14
cpu MHz             : 1599.900
cache size          : 12288 KB
physical id         : 1
siblings            : 12
core id             : 0
cpu cores           : 6
apicid              : 32
initial apicid      : 32
fpu                 : yes
fpu_exception       : yes
cpuid level         : 11
wp                  : yes
flags               : fpu vme de pse tsc msr pae mce cx8 apic sep mtrr pge mca cmov pat pse36 clflush
dts acpi mmx fxsr sse sse2 ss ht tm pbe syscall nx pdpe1gb rdtscp lm constant_tsc arch_perfmon pebs bts
rep_good nopl xtopology nonstop_tsc cpuid aperfmperf pni pclmulqdq dtes64 monitor ds_cpl vmx smx est tm2
ssse3 cx16 xtpr pdcm pcid dca sse4_1 sse4_2 popcnt aes lahf_lm epb pti tpr_shadow vnmi flexpriority ept
vpid dtherm ida arat
bugs                : cpu_meltdown spectre_v1 spectre_v2 spec_store_bypass l1tf
bogomips            : 4533.89
clflush size        : 64
cache_alignment     : 64
address sizes       : 40 bits physical, 48 bits virtual
```

第 2 步，使用命令“apt-get install qemu-kvm”安装 qemu-kvm 组件，注意 Ubuntu 操作系统使用的命令与 CentOS 操作系统的不同。

```
root@kvm-server:/home/admin# apt-get install qemu-kvm
Reading package lists... Done
Building dependency tree
Reading state information... Done
The following additional packages will be installed:
  cpu-checker dconf-gsettings-backend dconf-service fontconfig fontconfig-config
……（省略）
Suggested packages:
  gvfs libasound2-plugins alsa-utils libdv-bin oss-compat libvisual-0.4-plugins
  gstreamer1.0-tools jackd2 opus-tools pulseaudio libraw1394-doc speex samba vde2 sgabios
  ovmf debootstrap sharutils-doc bsd-mailx | mailx
The following NEW packages will be installed:
  cpu-checker dconf-gsettings-backend dconf-service fontconfig fontconfig-config
  fonts-dejavu-core glib-networking glib-networking-common glib-networking-services
```

```
……(省略)
0 upgraded, 108 newly installed, 0 to remove and 9 not upgraded.
Need to get 27.9 MB of archives.
After this operation, 107 MB of additional disk space will be used.
Do you want to continue? [Y/n] y  #输入y确认安装
Get:1 http://mirrors.163.com/ubuntu bionic/main amd64 fonts-dejavu-core all 2.37-1 [1,041 kB]
……(省略)
Setting up gstreamer1.0-x:amd64 (1.14.1-1ubuntu1~ubuntu18.04.1) ...
Processing triggers for libc-bin (2.27-3ubuntu1) ...
```

第3步，使用命令“apt-get install libvirt-bin virtinst virt-manager”安装管理工具。

```
root@kvm-server:/home/admin# apt-get install libvirt-bin virtinst virt-manager
Reading package lists... Done
Building dependency tree
Reading state information... Done
The following additional packages will be installed:
  adwaita-icon-theme at-spi2-core augeas-lenses bridge-utils genisoimage
  gir1.2-appindicator3-0.1 gir1.2-atk-1.0 gir1.2-freedesktop gir1.2-gdkpixbuf-2.0
……(省略)
0 upgraded, 129 newly installed, 0 to remove and 9 not upgraded.
Need to get 23.9 MB of archives.
After this operation, 119 MB of additional disk space will be used.
Do you want to continue? [Y/n] y  #输入y确认安装
Get:1 http://mirrors.163.com/ubuntu bionic/main amd64 libdbusmenu-glib4 amd64 16.04.1+18.04.20171206-
0ubuntu1 [41.4 kB]
Get:2 http://mirrors.163.com/ubuntu bionic/main amd64 libatk1.0-data all 2.28.1-1 [2,992 B]
……(省略)
Processing triggers for libgdk-pixbuf2.0-0:amd64 (2.36.11-2) ...
```

至此，CentOS和Ubuntu操作系统的KVM部署完成。通过命令的部署并不复杂，主要是确保YUM源能正常访问，同时确保下载过程中不要出现问题，否则也可能导致部署出现问题。

3.2 使用命令行部署虚拟机

在正式创建虚拟机前还需要做一些准备工作，如需要把Linux的ISO安装文件上传到物理服务器、创建虚拟机硬盘等。当然还要安装虚拟网络控制台（Virtual Network Console，VNC）客户端工具，用于连接到KVM虚拟机进行操作。本节实战操作将创建两台Linux虚拟机，其中一台使用纯命令方式创建，另一台配合使用VNC方式创建。

3.2.1 使用纯命令安装Linux虚拟机

第1步，使用WinSCP工具或其他工具连接到KVM主机，如图3-2-1所示。

第2步，上传CentOS-7-x86_64-Minimal-1708.iso文件到vm-iso目录，如图3-2-2所示。

第3步，使用命令“chmod”修改vm-iso目录权限，如果不修改目录权限，后续在安装虚拟机过程中挂载ISO文件可能会出现问题。

```
[root@localhost /]# chmod 777 vm-iso
```

第4步，使用命令“qemu-img”创建虚拟机硬盘文件并查看硬盘信息，如果未创建磁盘文件，在安装虚拟机过程中可以创建。

```
[root@localhost vm-centos7]# qemu-img create -f raw centos7.raw 10G
Formatting 'centos7.raw', fmt=raw size=10737418240
[root@localhost vm-centos7]# qemu-img info centos7.raw
```

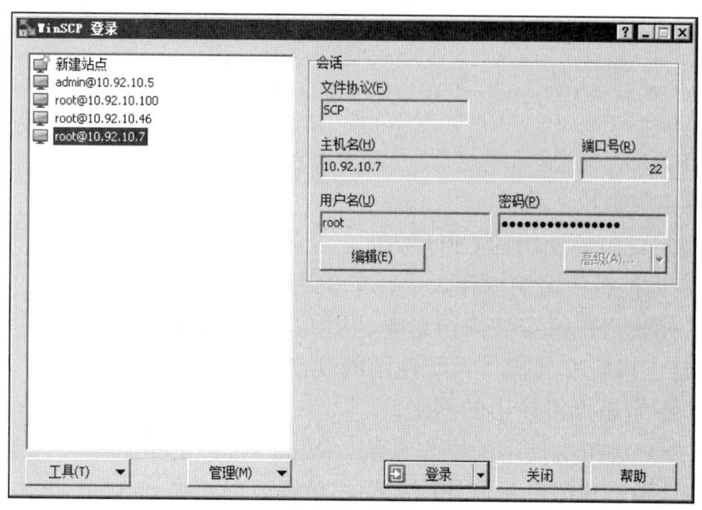

图 3-2-1

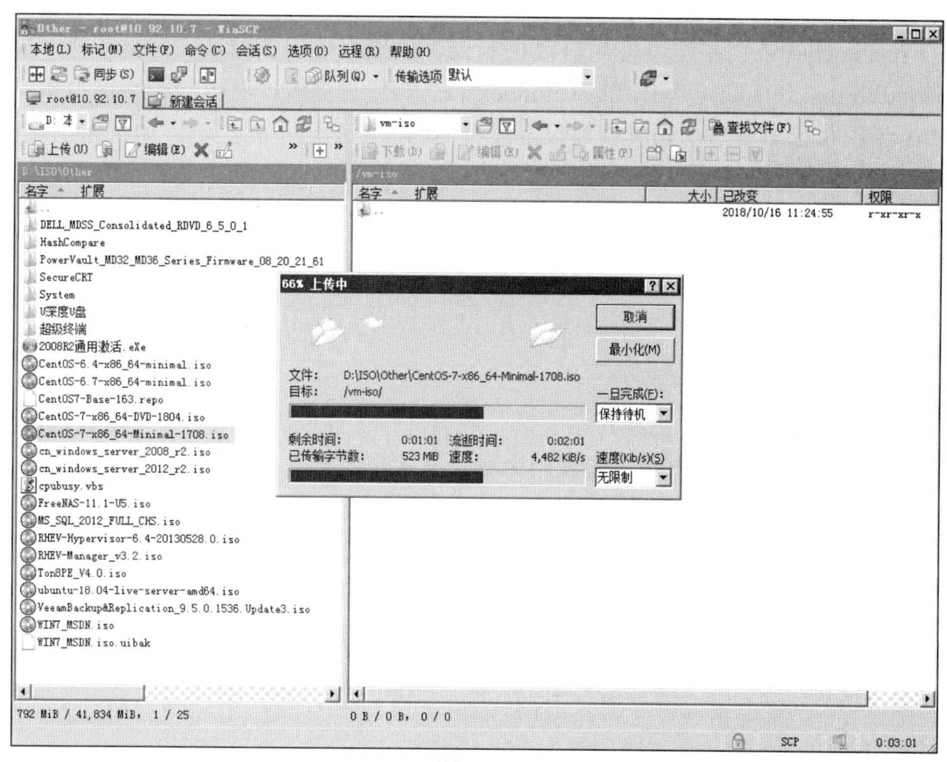

图 3-2-2

```
image: centos7.raw
file format: raw
virtual size: 10G (10737418240 bytes)
disk size: 0
```

参数解释如下。

1）raw：虚拟机硬盘使用的格式。

2）10G：虚拟机硬盘使用的容量。

第5步，使用命令“virt-install”创建虚拟机并安装Linux操作系统。

```
[root@localhost vm-centos7]# virt-install  --name centos7 --vcpus 1 --ram 2048 --location=/vm-iso/CentOS-7-x86_64-Minimal-1708.iso --disk path=/vm-centos7/centos7.raw,size=10,format=raw  --network bridge=virbr0 --os-type=linux --os-variant=rhel7 --extra-args='console=ttyS0' --noautoconsole --force
```

参数解释如下。

1）name：指定虚拟机名称。

2）vcpus：指定虚拟机使用的CPU数量。

3）ram：指定虚拟机使用的内存大小，单位为MB。

4）location：指定虚拟机安装操作系统调用ISO文件的路径。

5）disk：指定虚拟机安装操作系统使用的硬盘。

6）network：指定虚拟机使用的网络。

7）os-type：指定虚拟机使用的操作系统，Windows或Linux。

8）os-variant：指定虚拟机具体使用的操作系统版本，如rhel7代表Red Hat Enterprise Linux 7。

9）extra-args：指定安装时控制的工具，注意不能与vnc混用。

10）noautoconsole：指定不使用自动控制台，如果不使用该参数，可能出现“ERROR unsupported format character '0xffffffe7) at index xx”等错误提示，但不会影响虚拟机的安装。

第6步，执行命令后，如果参数设定没有问题，会出现下列提示。如果未出现“开始安装”或者其他提示，请检查参数设定是否正确。

```
开始安装......
搜索文件    .treeinfo......            |   354 B     00:00:00
搜索文件    vmlinuz......              |   5.6 MB    00:00:00
搜索文件    initrd.img......           |   46 MB     00:00:00
域安装仍在进行。您可以重新连接
到控制台以便完成安装进程。
[root@localhost vm-centos7]#
```

第7步，虚拟机centos7已经开始安装。如何进行安装控制操作呢？可以通过命令行和VNC工具连接到虚拟机进行控制。先使用命令行进行安装操作，而后使用命令“virsh console centos7”连接到虚拟机控制台，控制台是纯命令操作界面。

注意：带[!]的参数是需要配置的，配置完成后按“b”键开始安装操作系统。如果参数没有设定正确，按“b”键不会安装操作系统，需要继续设定参数。

```
Starting installer, one moment...
anaconda 21.48.22.121-1 for CentOS 7 started.
 * installation log files are stored in /tmp during the installation
 * shell is available on TTY2
 * when reporting a bug add logs from /tmp as separate text/plain attachments
04:44:07 Not asking for VNC because we don't have a network
===============================================================================
Installation
 1) [x] Language settings                 2)  [!] Time settings
       (English (United States))              (Timezone is not set.)
 3) [!] Installation source               4)  [!] Software selection
       (Processing...)                        (Processing...)
 5) [!] Installation Destination          6)  [x] Kdump
       (No disks selected)                    (Kdump is enabled)
 7) [ ] Network configuration             8)  [!] Root password
       (Not connected)                        (Password is not set.)
```

```
 9) [!] User creation
        (No user will be created)
  Please make your choice from above ['q' to quit | 'b' to begin installation |
  'r' to refresh]:
[anaconda] 1:main* 2:shell  3:log  4:storage-lo> Switch tab: Alt+Tab | Help: F1
```

参数解释如下。

1）Language settings：设定安装语言。

2）Time settings：设定时区。

3）Installation source：设定安装源。

4）Software selection：选择安装的软件包。

5）Installation Destination：选择安装目的硬盘。

6）Kdump：内核崩溃转储机制，使用默认配置即可。

7）Network configuration：设定网络。

8）Root password：设定 root 用户密码。

9）User creation：创建新的用户。

第 8 步，输入数字 2 进入 Time settings，可以选择手动设定时区，也可以配置 NTP 服务器，根据实际情况进行设定即可。

```
Time settings
Timezone: not set
NTP servers:not configured
 1)  Set timezone  #设置时区
 2)  Configure NTP servers  #配置 NTP 服务器
  Please make your choice from above ['q' to quit | 'c' to continue |
  'r' to refresh]: 1
```

第 9 步，选择需要设定的时区，输入数字 2 选择 Asia。

```
Timezone settings
Available regions
 1)  Europe               6)  Pacific             10)  Arctic
 2)  Asia                 7)  Australia           11)  US
 3)  America              8)  Atlantic            12)  Etc
 4)  Africa               9)  Indian
 5)  Antarctica
Please select the timezone.
Use numbers or type names directly [b to region list, q to quit]:2
```

第 10 步，选择时区所在城市，输入数字 64 选择 Shanghai。

```
1)  Aden                  29)  Hong_Kong           56)  Pontianak
2)  Almaty                30)  Hovd                57)  Pyongyang
……（省略）
9)  Baghdad               37)  Karachi             64)  Shanghai
Please select the timezone.
Use numbers or type names directly [b to region list, q to quit]:64
```

第 11 步，时区配置完成后回到主配置界面，可以看到 Time settings 选项由原来未配的 [!] 状态变为[x]状态，说明已配置，同时下面会显示配置的相关参数“Asia/Shanghai timezone”。

```
Installation
 1) [x] Language settings                 2) [x] Time settings
        (English (United States))               (Asia/Shanghai timezone)
 3) [x] Installation source               4) [x] Software selection
        (Local media)                           (Minimal Install)
```

```
    5) [!] Installation Destination          6) [x] Kdump
         (No disks selected)                      (Kdump is enabled)
    7) [ ] Network configuration             8) [!] Root password
         (Not connected)                          (Password is not set.)
    9) [!] User creation
         (No user will be created)
     Please make your choice from above ['q' to quit | 'b' to begin installation |
     'r' to refresh]:
```

第 12 步，其他配置选项的操作方式与 Time settings 基本一致，配置完成后按“b”键，开始操作系统的安装，安装完成后会重启。

```
  Please make your choice from above ['q' to quit | 'b' to begin installation |
  'r' to refresh]: b
Progress
Setting up the installation environment
Creating disklabel on /dev/vda
Creating xfs on /dev/vda1
Creating lvmpv on /dev/vda2
Creating swap on /dev/mapper/centos-swap
……（省略）
Installing iwl6000-firmware (299/299)
Performing post-installation setup tasks
Installing boot loader
Performing post-installation setup tasks
Configuring installed system
Writing network configuration
.reating users
Running post-installation scripts
Use of this product is subject to the license agreement found at /usr/share/centos-release/EULA
Installation complete.  Press return to quit
dracut Warning: Killing all remaining processes
Rebooting.
[ 1728.665719] Restarting system.
[root@localhost vm-centos7]#
```

第 13 步，使用命令“virsh start centos7”启动虚拟机，再使用“virsh console centos7”命令连接到虚拟机控制台。

```
[root@localhost vm-centos7]# virsh start centos7
域 centos7 已开始
[root@localhost vm-centos7]# virsh list
 Id    名称                          状态
----------------------------------------------------
 10    centos7                       running
[root@localhost vm-centos7]# virsh console centos7
连接到域 centos7
换码符为 ^]
CentOS Linux 7 (Core)
Kernel 3.10.0-693.el7.x86_64 on an x86_64
kvm-centos login: root     #登录到虚拟机
[   24.879010] random: crng init done
Password:
[root@kvm-centos ~]# uname -a    #查看 Linux 版本信息
Linux kvm-centos 3.10.0-693.el7.x86_64 #1 SMP Tue Aug 22 21:09:27 UTC 2017 x86_64 x86_64 x86_64 GNU/Linux
```

第 14 步，使用命令“ip addr”查看虚拟机 IP 地址，注意 IP 地址目前为自动获取的地址，可以通过 KVM 主机访问外部网络。使用快捷键 ctrl+]可以从虚拟机控制台退回到 KVM 主机命令行界面。

```
[root@kvm-centos ~]# ip addr
1: lo: <LOOPBACK,UP,LOWER_UP> mtu 65536 qdisc noqueue state UNKNOWN qlen 1
    link/loopback 00:00:00:00:00:00 brd 00:00:00:00:00:00
    inet 127.0.0.1/8 scope host lo
      valid_lft forever preferred_lft forever
    inet6 ::1/128 scope host
      valid_lft forever preferred_lft forever
2: eth0: <BROADCAST,MULTICAST,UP,LOWER_UP> mtu 1500 qdisc pfifo_fast state UP qlen 1000
    link/ether 52:54:00:43:5c:3b brd ff:ff:ff:ff:ff:ff
    inet 192.168.122.229/24 brd 192.168.122.255 scope global dynamic eth0
      valid_lft 3547sec preferred_lft 3547sec
    inet6 fe80::583e:4057:f346:a0a/64 scope link
      valid_lft forever preferred_lft forever
[root@kvm-centos ~]# ping 119.6.6.6
PING 119.6.6.6 (119.6.6.6) 56(84) bytes of data.
64 bytes from 119.6.6.6: icmp_seq=1 ttl=250 time=6.19 ms
64 bytes from 119.6.6.6: icmp_seq=2 ttl=250 time=6.52 ms
……(省略)
--- 119.6.6.6 ping statistics ---
5 packets transmitted, 5 received, 0% packet loss, time 4003ms
rtt min/avg/max/mdev = 4.580/6.965/11.150/2.208 ms
[root@kvm-centos ~]#   # 按 ctrl+]从虚拟机退出
[root@localhost vm-centos7]#   #退出到 KVM 主机控制台
```

至此，在 KVM 主机上使用纯命令方式成功部署第一台 Linux 虚拟机，虚拟机已经能够正常使用。整体来看，使用的命令并不多，注意使用命令 virt-install 创建虚拟机时参数的设定。

3.2.2 使用 VNC 安装 Linux 虚拟机

除了使用纯命令方式安装虚拟机外，也可以配置启用 VNC Server，然后通过 VNC Viewer 工具连接到虚拟机更加直观地监控虚拟机的安装过程。使用 VNC Viewer 连接虚拟机需要先在 KVM 主机上配置 VNC Server。

第 1 步，使用命令“yum install tigervnc tigervnc-server”在 KVM 主机上安装 VNC Server。

```
[root@localhost /]# yum install tigervnc tigervnc-server
已加载插件: fastestmirror, langpacks
Loading mirror speeds from cached hostfile
 * base: mirrors.aliyun.com
 * extras: mirrors.aliyun.com
 * updates: mirrors.aliyun.com
正在解决依赖关系
--> 正在检查事务
---> 软件包 tigervnc.x86_64.0.1.8.0-5.el7 将被安装
--> 正在处理依赖关系 tigervnc-icons，它被软件包 tigervnc-1.8.0-5.el7.x86_64 需要
……(省略)
--> 解决依赖关系完成
依赖关系解决
================================================================================
Package               架构         版本                源                  大小
================================================================================
正在安装:
 tigervnc             x86_64       1.8.0-5.el7         base                239 k
 tigervnc-server      x86_64       1.8.0-5.el7         base                214 k
为依赖而安装:
 fltk                 x86_64       1.3.4-1.el7         base                560 k
```

```
 mesa-libGLU        x86_64        9.0.0-4.el7         base              196 k
 tigervnc-icons     noarch        1.8.0-5.el7         base              37 k
事务概要
=============================================================================
安装  2 软件包 (+3 依赖软件包)
总下载量：1.2 M
安装大小：3.1 M
Is this ok [y/d/N]: y
Downloading packages:
(1/5): mesa-libGLU-9.0.0-4.el7.x86_64.rpm           | 196 kB  00:00:06
(2/5): tigervnc-1.8.0-5.el7.x86_64.rpm              | 239 kB  00:00:00
……（省略）
-----------------------------------------------------------------------------------------------
总计                                          152 kB/s  | 1.2 MB  00:08
Running transaction check
Running transaction test
Transaction test succeeded
Running transaction
  正在安装：mesa-libGLU-9.0.0-4.el7.x86_64            1/5
……（省略）
已安装：
  tigervnc.x86_64 0:1.8.0-5.el7             tigervnc-server.x86_64 0:1.8.0-5.el7
作为依赖被安装：
  fltk.x86_64 0:1.3.4-1.el7                 mesa-libGLU.x86_64 0:9.0.0-4.el7
  tigervnc-icons.noarch 0:1.8.0-5.el7
完毕！
```

第2步，使用命令“cp /lib/systemd/system/vncserver@.service /lib/systemd/system/vncserver@:1.service”复制vncserver@.service文件，生成新的文件vncserver@:1.service。

```
[root@localhost /]#
cp /lib/systemd/system/vncserver@.service /lib/systemd/system/vncserver@: 1.service
```

第3步，使用命令“vi /lib/systemd/system/vncserver@:1.service”编辑复制生成的文件，注意调整路径和root权限的配置。

```
[root@localhost /]# vi /lib/systemd/system/vncserver@:1.service
[Unit]
Description=Remote desktop service (VNC)
After=syslog.target network.target
[Service]
Type=forking
# Clean any existing files in /tmp/.X11-unix environment
ExecStartPre=/bin/sh -c '/usr/bin/vncserver -kill %i > /dev/null 2>&1 || :'
ExecStart=/sbin/runuser -l root -c "/usr/bin/vncserver %i"
PIDFile=/root/.vnc/%H%i.pid
ExecStop=/bin/sh -c '/usr/bin/vncserver -kill %i > /dev/null 2>&1 || :'
[Install]
WantedBy=multi-user.target
```

第4步，使用命令“vncpasswd”配置连接密码。

```
[root@localhost ~]# vncpasswd
Password:
Verify:
Would you like to enter a view-only password (y/n)? y
Password:
Verify:
[root@localhost ~]#
```

第5步，使用命令“systemctl start vncserver@:1.service”启动服务，启动后使用命令

“systemctl status vncserver@:1.service”查看服务状态是否处于“active (running)”状态。

```
[root@localhost ~]# systemctl start vncserver@:1.service
[root@localhost ~]# systemctl status vncserver@:1.service
vncserver@:1.service - Remote desktop service (VNC)
   Loaded: loaded (/usr/lib/systemd/system/vncserver@:1.service; disabled; vendor preset: disabled)
   Active: active (running) since 二 2018-10-16 23:38:33 CST; 14h ago
  Process: 5015 ExecStart=/sbin/runuser -l root -c /usr/bin/vncserver %i (code=exited, status=0/
SUCCESS)
  Process: 5008 ExecStartPre=/bin/sh -c /usr/bin/vncserver -kill %i > /dev/null 2>&1 || : (code=exited,
status=0/SUCCESS)
 Main PID: 2671 (Xvnc)
   CGroup: /system.slice/system-vncserver.slice/vncserver@:1.service
           ‣ 2671 /usr/bin/Xvnc :1 -auth /root/.Xauthority -desktop localhost.localdomain:1 (root) -fp
catalogue:/e...
```

第 6 步，使用命令“vncserver”启动 VNC Server，其中 root 代表使用 root 用户进行连接。

```
[root@localhost ~]# vncserver
New 'localhost.localdomain:4 (root)' desktop is localhost.localdomain:4
Starting applications specified in /root/.vnc/xstartup
Log file is /root/.vnc/localhost.localdomain:4.log
```

第 7 步，使用 VNC Viewer 工具连接 KVM 主机，如图 3-2-3 所示，单击“Connect”按钮。

第 8 步，出现未加密的连接提示，如图 3-2-4 所示，单击“Continue”按钮。

第 9 步，输入连接密码，如图 3-2-5 所示，单击“OK”按钮。

第 10 步，使用 VNC Viewer 工具成功连接到 KVM 主机，如图 3-2-6 所示。

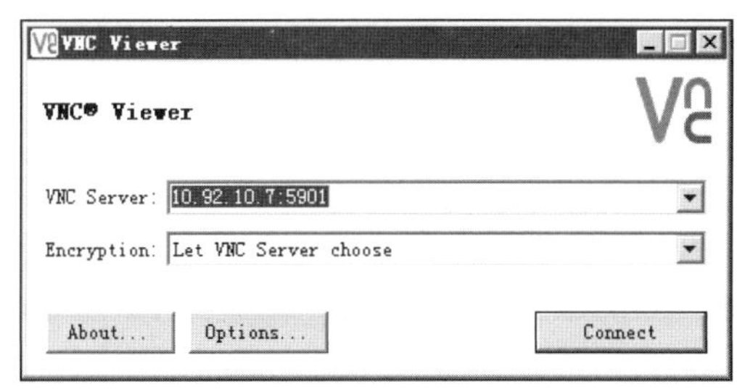

图 3-2-3

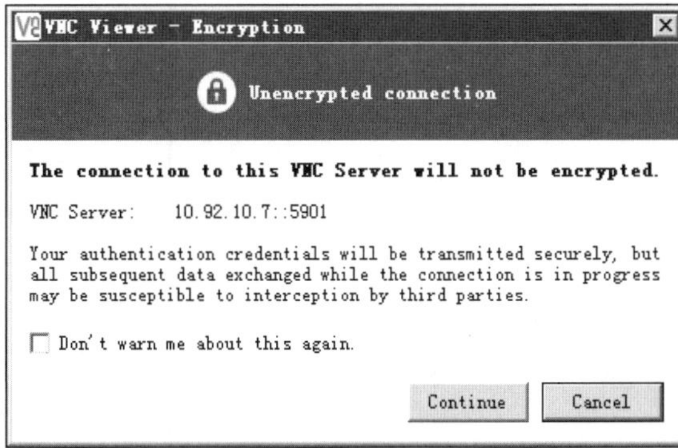

图 3-2-4

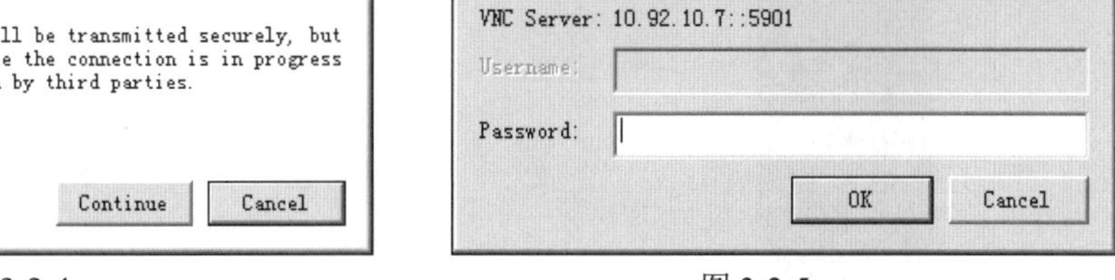

图 3-2-5

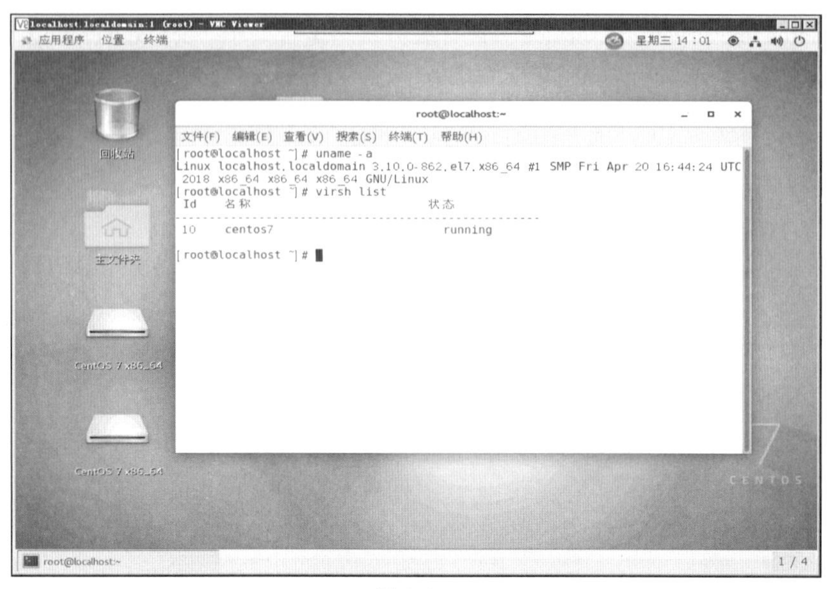

图 3-2-6

第 11 步，使用命令“virt-install”安装虚拟机，详细的参数解释参考前文内容，注意使用 vnc 参数后不再使用 extra-args 控制参数，否则 VNC Viewer 工具不能操作。

```
[root@localhost ~]# virt-install  --name centos7-02 --vcpus 1 --ram 2048 --location=/vm-iso/CentOS-7-x86_64-Minimal-1708.iso --disk path=/vm-centos7/centos7-02.raw,size=10,format=raw  --network bridge=virbr0 --os-type=linux --os-variant=rhel7  --vnc --vncport=-1 --vnclisten=0.0.0.0 --noautoconsole --force
开始安装......
搜索文件    .treeinfo......            |    354 B     00:00:00
搜索文件    vmlinuz......              |    5.6 MB    00:00:00
搜索文件    initrd.img......           |    46 MB     00:00:00
正在分配    'centos7-02.raw'           |    10 GB     00:00:00
域安装仍在进行。您可以重新连接
到控制台以便完成安装进程。
```

第 12 步，使用 VNC Viewer 工具连接到 centos7-02 虚拟机，如图 3-2-7 所示。

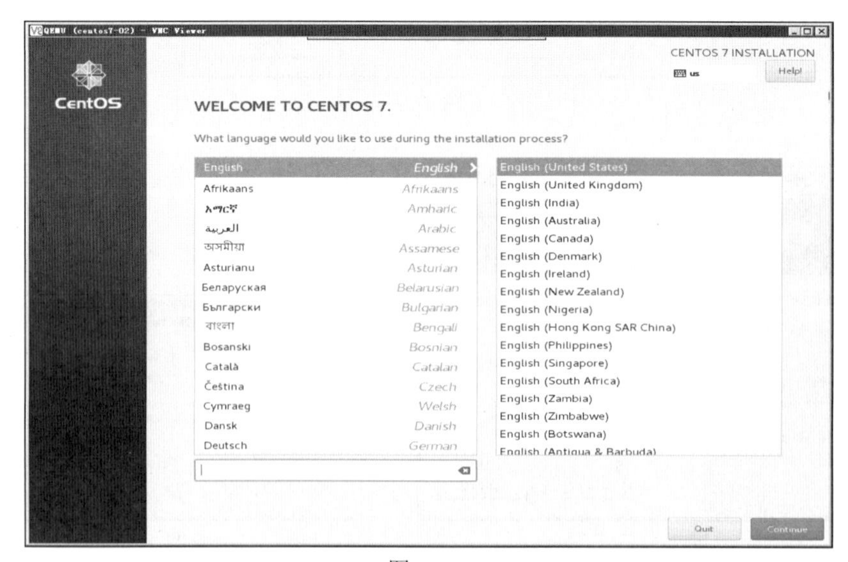

图 3-2-7

第 13 步，完成虚拟机安装，需要重启虚拟机，如图 3-2-8 所示。

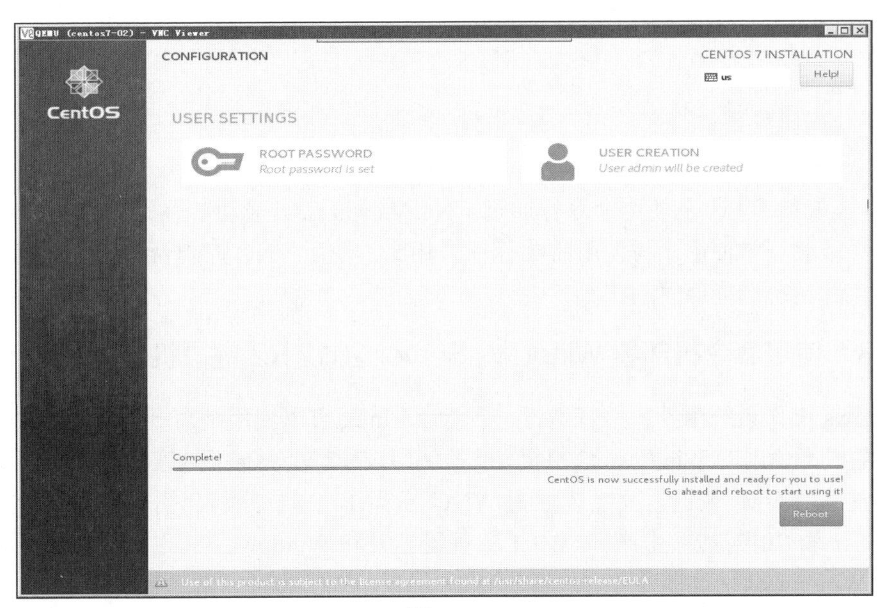

图 3-2-8

第 14 步，重启虚拟机会断开 VNC Viewer 工具的连接，虚拟机处于“关闭”状态，使用命令“virsh start centos7-02”启动虚拟机。

```
[root@localhost ~]# virsh start centos7-02
域 centos7-02 已开始
```

第 15 步，使用 VNC Viewer 工具连接到虚拟机，使用 root 用户登录，查看 Linux 版本并测试网络连通性是否正常，如图 3-2-9 所示。

```
QEMU (centos7-02) - VNC Viewer
CentOS Linux 7 (Core)
Kernel 3.10.0-693.el7.x86_64 on an x86_64

localhost login: root
Password:
Last login: Wed Oct 17 02:59:23 on tty1
[root@localhost ~]# uname -a
Linux localhost.localdomain 3.10.0-693.el7.x86_64 #1 SMP Tue Aug 22 21:09:27 UTC 2017 x86_64 x86_64 x86_64 GNU/Linux
[root@localhost ~]# ip addr
1: lo: <LOOPBACK,UP,LOWER_UP> mtu 65536 qdisc noqueue state UNKNOWN qlen 1
    link/loopback 00:00:00:00:00:00 brd 00:00:00:00:00:00
    inet 127.0.0.1/8 scope host lo
       valid_lft forever preferred_lft forever
    inet6 ::1/128 scope host
       valid_lft forever preferred_lft forever
2: eth0: <BROADCAST,MULTICAST,UP,LOWER_UP> mtu 1500 qdisc pfifo_fast state UP qlen 1000
    link/ether 52:54:00:a9:8f:31 brd ff:ff:ff:ff:ff:ff
    inet 192.168.122.181/24 brd 192.168.122.255 scope global dynamic eth0
       valid_lft 3523sec preferred_lft 3523sec
    inet6 fe80::dd4c:288a:4f38:4d26/64 scope link
       valid_lft forever preferred_lft forever
[root@localhost ~]# ping 61.139.2.69
PING 61.139.2.69 (61.139.2.69) 56(84) bytes of data.
64 bytes from 61.139.2.69: icmp_seq=1 ttl=54 time=5.54 ms
64 bytes from 61.139.2.69: icmp_seq=2 ttl=54 time=6.69 ms
64 bytes from 61.139.2.69: icmp_seq=3 ttl=54 time=5.87 ms
64 bytes from 61.139.2.69: icmp_seq=4 ttl=54 time=4.95 ms
64 bytes from 61.139.2.69: icmp_seq=5 ttl=54 time=6.21 ms
64 bytes from 61.139.2.69: icmp_seq=6 ttl=54 time=4.31 ms
64 bytes from 61.139.2.69: icmp_seq=7 ttl=54 time=6.75 ms
64 bytes from 61.139.2.69: icmp_seq=8 ttl=54 time=6.39 ms
64 bytes from 61.139.2.69: icmp_seq=9 ttl=54 time=4.80 ms
64 bytes from 61.139.2.69: icmp_seq=10 ttl=54 time=5.05 ms
^C
--- 61.139.2.69 ping statistics ---
10 packets transmitted, 10 received, 0% packet loss, time 9014ms
rtt min/avg/max/mdev = 4.310/5.660/6.758/0.817 ms
[root@localhost ~]# _
```

图 3-2-9

第16步，使用命令“virsh list --all”查看KVM主机运行的虚拟机情况，可以看到新创建的两台Linux虚拟机均处于“运行”状态。

```
[root@localhost ~]# virsh list --all
 Id    名称                         状态
----------------------------------------------------------------------------------------
 2     centos7-02                   running
 3     centos7                      running
```

至此，在KVM主机上使用命令行通过VNC Viewer工具成功部署第二台Linux虚拟机，虚拟机已经能够正常使用。与使用命令行部署相比，使用VNC控制台更直观，读者可以根据自己的实际情况选择部署的工具。

3.2.3 使用命令行部署Windows Server 2012 R2虚拟机

Windows操作系统由于一些特殊性，不能像Linux可以使用纯命令界面进行部署。对于在KVM主机上使用的Windows虚拟机，推荐使用VNC Viewer工具连接到虚拟机进行后续的安装，注意KVM主机需要安装配置VNC Server。本节分别介绍如何使用命令行部署Windows Server 2012 R2（生产环境用于服务器端）和Windows 7（生产环境用于虚拟桌面端）两种Windows虚拟机。

第1步，使用命令“virt-install”安装虚拟机，详细的参数解释如下。注意安装Windows虚拟机使用了boot引导参数和cdrom路径。

```
    root@kvm-server:/home/admin# virt-install  --name win2012 --vcpus 2 --ram 4096 --boot cdrom --cdrom=/vm-iso/cn_windows_server_2012_r2.iso  --disk  path=/vm-win/win2012.raw,size=20,format=raw   --network bridge=virbr0 --os-type=windows  --vnc --vncport=5900 --vnclisten=0.0.0.0 --noautoconsole --force
    Starting install...
    Allocating 'win2012.raw'
    |  20 GB  00:00:00
    Domain installation still in progress. You can reconnect to
    the console to complete the installation process.
```

参数解释如下。

1）name：指定虚拟机名称。

2）vcpus：指定虚拟机使用的CPU数量。

3）ram：指定虚拟机使用的内存大小，单位为MB。

4）disk：指定虚拟机安装操作系统使用的硬盘。

5）network：指定虚拟机使用的网络。

6）os-type：指定虚拟机使用的操作系统。

7）vnc：指定安装使用VNC进行控制。

8）vncport：指定VNC使用的端口。

9）vnclisten=0.0.0.0：指定VNC监听的IP地址。

10）noautoconsole：指定不使用自动控制台，如果不使用该参数，可能出现错误提示，但不会影响虚拟机的安装。

第2步，使用VNC Viewer工具连接到虚拟机，如图3-2-10所示，安装过程与在物理服务器上安装Windows操作系统一样，系统开始载入安装文件。

第3步，进入Windows Server 2012 R2操作系统安装界面，如图3-2-11所示。

图 3-2-10

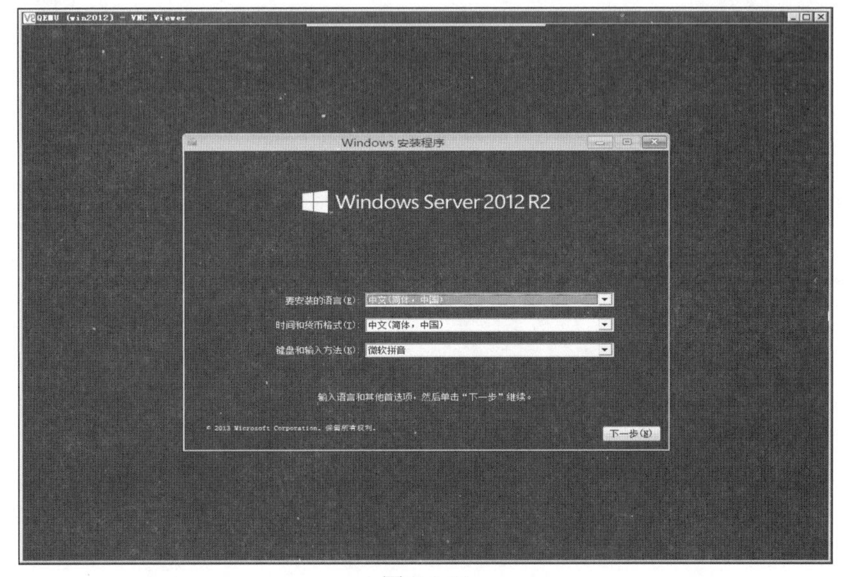

图 3-2-11

第 4 步，开始安装 Windows Server 2012 R2 操作系统，如图 3-2-12 所示。

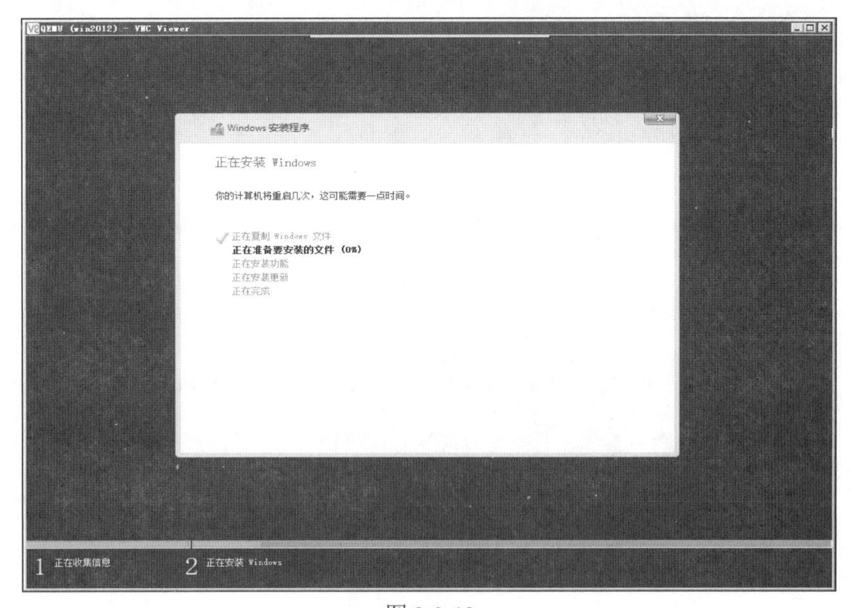

图 3-2-12

第 5 步，完成 Windows Server 2012 R2 操作系统安装，查看计算机的基本信息，如图 3-2-13 所示。

图 3-2-13

第 6 步，查看 Windows Server 2012 R2 操作系统网络的连通性，通过图 3-2-14 可以看到，网络连接正常。

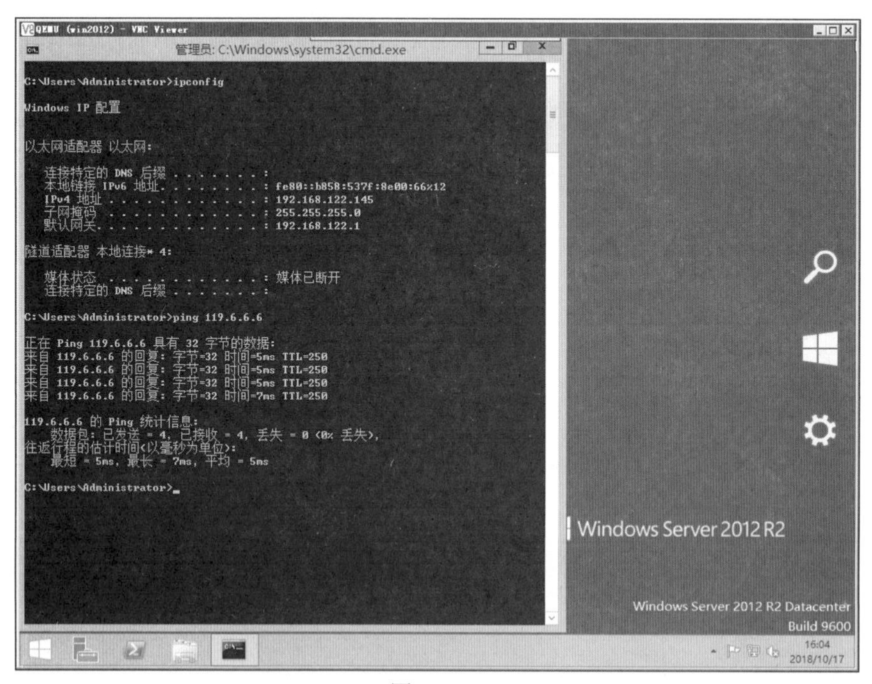

图 3-2-14

第 7 步，使用命令“virsh list”查看虚拟机的运行状态，目前为 running（运行）状态。

```
root@kvm-server:/home/admin# virsh list
 Id    Name                           State
----------------------------------------------------------------------------------------------------
 8     win2012                        running
```

3.2.4 使用命令行部署 Windows 7 虚拟机

使用命令行部署 Windows 7 虚拟机和部署 Windows Server 2012 R2 虚拟机基本一致，主要注意在安装过程中 Windows 7 驱动程序的加载问题。

第 1 步，使用命令“virt-install”安装虚拟机。注意安装 Windows 虚拟机使用了 boot 引导参数和 cdrom 路径。

```
root@kvm-server:/home/admin# virt-install  --name win7 --vcpus 2 --ram 4096 --boot cdrom --cdrom=/vm-iso/WIN7_MSDN.iso --disk path=/vm-win/win7.raw,size=20,format=raw  --network bridge=virbr0 --os-type=windows  --vnc --vncport=5901 --vnclisten=0.0.0.0 --noautoconsole --force
Starting install...
Allocating 'win7.raw'
|  20 GB  00:00:00
Domain installation still in progress. You can reconnect to
the console to complete the installation process.
```

第 2 步，使用 VNC Viewer 工具连接到虚拟机，安装过程与安装 Windows Server 2012 R2 操作系统类似，需要注意的是，使用早期的 KVM 版本安装 Windows 7 操作系统的过程中，会出现无法识别硬盘的问题，需要手动加载 KVM 虚拟化专用的 virtio 驱动程序。作者使用新的版本测试了 CentOS 和 Ubuntu 操作系统，均无须再加载驱动程序，可直接识别到硬盘，如图 3-2-15 所示。

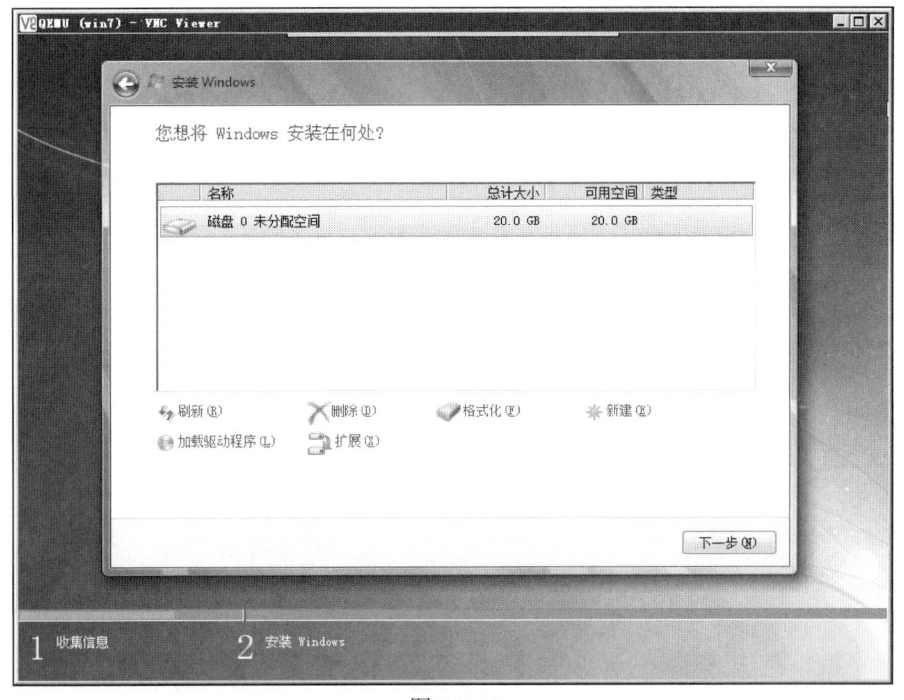

图 3-2-15

第 3 步，开始安装 Windows 7 操作系统，如图 3-2-16 所示。

图 3-2-16

第 4 步，完成 Windows 7 操作系统的安装，查看计算机的基本信息，通过图 3-2-17 可以看到虚拟机使用的硬盘为 QEMU 虚拟出来的硬盘。

图 3-2-17

第 5 步，查看 Windows 7 操作系统网络的连通性，通过图 3-2-18 可以看到，网络连接正常。

图 3-2-18

第 6 步，使用命令“virsh list”查看虚拟机的运行状态，目前为 running（运行）状态。

```
root@kvm-server:/vm-win# virsh list
 Id    Name                           State
----------------------------------------------------------------------------------------------------
 8     win2012                        running
 12    win7                           running
```

3.2.5　部署 Windows 虚拟机常见问题

KVM 对于 Linux 虚拟机的支持是非常好的，但对于 Windows 虚拟机的支持存在一些问题，本节针对部署 Windows 虚拟机的一些常见问题进行解释说明。

1．部署过程中不能识别硬盘问题

用早期的 KVM 版本安装 Windows 操作系统的过程中，会出现无法识别硬盘的问题，需要手动加载 KVM 虚拟化专用的 virtio 驱动程序。如果读者遇到这样的问题，可以下载专用的 virtio 驱动程序解决，访问 https://fedorapeople.org/groups/virt/virtio-win/direct-downloads/archive-virtio/virtio-win-0.1.160-1/进行下载，如图 3-2-19 所示。

2．部署过程中鼠标异常问题

在使用 VNC Viewer 连接虚拟机部署 Windows 操作系统的时候，可能会出现鼠标异常的情况，特别是在早期的 KVM 版本中经常出现，一般可以通过添加 usb 鼠标来解决，修改虚拟机 xml 配置文件即可。

```
<input type='tablet' bus='usb' />
```

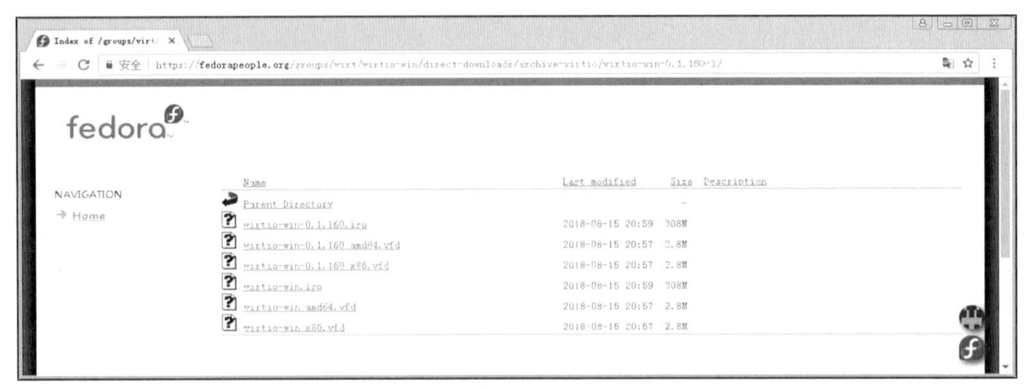

图 3-2-19

3. VNC Viewer 闪退问题

在使用 VNC Viewer 连接虚拟机特别是 Windows 虚拟机时，可能会出现 VNC Viewer 闪退问题，多数是由于 VNC Viewer 工具 network speed 配置不当。取消勾选"Adapt to network speed(recommended)"，移动滑块到最右边"Best quality"即可，如图 3-2-20 所示。

至此，在 KVM 主机成功部署 Windows Server 2012 R2 和 Windows 7 两台虚拟机。整体来说，做好准备工作，使用命令行创建虚拟机没有多大难度。创建失败的原因经常是不少人没有系统地看完步骤就开始操作，结果过程中问题百出。强烈建议初学者一步一步地认真操作。

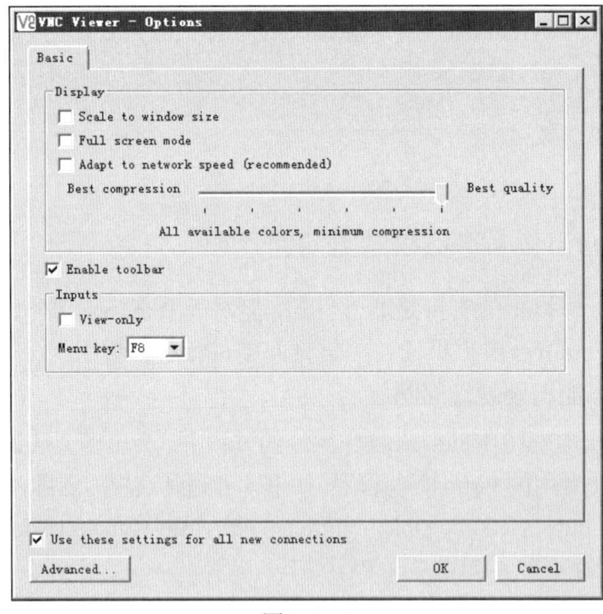

图 3-2-20

3.2.6 常用 virsh 命令总结

为便于读者安装部署虚拟机，现将常用 virsh 命令进行总结，如表 3-2-1 所示。当然 virsh 命令不仅仅是这些，读者可以通过系统进行查看。

表 3-2-1 常用 virsh 命令

命令	解释
virsh list	列出正在运行的虚拟机，使用--all 参数列出所有虚拟机
virsh start 虚拟机名	启动虚拟机
virsh autostart 虚拟机名	开机自动启动虚拟机
virsh shutdown 虚拟机名	关闭虚拟机
virsh console 虚拟机名	使用控制台连接到虚拟机
virsh destroy 虚拟机名	强制关闭虚拟机
virsh suspend 虚拟机名	挂起虚拟机
virsh resumed 虚拟机名	虚拟机从挂起状态恢复
virsh edit 虚拟机名	修改虚拟机配置文件
virsh dumpxml 虚拟机名	查看虚拟机配置文件
virsh undefine 虚拟机名	删除虚拟机配置文件

3.3 使用 GUI 部署虚拟机

前文介绍了如何使用纯命令方式部署虚拟机，相信读者应该掌握了。但在实际使用过程中，特别是对习惯使用 Windows 操作系统 GUI 的技术人员来说，还是更愿意使用 GUI 来创建和管理虚拟机。

KVM 虚拟化也提供了创建和管理虚拟机的 GUI，就是 virt-manager 管理工具。在前文中已经介绍过如何安装该管理工具，本节介绍如何使用 virt-manager 工具创建 Linux 虚拟机和 Windows 虚拟机。

3.3.1 使用 GUI 部署 Linux 虚拟机

第 1 步，使用 VNC Viewer 工具连接到 KVM 主机，打开终端窗口，输入命令“virt-manager”，如图 3-3-1 所示。

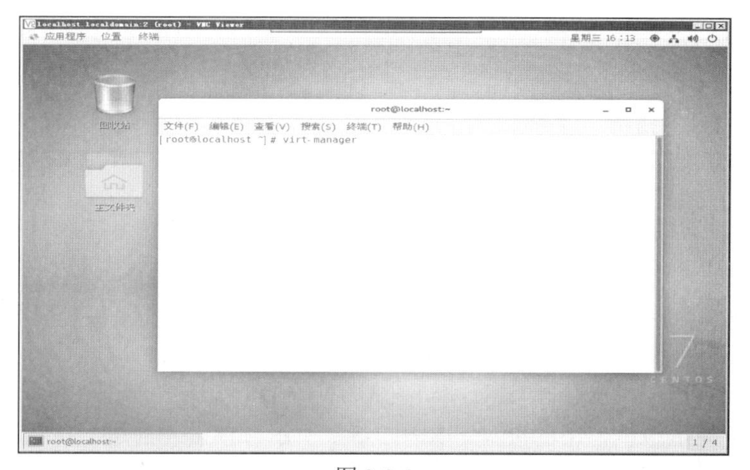

图 3-3-1

第 2 步，如果正确安装了 virt-manager 管理工具，会打开“虚拟系统管理器”窗口，如图 3-3-2 所示，可看到正在运行之前创建的 2 台虚拟机。

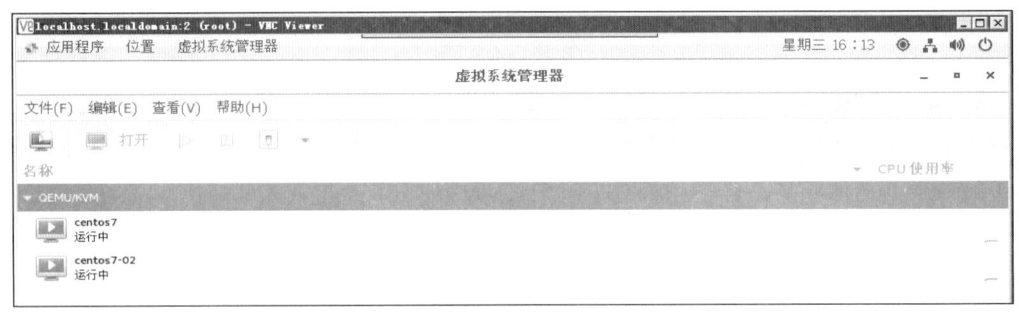

图 3-3-2

第 3 步，在“文件”菜单中选择“新建虚拟机”，打开新建虚拟机向导，如图 3-3-3 所示，单击“前进”按钮。

图 3-3-3

第 4 步，选择存储卷，可以理解为选择操作系统的 ISO 文件，如图 3-3-4 所示，单击“选择卷”按钮。

第 5 步，选择后管理器会自动侦测操作系统，操作系统类型为 Linux，版本为 CentOS 7.0，如图 3-3-5 所示，单击“前进”按钮。

第 6 步，配置虚拟机内存和 CPU，如图 3-3-6 所示，单击“前进”按钮。

第 7 步，配置虚拟机使用的硬盘，如图 3-3-7 所示，单击“前进”按钮。

第 8 步，配置虚拟机名称和网络，如图 3-3-8 所示，单击“完成”按钮。

localhost.localdomain:2 (root) - VNC Viewer
应用程序 位置 虚拟系统管理器
星期三 16:14
虚拟系统管理器
文件(F) 编辑(E) 查看(V) 帮助(H)
打开
名称
CPU 使用率
QEMU/KVM
centos7
运行中
centos7-02
运行中
选择存储卷
default
文件系统目录
vm-centos7
文件系统目录
vm-iso
文件系统目录
vm-ubuntu
文件系统目录
大小： 43.14 GiB Free / 6.84 GiB In Use
位置： /vm-iso
卷
卷 大小 格式 用于
CentOS-7-x86_64-Minimal-1708.iso 792.00 MiB iso
ubuntu-18.04-live-server-amd64.iso 806.00 MiB iso
本地浏览(B) 取消(C) 选择卷
[root@localhost:~]
虚拟系统管理器
1 / 4

图 3-3-4

localhost.localdomain:2 (root) - VNC Viewer
应用程序 位置 虚拟系统管理器
星期三 16:14
虚拟系统管理器
文件(F) 编辑(E) 查看(V) 帮助(H)
打开
名称
CPU 使用率
QEMU/KVM
centos7
运行中
centos7-02
运行中
新建虚拟机
生成新虚拟机
5 的步骤 2
定位您的安装介质
使用 CD-ROM 或 DVD：
没有探测到介质 (/dev/sr0)
使用 ISO 映像：
/vm-iso/CentOS-7-x86_64-Minimal-1
浏览(W)...
根据安装介质自动侦测操作系统(U)
操作系统类型(T)： Linux
版本(V)： CentOS 7.0
取消(C) 后退(B) 前进(F)
[root@localhost:~]
虚拟系统管理器
1 / 4

图 3-3-5

localhost.localdomain:2 (root) - VNC Viewer
应用程序 位置 虚拟系统管理器
星期三 16：14
虚拟系统管理器
文件(F) 编辑(E) 查看(V) 帮助(H)
名称
CPU 使用率
QEMU/KVM
centos7
运行中
centos7-02
运行中
新建虚拟机
生成新虚拟机
选择内存和 CPU 设置
内存(M)，RAM： 1024
主机中最多有 65525 MIB 可用
CPU： 1
最多有 24 可用
取消(C) 后退(B) 前进(F)
[root@localhost:~]
虚拟系统管理器
1 / 4

图 3-3-6

localhost.localdomain:2 (root) - VNC Viewer
应用程序 位置 虚拟系统管理器
星期三 16：14
虚拟系统管理器
文件(F) 编辑(E) 查看(V) 帮助(H)
名称
CPU 使用率
QEMU/KVM
centos7
运行中
centos7-02
运行中
新建虚拟机
生成新虚拟机
为虚拟机启用存储(E)
为虚拟机创建磁盘镜像(R)
9.0
GiB
43.1 GiB available in the default location
选择或创建自定义存储(S)
管理(M)...
取消(C) 后退(B) 前进(F)
[root@localhost:~]
虚拟系统管理器
1 / 4

图 3-3-7

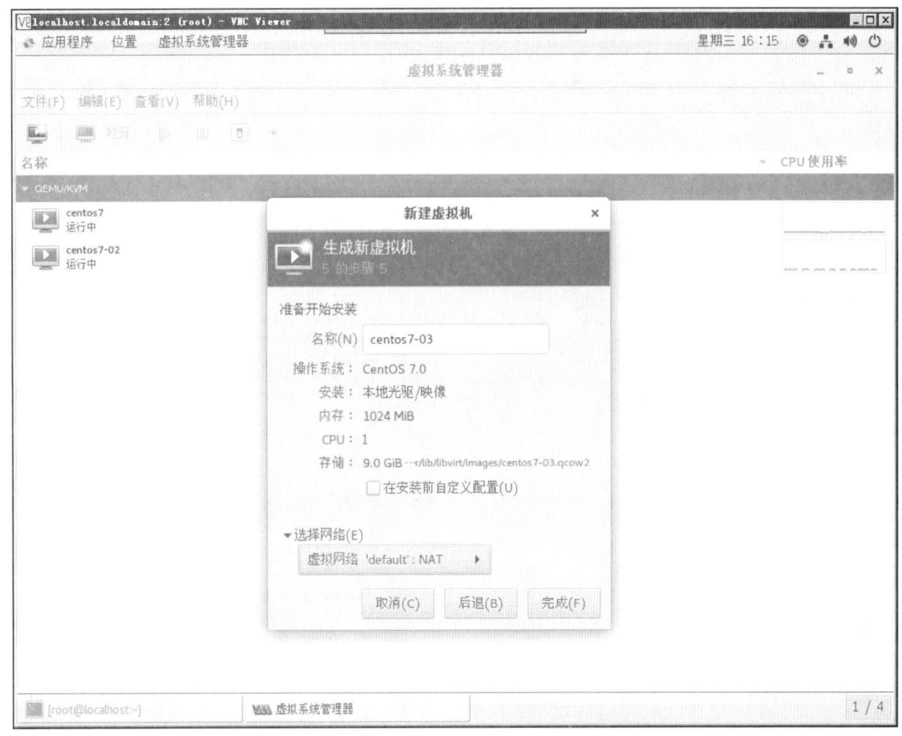

图 3-3-8

第 9 步，开始为虚拟机安装操作系统，如图 3-3-9 所示。

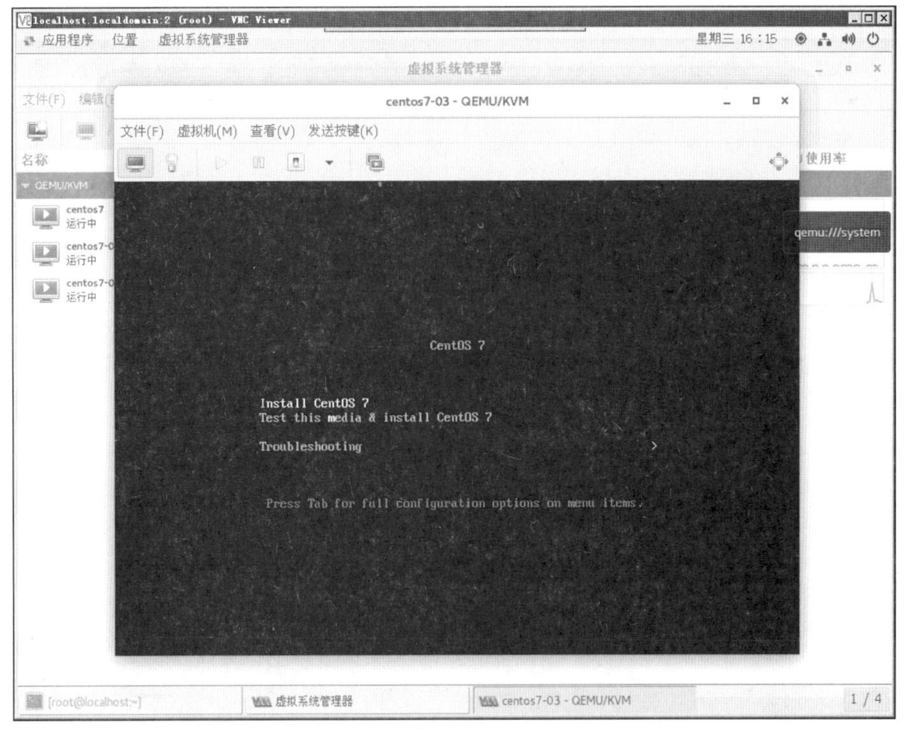

图 3-3-9

第 10 步，进入操作系统安装界面，如图 3-3-10 所示。

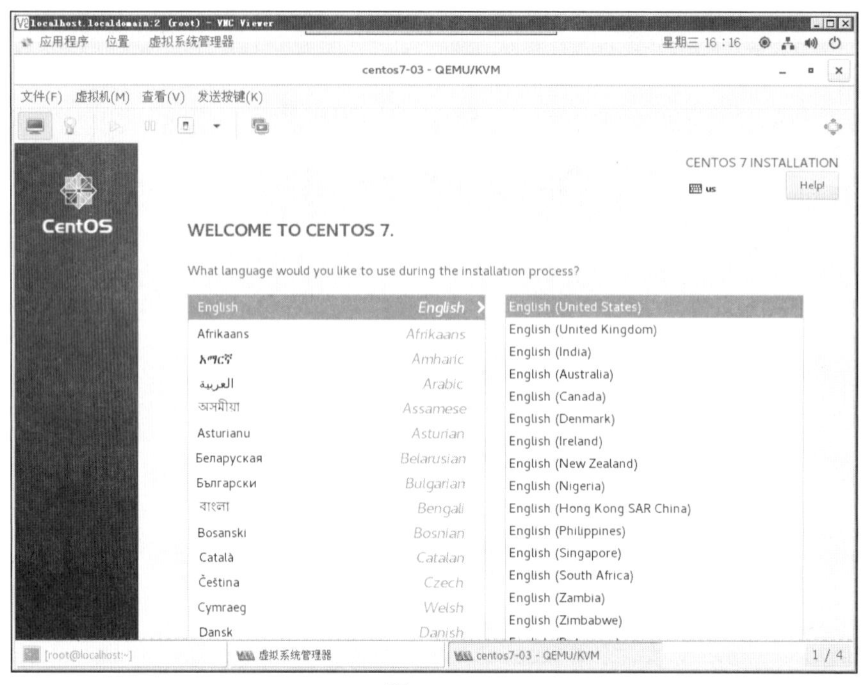

图 3-3-10

第 11 步，切换到“虚拟系统管理器”，可以看到名为 centos7-03 的虚拟机处于“运行中”状态，如图 3-3-11 所示。

图 3-3-11

第 12 步，使用“虚拟系统管理器”可以对虚拟机进行日常管理操作，如图 3-3-12 所示。

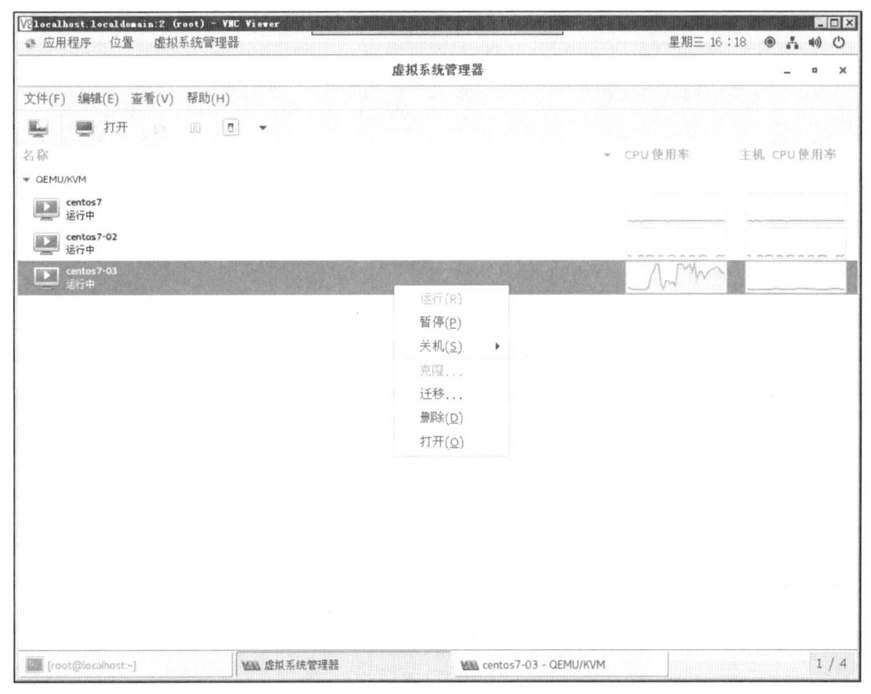

图 3-3-12

第 13 步，继续安装虚拟机操作系统，通过图 3-3-13 可以看出，虚拟机硬盘使用的是虚拟设备 Virtio Block Device，说明虚拟机运行在 KVM 环境下。

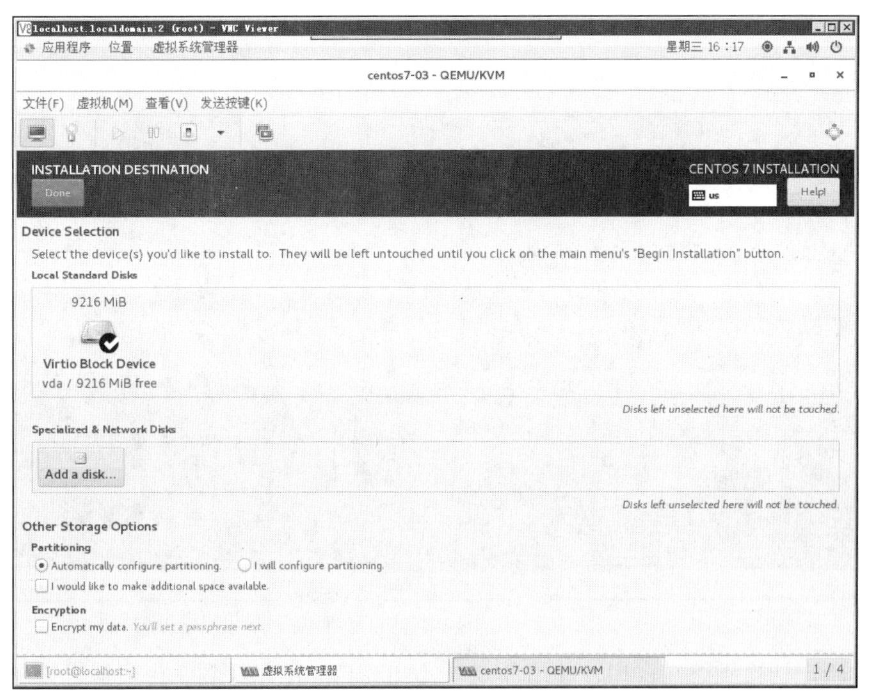

图 3-3-13

第 14 步，完成虚拟机操作系统的安装，如图 3-3-14 所示，需要重启虚拟机。

图 3-3-14

第 15 步，虚拟机重启完成后，使用 root 用户名和密码登录操作系统，查看 Linux 版本信息和网络的连通性，如图 3-3-15 所示，网络连接正常。

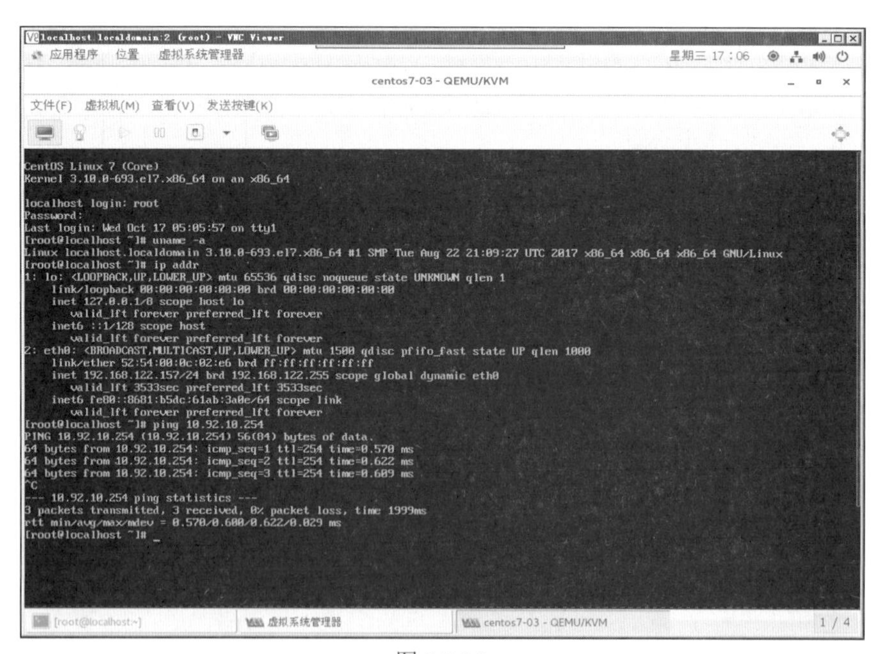

图 3-3-15

第 16 步，切换到终端窗口，使用命令“virsh list”，可以看到 3 台虚拟机处于“运行”状态，如图 3-3-16 所示。

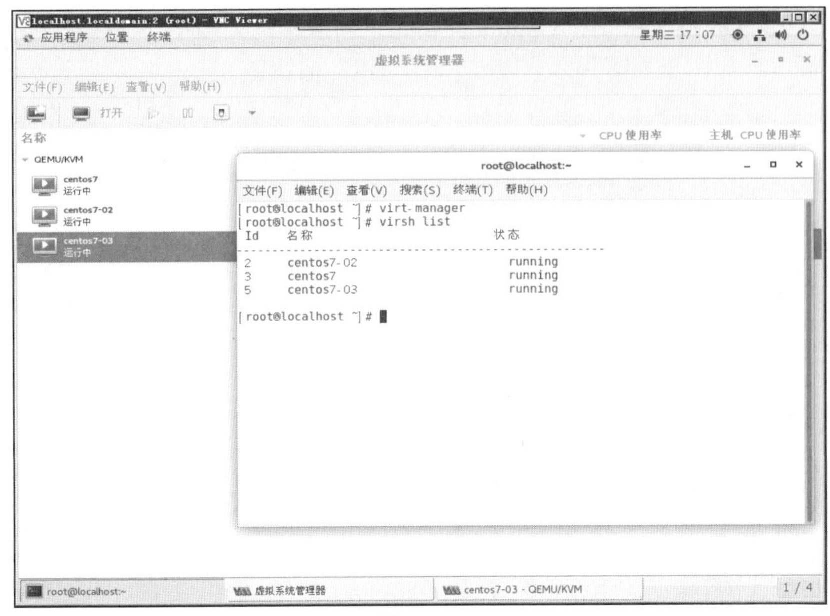

图 3-3-16

至此，使用 GUI 部署 Linux 虚拟机完成，相比使用命令行方式更简洁明了，缺点是需要安装 GUI 进行操作。对于不愿意使用命令行的读者来说，使用 GUI 部署虚拟机也是一种选择。

3.3.2 使用 GUI 部署 Windows 虚拟机

使用 GUI 部署 Windows 虚拟机过程与部署 Linux 虚拟机过程相似，主要不同在于 ISO 镜像的选择。

第 1 步，进行 Windows 7 操作系统安装，如图 3-3-17 所示，单击“下一步”按钮。

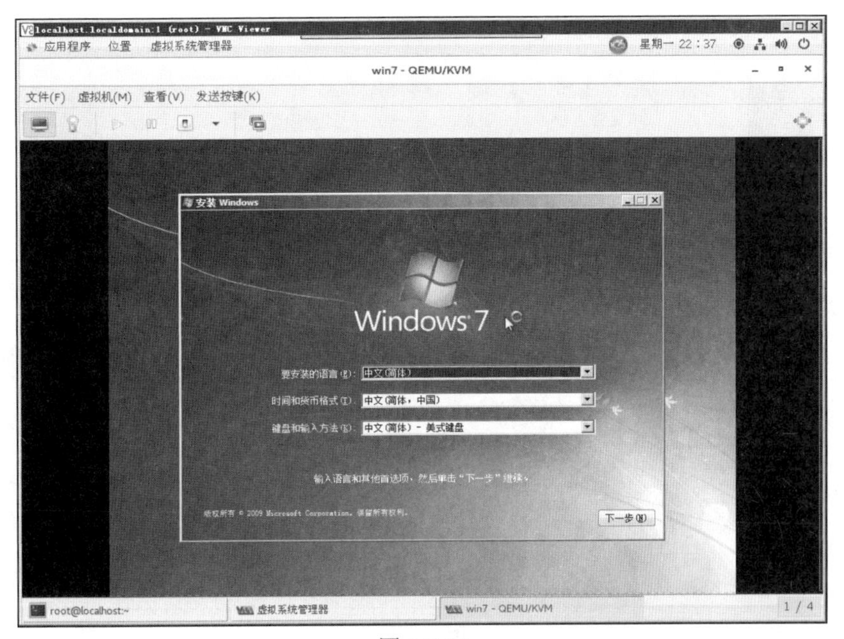

图 3-3-17

第 2 步，CentOS 操作系统和 Ubuntu 操作系统新版本均无须手动加载 virtio 驱动程序，可直接识别硬盘，如图 3-3-18 所示，单击“下一步”按钮。

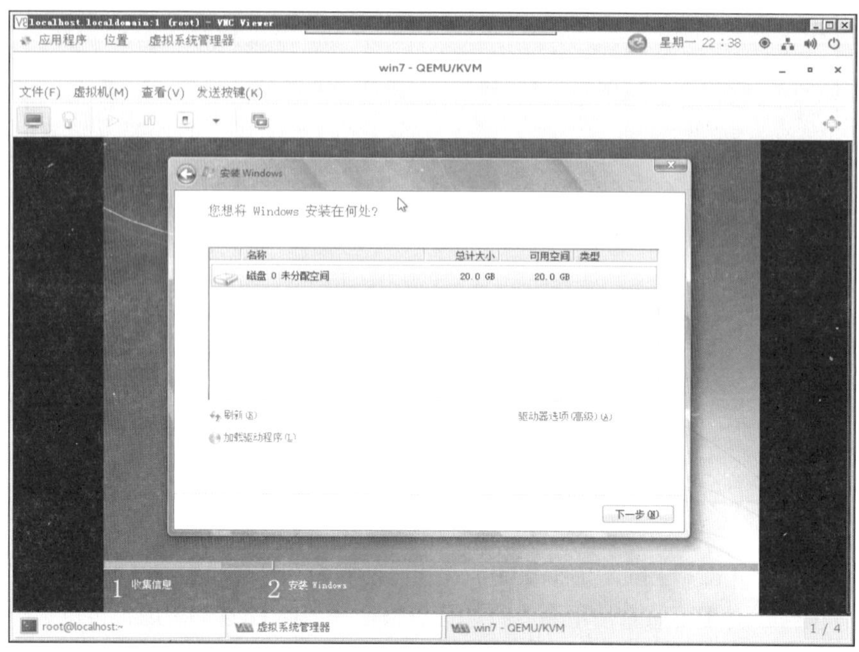

图 3-3-18

第 3 步，完成 Windows 7 操作系统的安装，查看计算机的基本信息，如图 3-3-19 所示，其中硬盘使用 QEMU 虚拟的硬盘。

图 3-3-19

第 4 步，查看 Windows 7 操作系统网络的连通性，通过图 3-3-20 可以看到，网络连接正常。

图 3-3-20

第 5 步，切换到终端窗口，使用命令“virsh list--all”可以看到 3 台虚拟机正在运行状态，1 台虚拟机处于关闭状态，如图 3-3-21 所示。

图 3-3-21

至此，使用命令行和 GUI 两种模式部署 Linux 虚拟机和 Windows 虚拟机完成。使用命令行方式效率相对较高，但是需要注意命令参数的配置；使用 GUI 模式不需要使用复杂的参数，使用向导方式即可完成。两种方式都可用于生产环境部署虚拟机，至于如何选择取决于读者。

3.4 使用模板部署虚拟机

前文介绍了使用命令行和 GUI 两种模式部署 Linux 虚拟机和 Windows 虚拟机。在生产环境中，一台一台地安装显然不符合实际需求。有没有其他方式可快速部署虚拟机呢？答案是肯定的，可以通过镜像的方式来部署虚拟机，也称为通过模板方式部署虚拟机。本节介绍如何使用镜像部署虚拟机。

3.4.1 理解 KVM 虚拟机硬盘镜像格式

在使用镜像部署虚拟机前，需要简单了解虚拟机硬盘镜像的两种格式。KVM 虚拟机使用两种硬盘镜像格式，一种是前文曾提及的 RAW 格式，另一种是 QCOW2 格式。这里先进行简单介绍，后文会进行详细介绍。

RAW 格式：KVM 虚拟机原始的硬盘镜像格式，性能好但功能简单。使用 RAW 格式会占用大量的存储空间，如创建了 10GB 空间直接使用 10GB 空间，不管安装的操作系统究竟使用了多少。

QCOW2 格式：新型的 KVM 虚拟机硬盘镜像格式，性能一般但功能多。使用 QCOW2 格式可减少实际使用空间，类似于精简置备，如创建了 10GB 空间，操作系统只使用了 5GB 空间，那么实际使用空间就是 5GB。

3.4.2 Backing file 的作用

了解 RAW 和 QCOW2 两种格式后，还需要了解了一下 KVM 中的 Backing file，可以翻译为后备镜像。后备镜像简单来说就是多台虚拟机共用一个镜像文件。我们以场景应用的方式来解释 Backing file 的作用。

场景：如前文所述，现已安装好一台 centos7 虚拟机，使用 RAW 格式 10GB，假设实际使用 5GB，现在想快速再创建 20 台，注意，是快速创建，不是一台一台重新安装。

方案 1：复制安装好的 centos7 虚拟机 RAW 硬盘，修改虚拟机配置，20 台虚拟机共有 200GB 使用空间。

方案 2：将 centos7 虚拟机 RAW 硬盘作为 Backing file，再创建 20 份 QCOW2 格式的镜像，修改虚拟机配置文件，20 台虚拟机原始使用 10GB 空间，后续数据写入时再增加空间。

很明显，使用方案 1 复制文件需要大量时间，使用方案 2 不需要大量复制数据就可以部署虚拟机，同时减少了空间的使用，这就是 Backing flie 的作用。

3.4.3 复制 Linux 虚拟机硬盘镜像创建虚拟机

前文已经讲解过多台 Linux 虚拟机的安装，现将 centos7 虚拟机作为模板，通过复制虚拟机硬盘镜像的方式来创建虚拟机。

第 1 步，使用命令“virsh list --all”查看 KVM 主机上的虚拟机信息，centos7 虚拟机处于关闭状态。

```
root@kvm-server:/vm-linux# virsh list --all
 Id   Name                    State
--------------------------------------------------------------------------------------------
 8    win2012                 running
 12   win7                    running
 -    centos7                 shut off
 -    centos7-02              shut off
 -    centos7-03              shut off
```

第 2 步，使用命令“ls”查看虚拟机硬盘镜像文件，可见其使用的是 RAW 格式。

```
root@kvm-server:/vm-linux# ls
centos7.raw
```

第 3 步，使用命令“qemu-img create -f raw -b centos7.raw centos7-04.raw”复制虚拟机硬盘镜像文件创建新的虚拟机硬盘镜像，会出现错误提示，说明 backing flie 不能使用 RAW 格式。

```
root@kvm-server:/vm-linux# qemu-img create -f raw -b centos7.raw centos7-04.raw
qemu-img: centos7-04.raw: Backing file not supported for file format 'raw'
```

第 4 步，调整参数，将 RAW 格式调整为 QCOW2 格式，再使用命令“qemu-img create -f qcow2 -b centos7.raw centos7-04.raw”复制虚拟机硬盘镜像文件创建新的虚拟机硬盘镜像。

```
root@kvm-server:/vm-linux# qemu-img create -f qcow2 -b centos7.raw centos7-04.raw
Formatting 'centos7-04.raw', fmt=qcow2 size=10737418240 backing_file=centos7.raw cluster_size=65536
lazy_refcounts=off refcount_bits=16
```

第 5 步，使用命令“ls”查看新创建的虚拟机硬盘镜像文件 centos7-04.raw，注意硬盘镜像文件是 RAW 格式。

```
root@kvm-server:/vm-linux# ls
centos7-04.raw  centos7.raw
```

第 6 步，使用命令“qemu-img info centos7-04.raw”查看硬盘镜像文件信息，从文件格式可以看到，新创建的硬盘镜像文件格式为 QCOW2。

```
root@kvm-server:/vm-linux# qemu-img info centos7-04.raw
image: centos7-04.raw
file format: qcow2
virtual size: 10G (10737418240 bytes)
disk size: 196K
cluster_size: 65536
backing file: centos7.raw
Format specific information:
    compat: 1.1
    lazy refcounts: false
    refcount bits: 16
    corrupt: false
```

第 7 步，既然新创建的硬盘镜像文件格式为 QCOW2，可以在实际使用中直接创建，那么使用命令“rm -rf centos7-04.raw”删除创建的硬盘镜像文件后重新创建。

```
root@kvm-server:/vm-linux# rm -rf centos7-04.raw
```

第 8 步，使用命令“qemu-img create -f qcow2 -b centos7.raw centos7-04.qcow2”创建新的虚拟机硬盘镜像，注意格式为 QCOW2。

```
root@kvm-server:/vm-linux# qemu-img create -f qcow2 -b centos7.raw centos7-04.qcow2
```

```
    Formatting    'centos7-04.qcow2',    fmt=qcow2    size=10737418240    backing_file=centos7.raw
cluster_size=65536 lazy_refcounts=off refcount_bits=16
```

第9步，使用命令“ls”查看新创建的虚拟机硬盘镜像文件centos7-04.qcow2，注意硬盘镜像文件是QCOW2格式。

```
root@kvm-server:/vm-linux# ls
centos7-04.qcow2  centos7.raw
```

第10步，使用命令“qemu-img info centos7-04.qcow2”查看硬盘镜像文件信息，硬盘镜像文件格式为QCOW2。

```
root@kvm-server:/vm-linux# qemu-img info centos7-04.qcow2
image: centos7-04.qcow2
file format: qcow2
virtual size: 10G (10737418240 bytes)
disk size: 196K
cluster_size: 65536
backing file: centos7.raw
Format specific information:
   compat: 1.1
   lazy refcounts: false
   refcount bits: 16
corrupt: false
```

至此，通过复制虚拟机硬盘镜像文件创建新的虚拟机硬盘镜像文件完成，但是虚拟机还不能使用，还需要对虚拟机的配置文件进行一些操作才能使用。

第11步，使用命令“virsh list --all”查看虚拟机信息，并没有centos7-04虚拟机，因为还未注册该虚拟机。

```
root@kvm-server:/vm-linux# virsh list --all
 Id    Name                           State
----------------------------------------------------------------------------------------------
 8     win2012                        running
 12    win7                           running
 -     centos7                        shut off
 -     centos7-02                     shut off
 -     centos7-03                     shut off
```

第12步，使用命令“ls /etc/libvirt/qemu/”查看现有虚拟机的配置文件，虚拟机配置文件就是.xml文件。注意，KVM主机虚拟机配置文件位于/etc/libvirt/qemu。

```
root@kvm-server:/vm-linux# ls /etc/libvirt/qemu/
centos7-02.xml  centos7-03.xml  centos7.xml  networks  win2012.xml  win7.xml
```

第13步，使用命令“cp /etc/libvirt/qemu/centos7.xml /etc/libvirt/qemu/centos7-04.xml”复制原centos7虚拟机配置文件，重新命名为centos7-04.xml。

```
root@kvm-server:/vm-linux#cp /etc/libvirt/qemu/centos7.xml /etc/libvirt/qemu/centos7-04.xml
```

第14步，使用命令“vi /etc/libvirt/qemu/centos7-04.xml”编辑配置文件，因为复制的是centos7的配置文件，虚拟机uuid、硬盘镜像路径以及MAC地址等均需要修改，否则新创建的虚拟机无法启动，虚拟机uuid和MAC地址推荐删除后重新生成。

```
root@kvm-server:/vm-linux# vi /etc/libvirt/qemu/centos7-04.xml
<domain type='kvm'>
  <name>centos7</name>  #虚拟机名
  <uuid>632d4cc5-2146-4222-826e-08fae04d847b</uuid>  #虚拟机uuid，推荐删除后重新生成
  <memory unit='KiB'>2097152</memory>   #虚拟机内存
```

```
  <currentMemory unit='KiB'>2097152</currentMemory>
  <vcpu placement='static'>1</vcpu>  #虚拟机 VCPU
<devices>
    <emulator>/usr/bin/kvm-spice</emulator>
    <disk type='file' device='disk'>
      <driver name='qemu' type='raw'/>    #虚拟机硬盘镜像格式，注意格式的修改
      <source file='/vm-ubuntu/ubuntu.raw'/>  #虚拟机硬盘镜像路径，注意路径的修改
      <target dev='vda' bus='virtio'/>
      <address type='pci' domain='0x0000' bus='0x00' slot='0x07' function='0x0'/>
    </disk>
<interface type='bridge'>
      <mac address='52:54:00:ad:3f:69'/>  #虚拟机 MAC 地址，推荐删除后重新生成
      <source bridge='virbr0'/>
      <model type='virtio'/>
      <address type='pci' domain='0x0000' bus='0x00' slot='0x03' function='0x0'/>
```

第 15 步，修改完配置文件后，使用命令“virsh define /etc/libvirt/qemu/centos7-04.xml”注册虚拟机。

```
root@kvm-server:/vm-linux# virsh define /etc/libvirt/qemu/centos7-04.xml
Domain centos7-04 defined from /etc/libvirt/qemu/centos7-04.xml
```

第 16 步，注册完成后，使用命令“virsh list --all”查看虚拟机是否注册成功，注册成功后 centos7-04 虚拟机处于关闭状态。

```
root@kvm-server:/vm-linux# virsh list --all
 Id    Name                           State
----------------------------------------------------
 8     win2012                        running
 12    win7                           running
 -     centos7                        shut off
 -     centos7-02                     shut off
 -     centos7-03                     shut off
 -     centos7-04                     shut off
```

第 17 步，使用命令“virsh start centos7-04”启动虚拟机。

```
root@kvm-server:/vm-linux# virsh start centos7-04
Domain centos7-04 started
```

第 18 步，使用命令“virsh console centos7-04”连接到虚拟机控制台。

```
root@kvm-server:/vm-linux# virsh console centos7-04
Connected to domain centos7-04
Escape character is ^]
```

第 19 步，如果虚拟机配置文件正确，虚拟机会正常启动，并出现启动自检等信息；如果配置文件存在问题，虚拟机控制台不会有其他显示，请关闭虚拟机电源后检查配置文件。

```
Welcome to CentOS Linux 7 (Core)!

[  OK  ] Stopped Switch Root.
[  OK  ] Stopped Journal Service.
         Starting Journal Service...
[  OK  ] Set up automount Arbitrary Executab...ats File System Automount Point.
[  OK  ] Listening on udev Control Socket.
[  OK  ] Created slice system-serial\x2dgetty.slice.
[  OK  ] Reached target Remote File Systems.
         Mounting Debug File System...
[  OK  ] Listening on /dev/initctl Compatibility Named Pipe.
[  OK  ] Created slice User and Session Slice.
[  OK  ] Reached target Slices.
```

```
[  OK  ] Created slice system-getty.slice.
[  OK  ] Listening on Delayed Shutdown Socket.
         Mounting Huge Pages File System...
[  OK  ] Stopped File System Check on Root Device.
……（省略）
[    7.990611] Netfilter messages via NETLINK v0.30.
[    7.998664] ip_set: protocol 6

CentOS Linux 7 (Core)
Kernel 3.10.0-693.el7.x86_64 on an x86_64
kvm-centos login: root
Password:
Last login: Tue Oct 16 17:34:45 on tty1
```

第20步，使用root用户名和密码登录虚拟机，使用命令“ip addr”查看IP地址，可以看到MAC地址不同于原centos7虚拟机的MAC地址，生成了新的MAC地址。

```
[root@kvm-centos ~]# ip addr
1: lo: <LOOPBACK,UP,LOWER_UP> mtu 65536 qdisc noqueue state UNKNOWN qlen 1
    link/loopback 00:00:00:00:00:00 brd 00:00:00:00:00:00
    inet 127.0.0.1/8 scope host lo
       valid_lft forever preferred_lft forever
    inet6 ::1/128 scope host
       valid_lft forever preferred_lft forever
2: eth0: <BROADCAST,MULTICAST,UP,LOWER_UP> mtu 1500 qdisc pfifo_fast state UP qlen 1000
    link/ether 52:54:00:38:93:21 brd ff:ff:ff:ff:ff:ff
    inet 192.168.122.215/24 brd 192.168.122.255 scope global dynamic eth0
       valid_lft 3546sec preferred_lft 3546sec
    inet6 fe80::f0b2:dd33:c652:568a/64 scope link
       valid_lft forever preferred_lft forever
```

第21步，使用命令“ping”检查网络的连通性，访问网络正常。

```
[root@kvm-centos ~]# ping 119.6.6.6
PING 119.6.6.6 (119.6.6.6) 56(84) bytes of data.
64 bytes from 119.6.6.6: icmp_seq=1 ttl=250 time=5.32 ms
64 bytes from 119.6.6.6: icmp_seq=2 ttl=250 time=3.90 ms
64 bytes from 119.6.6.6: icmp_seq=3 ttl=250 time=4.45 ms
64 bytes from 119.6.6.6: icmp_seq=4 ttl=250 time=4.72 ms
____________________________________
--- 119.6.6.6 ping statistics ---
4 packets transmitted, 4 received, 0% packet loss, time 3005ms
rtt min/avg/max/mdev = 3.908/4.604/5.327/0.512 ms
```

第22步，使用命令“virsh list --all”查看虚拟机信息，centos7-04虚拟机处于运行状态。

```
root@kvm-server:/vm-linux# virsh list --all
 Id    Name                           State
----------------------------------------------------------------------------------------------
 8     win2012                        running
 12    win7                           running
 19    centos7-04                     running
 -     centos7                        shut off
 -     centos7-02                     shut off
 -     centos7-03                     shut off
```

第23步，尝试启动centos7虚拟机，使用命令“virsh start centos7”后会出现错误提示，提示无法启动，因为已将centos7虚拟机硬盘镜像作为backing file文件。

```
root@kvm-server:/vm-linux# virsh start centos7
error: Failed to start domain centos7
error: internal error: qemu unexpectedly closed the monitor: 2018-10-29T13:55:30.517655Z qemu-system-
```

```
x86_64:  -device  virtio-blk-pci,scsi=off,bus=pci.0,addr=0x7,drive=drive-virtio-disk0,id=virtio-disk0,
bootindex=1: Failed to get "write" lock
     Is another process using the image?
```

第 24 步，使用命令“vi /etc/libvirt/qemu/centos7-04.xml”查看虚拟机新的配置文件，注意重新生成的虚拟机 uuid 和 MAC 地址。

```
root@kvm-server:/vm-linux#vi  /etc/libvirt/qemu/centos7-04.xml
<domain type='kvm'>
  <name>centos7-04</name>
  <uuid>1f56b567-d667-414f-a4cd-74fa9749e2e3</uuid>  #重新生成的虚拟机 uuid
  <memory unit='KiB'>2097152</memory>
  <currentMemory unit='KiB'>2097152</currentMemory>
  <vcpu placement='static'>1</vcpu>
<interface type='bridge'>
    <mac address='52:54:00:38:93:21'/>  #重新生成的虚拟机 MAC 地址
    <source bridge='virbr0'/>
    <model type='virtio'/>
    <address type='pci' domain='0x0000' bus='0x00' slot='0x03' function='0x0'/>
```

至此，通过复制已安装 Linux 虚拟机硬盘镜像创建虚拟机完成，整体来说是比较简单的，重点是修改虚拟机配置文件。对于虚拟机配置文件的修改和调整，后文还会详细介绍。

3.4.4 复制 Windows 虚拟机硬盘镜像创建虚拟机

通过复制已安装 Windows 虚拟机硬盘镜像创建虚拟机的操作与 3.4.3 节所介绍的相似，下面继续进行操作介绍。

第 1 步，使用命令“qemu-img create -f qcow2 -b win7.raw win7-02.qcow2”创建新的虚拟机硬盘镜像，注意格式为 QCOW2。

```
root@kvm-server:/vm-win# qemu-img create -f qcow2 -b win7.raw win7-02.qcow2
Formatting 'win7-02.qcow2', fmt=qcow2 size=21474836480 backing_file=win7.raw cluster_size=65536
lazy_refcounts=off refcount_bits=16
```

第 2 步，使用命令“cp /etc/libvirt/qemu/win7.xml /etc/libvirt/qemu/win7-02.xml”复制原 win7 虚拟机配置文件，重新命名为 win7-02.xml。

```
root@kvm-server:/vm-win# cp /etc/libvirt/qemu/win7.xml /etc/libvirt/qemu/win7-02.xml
```

第 3 步，使用命令“vi /etc/libvirt/qemu/win7-02.xml”编辑配置文件，与 Linux 虚拟机类似，因为是复制的 win7 的配置文件，虚拟机 uuid、硬盘镜像路径以及 MAC 地址等需要修改，否则新创建的虚拟机无法启动，虚拟机 uuid 和 MAC 地址推荐删除后重新生成。

```
root@kvm-server:/vm-win# vi /etc/libvirt/qemu/win7-02.xml
<domain type='kvm'>
  <name>win7</name>
  <uuid>b052cc07-0a98-4102-9f7d-e1438725665b</uuid> #虚拟机 uuid，推荐删除后重新生成
  <memory unit='KiB'>4194304</memory>
  <currentMemory unit='KiB'>4194304</currentMemory>
  <vcpu placement='static'>2</vcpu>
<devices>
    <emulator>/usr/bin/kvm-spice</emulator>
    <disk type='file' device='disk'>
      <driver name='qemu' type='raw'/>  #虚拟机硬盘镜像格式，注意格式的修改
      <source file='/vm-win/win7.raw'/>  #虚拟机硬盘镜像路径，注意路径的修改
      <target dev='hda' bus='ide'/>
      <address type='drive' controller='0' bus='0' target='0' unit='0'/>
<interface type='bridge'>
```

```
    <mac address='52:54:00:d2:ef:25'/>  #虚拟机MAC地址，推荐删除后重新生成
    <source bridge='virbr0'/>
    <model type='rtl8139'/>
    <address type='pci' domain='0x0000' bus='0x00' slot='0x03' function='0x0'/>
  <graphics type='vnc' port='5901' autoport='no' listen='0.0.0.0'>  #虚拟机VNC连接端口
    <listen type='address' address='0.0.0.0'/>
```

第4步，修改完配置文件后，使用命令“virsh define /etc/libvirt/qemu/win7-02.xml”注册虚拟机。

```
root@kvm-server:/vm-win# virsh define /etc/libvirt/qemu/win7-02.xml
Domain win7-02 defined from /etc/libvirt/qemu/win7-02.xml
```

第5步，注册完成后，使用命令“virsh list --all”查看虚拟机是否注册成功，注册成功后win7-02虚拟机处于关闭状态。

```
root@kvm-server:/vm-win# virsh list --all
 Id    Name                           State
----------------------------------------------------------------------------------------------------
 8     win2012                        running
 19    centos7-04                     running
 -     centos7                        shut off
 -     centos7-02                     shut off
 -     centos7-03                     shut off
 -     win7                           shut off
 -     win7-02                        shut off
```

第6步，使用命令“virsh start win7-02”启动虚拟机。

```
root@kvm-server:/vm-win# virsh start win7-02
Domain win7-02 started
```

第7步，使用VNC Viewer连接到虚拟机控制台，如图3-4-1所示。

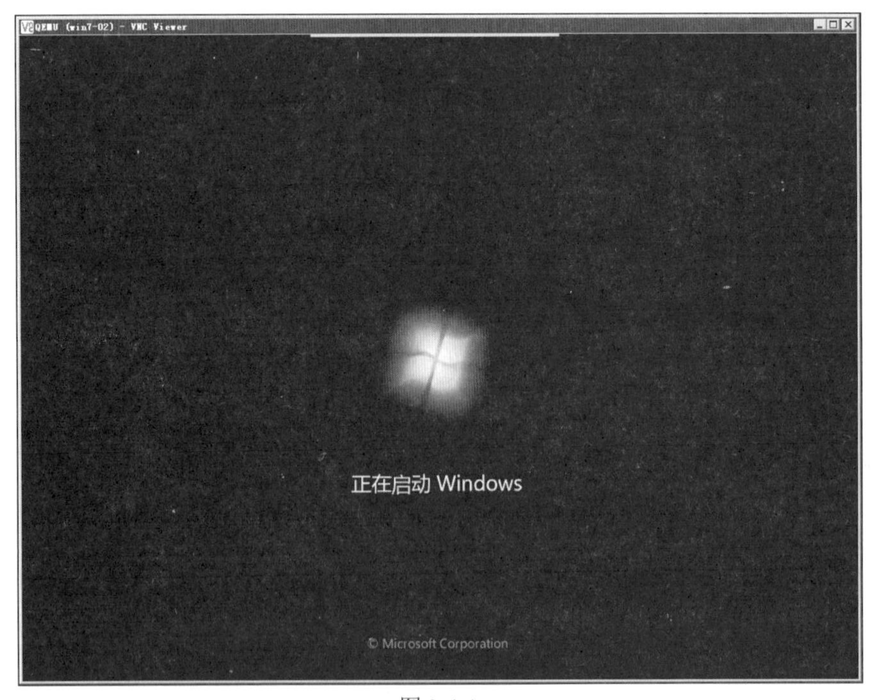

图3-4-1

第 8 步，新创建的虚拟机启动完成，检测到新的硬件需要重启虚拟机，如图 3-4-2 所示。

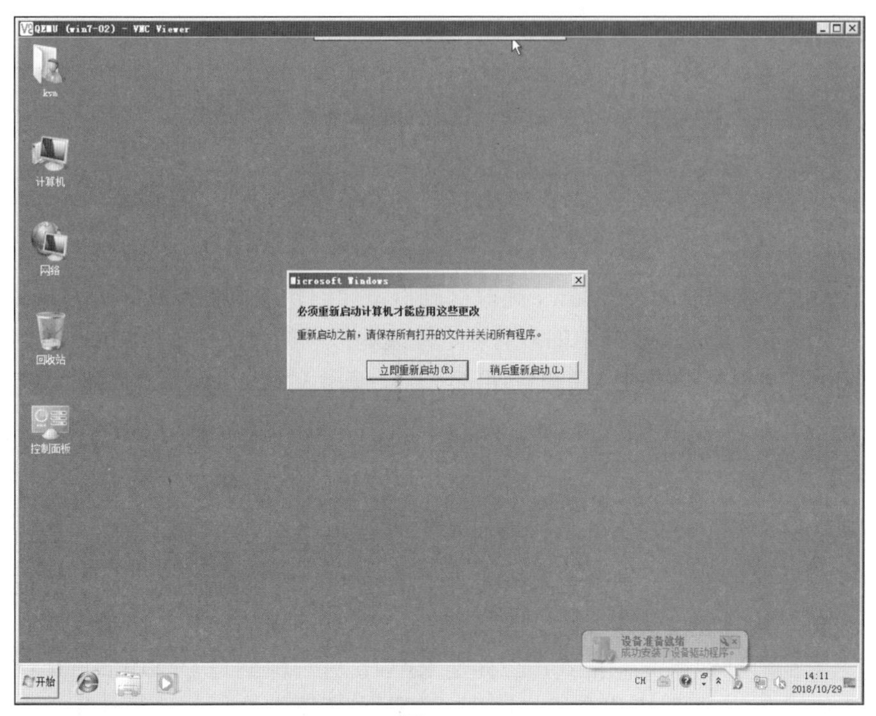

图 3-4-2

第 9 步，虚拟机重启完成，查看计算机的基本信息和网络的连通性，如图 3-4-3 所示。

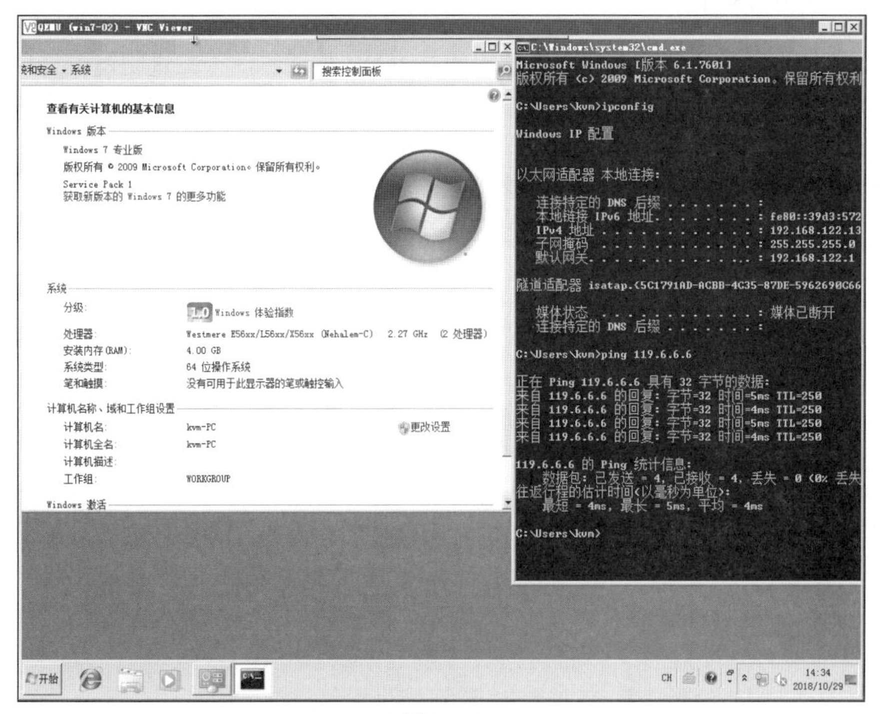

图 3-4-3

第10步，使用命令“virsh list --all”查看虚拟机信息，win7-02虚拟机处于运行状态。

```
root@kvm-server:/vm-win# virsh list --all
Id    Name                         State
----------------------------------------------------------------------------------------------
 8     win2012                     running
 19    centos7-04                  running
 21    win7-02                     running
 -     centos7                     shut off
 -     centos7-02                  shut off
 -     centos7-03                  shut off
 -     win7                        shut off
```

至此，通过复制已安装Windows虚拟机硬盘镜像创建虚拟机完成，与Linux虚拟机相关操作类似，整体来说是比较简单的，重点是修改虚拟机配置文件。与Linux虚拟机有差异的是，Windows虚拟机不能通过控制台连接，需要使用VNC Viewer工具进行连接。

3.5 虚拟机硬盘格式

在物理服务器中，会使用硬盘来安装操作系统和存放数据。虚拟机同样需要用硬盘来安装操作系统和存放数据，但在虚拟环境中，没有物理存在的硬盘，使用的是虚拟出来的硬盘，其实质就是硬盘镜像文件。硬盘镜像文件有不同的格式。一般来说，在KVM环境中会使用到两种镜像文件格式，一种是RAW，另一种就是QCOW2（QCOW是上一代，已经不使用），这两种格式有什么区别，在生产环境中该如何选择使用，本节将进行介绍。

3.5.1 RAW格式

KVM虚拟机原始的硬盘镜像文件使用的是RAW格式，属于二进制镜像文件，性能好但功能简单。使用RAW格式会占用大量的存储空间，一次性会把创建的空间全部使用，如创建了10GB空间就直接使用10GB空间，不管实际使用多少，同时RAW格式需要EXT3/EXT4文件系统的支持。使用命令“qemu-img create”可以创建镜像文件，如果不指定参数默认使用RAW格式创建，具体过程如下：

```
root@kvm-server:/vm-win# qemu-img create win10 20G  #创建名为win10镜像文件
Formatting 'win10', fmt=raw size=21474836480
root@kvm-server:/vm-win# ls  #查看新创建的win10镜像文件
win10  win2012.raw  win7-02.qcow2  win7.raw
root@kvm-server:/vm-win# qemu-img info win10  #查看新创建的win10镜像文件信息
image: win10
file format: raw
virtual size: 20G (21474836480 bytes)
disk size: 0
```

RAW格式支持稀疏文件特性，所谓的稀疏文件特性是指文件系统会把分配的空字节文件记录在元数据中，而不会实际占用硬盘空间。使用ls命令或du命令查看RAW镜像文件，会得出大小不一样的结果。

```
root@kvm-server:/vm-win# ls -lh win2012.raw  #使用ls命令查看镜像文件大小20GB
-rw------- 1 libvirt-qemu kvm 20G Nov  4 12:45 win2012.raw
root@kvm-server:/vm-win# du -h win2012.raw  #使用du命令查镜像文件实际使用7.5GB
7.5G    win2012.raw
```

整体来说，RAW 格式可以满足生产环境虚拟机的使用要求。但需要注意的是，RAW 格式不支持虚拟机快照，如果生产环境需要使用快照，可以通过命令将 RAW 格式转换为 QCOW2 格式，后文会进行介绍。

3.5.2 QCOW2 格式

QCOW2 格式是 QCOW 格式的升级版，QCOW 格式基本被 QCOW2 格式取代。QCOW2 格式是新的 KVM 虚拟机硬盘镜像文件格式，性能一般但功能较多，在使用镜像部署虚拟机时使用。使用 QCOW2 格式可以减少实际使用空间，类似于精简置备，如创建了 10GB 空间，操作系统只使用了 5GB 空间，那么实际使用空间就是 5GB。使用命令“qemu-img create”加参数可以创建 QCOW2 镜像文件，具体的创建过程如下：

```
root@kvm-server:/vm-win# qemu-img create -f qcow2 win10.qcow2 20G
Formatting 'win10.qcow2', fmt=qcow2 size=21474836480 cluster_size=65536 lazy_refcounts=off
refcount_bits=16
root@kvm-server:/vm-win# ls
win10  win10.qcow2  win2012.raw  win7-02.qcow2  win7.raw
root@kvm-server:/vm-win# qemu-img info win10.qcow2
image: win10.qcow2
file format: qcow2
virtual size: 20G (21474836480 bytes)
disk size: 196K
cluster_size: 65536
Format specific information:
   compat: 1.1
   lazy refcounts: false
   refcount bits: 16
   corrupt: false
root@kvm-server:/vm-win#
```

与 RAW 格式一样，QCOW2 格式也支持稀疏文件特性。使用 ls 命令或 du 命令查看 QCOW2 镜像文件会得出大小一样的结果。

```
root@kvm-server:/vm-win# ls -lh win10.qcow2
-rw-r--r-- 1 root root 193K Nov  4 13:08 win10.qcow2
root@kvm-server:/vm-win# du -h win10.qcow2
196K    win10.qcow2
```

由于 QCOW2 格式支持其他一些高级特性，如虚拟机快照、压缩以及加密等，因此 QCOW2 格式也是目前推荐使用的虚拟机硬盘镜像格式。

3.5.3 RAW/QCOW2 格式对比

了解了 RAW 格式和 QCOW2 格式后，下面介绍它们的对比，如表 3-5-1 所示。

表 3-5-1 RAW/QCOW2 格式对比

	RAW 格式	QCOW2 格式
空间占用	全部使用	实际使用
虚拟机快照	不支持	支持
镜像文件加密	不支持	支持

续表

	RAW 格式	QCOW2 格式
文件压缩	不支持	支持
格式转换	可直接转换	需先转换为 RAW 后再转换
文件删除后空间占用	空间变小	空间不变

对于生产环境来说，多数的选择是使用 QCOW2 格式，当然也可以使用 RAW 格式，后续也可以进行格式之间的转换。

3.5.4 RAW/QCOW2 格式常见操作

在生产环境中，会经常根据需要对两种格式进行操作，常见的操作有格式转换、增加硬盘容量等，本小节将介绍这些常见的操作。

1. 将 RAW 格式转换为 QCOW2 格式

第 1 步，进行格式转换前需要关闭虚拟机电源，使用命令“virsh list --all”查看虚拟机运行状态。

```
root@kvm-server:/vm-win# virsh list --all
 Id    Name                           State
----------------------------------------------------------------------------------------------------
 8     win2012                        running
 19    centos7-04                     running
 33    win7-02                        running
 -     centos7                        shut off
 -     centos7-02                     shut off
 -     centos7-03                     shut off
 -     win7                           shut off
 -     win7-kvm                       shut off
```

第 2 步，使用命令“ls”查看虚拟机硬盘文件。

```
root@kvm-server:/vm-win# ls
win10  win10.qcow2  win2012.raw  win7-02.qcow2  win7-kvm.raw  win7.raw
```

第 3 步，使用命令“qemu-img convert -f raw -O qcow2 win7-kvm.raw win7-kvm.qcow2”进行格式转换。

```
root@kvm-server:/vm-win# qemu-img convert -f raw -O qcow2 win7-kvm.raw win7-kvm.qcow2
```

第 4 步，使用命令“ls”查看通过转换生成的新文件，注意保留原虚拟机硬盘文件。

```
root@kvm-server:/vm-win# ls
win10  win10.qcow2  win2012.raw  win7-02.qcow2  win7-kvm.qcow2  win7-kvm.raw  win7.raw
```

第 5 步，使用命令“qemu-img info win7-kvm.qcow2”查看新文件的相关信息。

```
root@kvm-server:/vm-win# qemu-img info win7-kvm.qcow2
image: win7-kvm.qcow2
file format: qcow2
virtual size: 20G (21474836480 bytes)
disk size: 19G
cluster_size: 65536
Format specific information:
    compat: 1.1
    lazy refcounts: false
```

```
    refcount bits: 16
    corrupt: false
root@kvm-server:/vm-win# qemu-img info win7-kvm.raw
image: win7-kvm.raw
file format: raw
virtual size: 20G (21474836480 bytes)
disk size: 19G
```

第 6 步，使用命令“virsh edit win7-kvm”编辑虚拟机的配置文件，修改硬盘文件格式和路径。

```
root@kvm-server:/vm-win# virsh edit win7-kvm
    <disk type='file' device='disk'>
      <driver name='qemu' type='qcow2'/>
      <source file='/vm-win/win7-kvm.qcow2'/>
      <target dev='hda' bus='ide'/>
      <address type='drive' controller='0' bus='0' target='0' unit='0'/>
Domain win7-kvm XML configuration edited.
```

第 7 步，使用命令“virsh start win7-kvm”启动虚拟机进行验证。

```
root@kvm-server:/vm-win# virsh start win7-kvm
Domain win7-kvm started
```

第 8 步，虚拟机启动正常，说明硬盘格式转换正常，如图 3-5-1 所示。如果无法启动或启动出现错误提示，说明硬盘格式转换出现问题。

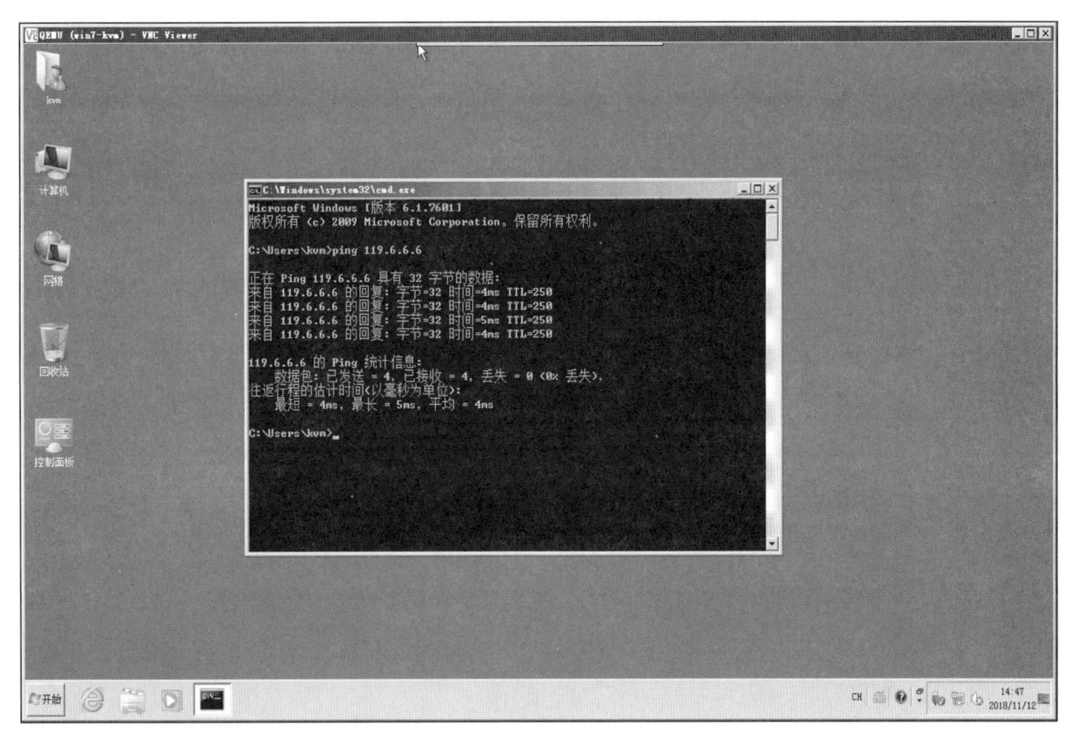

图 3-5-1

2. 虚拟机硬盘扩容

虚拟机硬盘扩容可以分为 Windows 操作系统和 Linux 操作系统两种情况，操作方式基本相同，下面分别进行介绍。

（1）Windows 虚拟机硬盘扩容

第 1 步，增加硬盘容量前需要关闭虚拟机电源，使用命令“virsh shutdown win7-kvm”关闭虚拟机电源。

```
root@kvm-server:/vm-win# virsh shutdown win7-kvm
Domain win7-kvm is being shutdown
```

第 2 步，使用命令“qemu-img resize win7-kvm.qcow2 +10G”增加 10GB 硬盘容量。

```
root@kvm-server:/vm-win# qemu-img resize win7-kvm.qcow2 +10G
Image resized.
```

第 3 步，使用命令“qemu-img info win7-kvm.qcow2”验证容量是否增加。

```
root@kvm-server:/vm-win# qemu-img info win7-kvm.qcow2
image: win7-kvm.qcow2
file format: qcow2
virtual size: 30G (32212254720 bytes)
disk size: 19G
cluster_size: 65536
Format specific information:
    compat: 1.1
    lazy refcounts: false
    refcount bits: 16
    corrupt: false
```

第 4 步，使用命令“virsh start win7-kvm”启动虚拟机，同时使用 VNC Viewer 连接到虚拟机进行查看，如图 3-5-2 所示，可以看到增加的 10GB 容量并没有被操作系统使用。

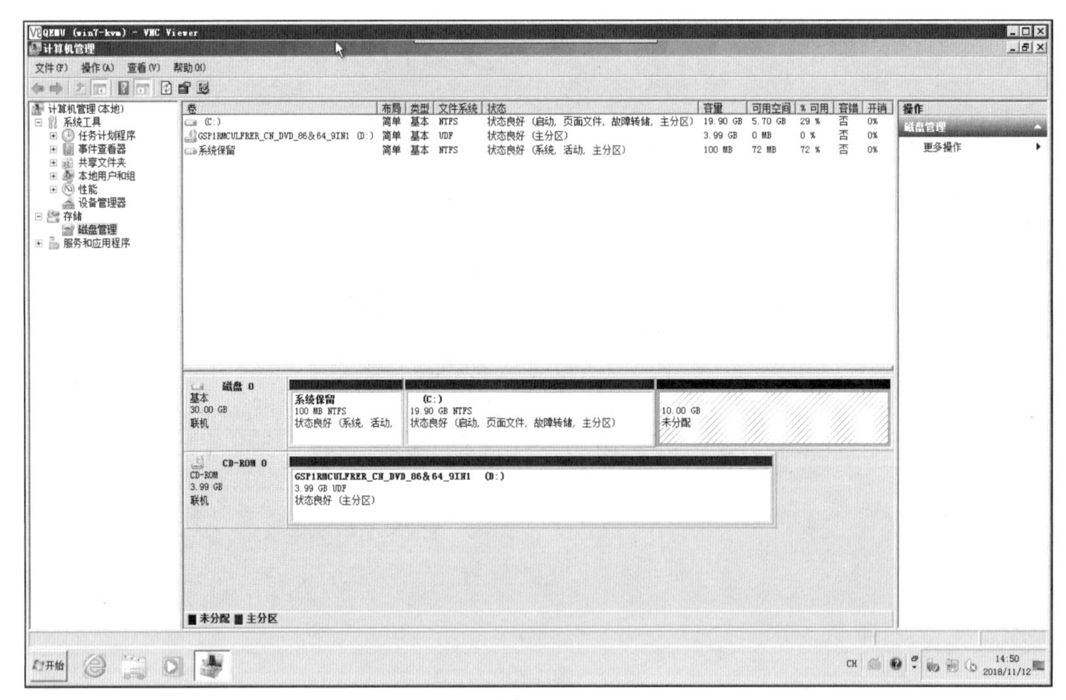

图 3-5-2

第 5 步，Windows 操作系统可以通过扩展卷的方式将新容量添加到 C 盘，扩容后 C 盘容量为 29.9GB，如图 3-5-3 所示。

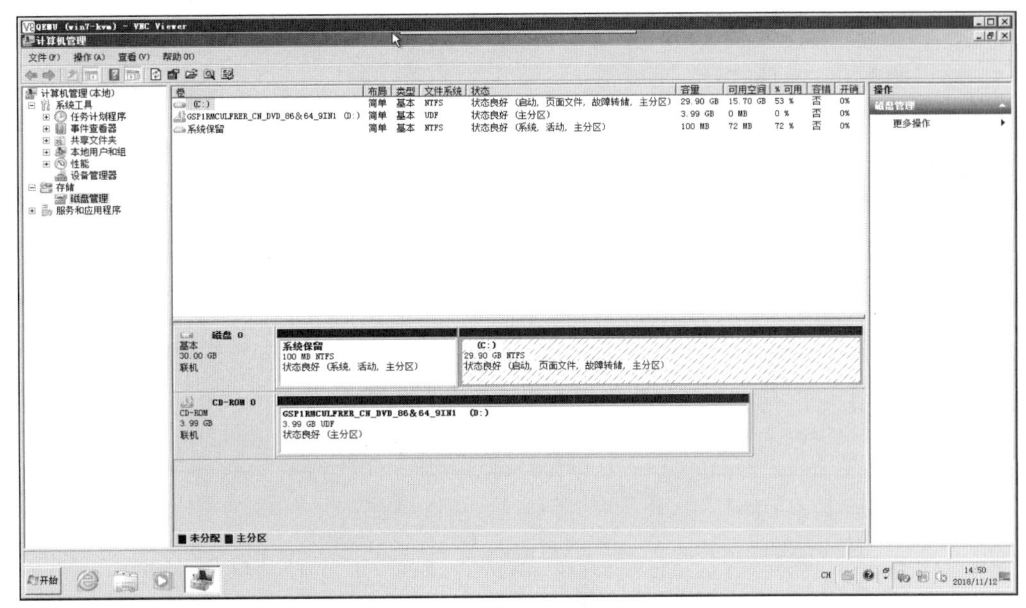

图 3-5-3

（2）Linux 虚拟机硬盘扩容

第 1 步，使用命令“fdisk -l”查看虚拟机硬盘信息，可以看到/dev/vda 容量为 16.1GB。

```
[root@kvm-centos ~]# fdisk -l
Disk /dev/vda: 16.1 GB, 16106127360 bytes, 31457280 sectors
Units = sectors of 1 * 512 = 512 bytes
Sector size (logical/physical): 512 bytes / 512 bytes
I/O size (minimum/optimal): 512 bytes / 512 bytes
Disk label type: dos
Disk identifier: 0x000043d2
   Device Boot      Start         End      Blocks   Id  System
/dev/vda1   *        2048     2099199     1048576   83  Linux
/dev/vda2         2099200     4196351     1048576   82  Linux swap / Solaris
/dev/vda3         4196352    20971519     8387584   83  Linux
```

第 2 步，使用命令“df -Th”查看各分区的容量情况，可以看到根分区容量为 8GB。

```
[root@kvm-centos ~]# df -Th
Filesystem      Type        Size     Used     Avail    Use%    Mounted on
/dev/vda3       xfs         8.0G     1.1G     7.0G     13%     /
devtmpfs        devtmpfs    911M     0        911M     0%      /dev
tmpfs           tmpfs       920M     0        920M     0%      /dev/shm
tmpfs           tmpfs       920M     8.5M     912M     1%      /run
tmpfs           tmpfs       920M     0        920M     0%      /sys/fs/cgroup
/dev/vda1       xfs         1014M    135M     880M     14%     /boot
tmpfs           tmpfs       184M     0        184M     0%      /run/user/0
```

第 3 步，关闭 Linux 虚拟机电源，在 KVM 主机上使用命令“qemu-img info centos7-04.qcow2”查看虚拟机硬盘信息，容量为 15GB，对容量进行调整。

```
root@kvm-server:/vm-linux# qemu-img info centos7-04.qcow2
image: centos7-04.qcow2
file format: qcow2
virtual size: 15G (16106127360 bytes)
disk size: 275M
cluster_size: 65536
backing file: centos7.raw
```

```
Format specific information:
    compat: 1.1
    lazy refcounts: false
    refcount bits: 16
corrupt: false
```

第 4 步，使用命令“qemu-img resize centos7-04.qcow2 +5G”增加 5GB 硬盘容量。

```
root@kvm-server:/vm-linux# qemu-img resize centos7-04.qcow2 +5G
Image resized.
root@kvm-server:/vm-linux# qemu-img info centos7-04.qcow2
image: centos7-04.qcow2
file format: qcow2
virtual size: 20G (21474836480 bytes)
disk size: 275M
cluster_size: 65536
backing file: centos7.raw
Format specific information:
    compat: 1.1
    lazy refcounts: false
    refcount bits: 16
corrupt: false
```

第 5 步，在 KVM 主机上使用命令“virsh start centos7-04”打开虚拟机电源。

```
root@kvm-server:/vm-linux# virsh start centos7-04
Domain centos7-04 started
```

第 6 步，使用命令“virsh console centos7-04”连接到虚拟机控制台。

```
root@kvm-server:/vm-linux# virsh console centos7-04
Connected to domain centos7-04
```

第 7 步，使用命令“fdisk -l”查看虚拟机硬盘信息，可以看到/dev/vda 容量已经变为 21.5GB，说明硬盘扩容成功。

```
[root@kvm-centos ~]# fdisk -l
Disk /dev/vda: 21.5 GB, 21474836480 bytes, 41943040 sectors
Units = sectors of 1 * 512 = 512 bytes
Sector size (logical/physical): 512 bytes / 512 bytes
I/O size (minimum/optimal): 512 bytes / 512 bytes
Disk label type: dos
Disk identifier: 0x000043d2
   Device Boot      Start         End      Blocks   Id  System
/dev/vda1   *        2048     2099199     1048576   83  Linux
/dev/vda2         2099200     4196351     1048576   82  Linux swap / Solaris
/dev/vda3         4196352    20971519     8387584   83  Linux
```

第 8 步，使用命令“df -Th”查看各分区的容量情况，可以看到根分区容量依旧为 8GB，说明硬盘扩容后的容量还未应用到分区。

```
[root@kvm-centos ~]# df -Th
Filesystem     Type      Size    Used   Avail   Use%    Mounted on
/dev/vda3      xfs       8.0G    1.1G   7.0G    13%     /
devtmpfs       devtmpfs  911M    0      911M    0%      /dev
tmpfs          tmpfs     920M    0      920M    0%      /dev/shm
tmpfs          tmpfs     920M    8.5M   912M    1%      /run
tmpfs          tmpfs     920M    0      920M    0%      /sys/fs/cgroup
/dev/vda1      xfs       1014M   135M   880M    14%     /boot
tmpfs          tmpfs     184M    0      184M    0%      /run/user/0
```

第 9 步，使用命令“fdisk /dev/vda”命令删除根分区进行重建。

```
[root@kvm-centos ~]# fdisk /dev/vda
Welcome to fdisk (util-linux 2.23.2).
Changes will remain in memory only, until you decide to write them.
Be careful before using the write command.

Command (m for help): m
Command action
   a       toggle a bootable flag
   b       edit bsd disklabel
   c       toggle the dos compatibility flag
   c       delete a partition
   g       create a new empty GPT partition table
   G       create an IRIX (SGI) partition table
   l       list known partition types
   m       print this menu
   n       add a new partition
   o       create a new empty DOS partition table
   p       print the partition table
   q       quit without saving changes
   s       create a new empty Sun disklabel
   t       change a partition's system id
   u       change display/entry units
   v       verify the partition table
   w       write table to disk and exit
   x       extra functionality (experts only)
```

第 10 步，在“Command”后输入“d”，删除根分区。

```
Command (m for help): d
Partition number (1-3, default 3):
Partition 3 is deleted
```

第 11 步，在“Command”后输入“p”，查看删除后的分区信息。

```
Command (m for help): p
Disk /dev/vda: 21.5 GB, 21474836480 bytes, 41943040 sectors
Units = sectors of 1 * 512 = 512 bytes
Sector size (logical/physical): 512 bytes / 512 bytes
I/O size (minimum/optimal): 512 bytes / 512 bytes
Disk label type: dos
Disk identifier: 0x000043d2
   Device Boot     Start       End          Blocks     Id    System
/dev/vda1          2048        2099199      1048576    83    Linux
/dev/vda2          2099200     4196351      1048576    82    Linux swap / Solaris
```

第 12 步，在“Command”后输入“n”，创建新的根分区，使用默认值即可。

```
Command (m for help): n
Partition type:
   p   primary (2 primary, 0 extended, 2 free)
   e   extended
Select (default p):
Using default response p
Partition number (3,4, default 3):
First sector (4196352-41943039, default 4196352):
Using default value 4196352
Last sector, +sectors or +size{K,M,G} (4196352-41943039, default 41943039):
Using default value 41943039
Partition 3 of type Linux and of size 18 GiB is set
```

第 13 步，在“Command”后输入“p”，查看新建根分区信息。

```
Command (m for help): p
```

```
Disk /dev/vda: 21.5 GB, 21474836480 bytes, 41943040 sectors
Units = sectors of 1 * 512 = 512 bytes
Sector size (logical/physical): 512 bytes / 512 bytes
I/O size (minimum/optimal): 512 bytes / 512 bytes
Disk label type: dos
Disk identifier: 0x000043d2
   Device Boot     Start        End         Blocks     Id   System
/dev/vda1          2048         2099199     1048576    83   Linux
/dev/vda2          2099200      4196351     1048576    82   Linux swap / Solaris
/dev/vda3          4196352      41943039    18873344   83   Linux
```

第14步，在“Command”后输入“w”，保存配置并退出。

```
Command (m for help): w
The partition table has been altered!
Calling ioctl() to re-read partition table.
WARNING: Re-reading the partition table failed with error 16: Device or resource busy.
The kernel still uses the old table. The new table will be used at
the next reboot or after you run partprobe(8) or kpartx(8)
Syncing disks.
```

第15步，使用命令“partprobe /dev/vda”激活容量。

```
[root@kvm-centos ~]# partprobe /dev/vda
```

第16步，使用命令“xfs_growfs /dev/vda3”刷新容量。

```
[root@kvm-centos ~]#xfs_growfs /dev/vda3
```

第17步，使用命令“df -Th”查看各分区的容量情况，可以看到根分区容量为18GB，说明扩容后硬盘容量成功应用到分区。

```
[root@kvm-centos ~]# df -Th
Filesystem    Type        Size     Used    Avail    Use%    Mounted on
/dev/vda3     xfs         18.0G    889M    18G      5%      /
devtmpfs      devtmpfs    911M     0       911M     0%      /dev
tmpfs         tmpfs       920M     0       920M     0%      /dev/shm
tmpfs         tmpfs       920M     8.5M    912M     1%      /run
tmpfs         tmpfs       920M     0       920M     0%      /sys/fs/cgroup
/dev/vda1     xfs         1014M    135M    880M     14%     /boot
tmpfs         tmpfs       184M     0       184M     0%      /run/user/0
```

3.6 虚拟机网络架构

无论是虚拟化还是传统架构，网络都是非常重要的环节，虚拟化的通信和对外提供服务都依赖于网络，在KVM环境下也提供了多种网络供选择使用。前文没有介绍网络，但细心的读者可能会发现，在KVM主机上创建的虚拟机获取的IP地址是“192.168”开头的地址，和KVM主机的IP地址并没有在同一网段，为什么还可以对外进行访问呢？本节将介绍在KVM环境下的网络配置。

3.6.1 KVM环境网络

在KVM环境下，虚拟机需要对外进行通信，通信就意味着网络连接的问题，KVM提供了内置和外部等多种网络支持方式。KVM网络分为NAT、Bridge、Route、Isolate这4种模式，在生产环境中，主要使用NAT和Bridge两种模式。

1. NAT 模式

NAT 模式是在安装 libvirt 后使用的默认模式，虚拟网络交换机以 NAT 模式运行。NAT 模式支持 KVM 主机与虚拟机相互访问，同时也支持虚拟机访问外部网络，但不支持外部网络访问虚拟机。简单来说，在 NAT 模式下虚拟机获取的 IP 地址是“192.168”开头的地址，属于 NAT 模式分配的 IP 地址，和 KVM 主机的 IP 地址并没有在同一网段，所以 KVM 主机可以访问虚拟机而外部网络不能访问。NAT 模式在生产环境中一般用于虚拟桌面。

（1）CentOS 7 操作系统安装 libvirt 默认创建的 virbr0 接口

```
[root@kvm-centos7 /]# ip addr
8: virbr0: <BROADCAST,MULTICAST,UP,LOWER_UP> mtu 1500 qdisc noqueue state UP group default qlen 1000
    link/ether 52:54:00:89:5e:bd brd ff:ff:ff:ff:ff:ff
    inet 192.168.122.1/24 brd 192.168.122.255 scope global virbr0
      valid_lft forever preferred_lft forever
```

（2）Ubuntu 18.04 操作系统安装 libvirt 默认创建的 virbr0 接口

```
root@kvm-server:/# ip addr
7: virbr0: <BROADCAST,MULTICAST,UP,LOWER_UP> mtu 1500 qdisc noqueue state UP group default qlen 1000
    link/ether 52:54:00:ed:52:28 brd ff:ff:ff:ff:ff:ff
    inet 192.168.122.1/24 brd 192.168.122.255 scope global virbr0
      valid_lft forever preferred_lft forever
```

2. Bridge 模式

Bridge 模式可以让虚拟机和 KVM 主机在同一网段，这样配置，外部网络就可以访问虚拟机提供的服务了。Bridge 模式的做法是新创建配置文件，将其关联到 KVM 主机物理网卡，通过这样的方式实现访问。Bridge 模式是生产环境中最常用的模式，用于虚拟机和外部网络的相互访问。

（1）查看 KVM 主机物理网卡 ens2f1 配置信息

```
[root@kvm-centos7 /]# cat /etc/sysconfig/network-scripts/ifcfg-ens2f1
TYPE=Ethernet
PROXY_METHOD=none
BROWSER_ONLY=no
DEFROUTE=yes
BOOTPROTO=static
IPV4_FAILURE_FATAL=no
IPV6INIT=yes
IPV6_AUTOCONF=yes
IPV6_DEFROUTE=yes
IPV6_FAILURE_FATAL=no
IPV6_ADDR_GEN_MODE=stable-privacy
NAME=ens2f1
UUID=05146ffb-e6a3-46e8-b541-229195f0da8d
DEVICE=ens2f1
ONBOOT=yes
BRIDGE=br0  #关联到 br0 配置文件
PADDR=10.92.10.122
NETMASK=255.255.255.0
GATEWAY=10.92.10.254
DNS1=119.6.6.6
```

（2）查看新创建的 br0 网卡配置信息

```
[root@kvm-centos7 /]# cat /etc/sysconfig/network-scripts/ifcfg-br0
TYPE=Bridge
```

```
PROXY_METHOD=none
BROWSER_ONLY=no
DEFROUTE=yes
BOOTPROTO=static
IPV4_FAILURE_FATAL=no
IPV6INIT=yes
IPV6_AUTOCONF=yes
IPV6_DEFROUTE=yes
IPV6_FAILURE_FATAL=no
IPV6_ADDR_GEN_MODE=stable-privacy
NAME=br0
UUID=05146ffb-e6a3-46e8-b541-229195f0da8d
DEVICE=br0
ONBOOT=yes
IPADDR=10.92.10.122
NETMASK=255.255.255.0
GATEWAY=10.92.10.254
DNS1=119.6.6.6
ZONE=public
```

（3）查看相关配置信息

```
[root@kvm-centos7 /]# ip addr
3: ens2f1: <BROADCAST,MULTICAST,UP,LOWER_UP> mtu 1500 qdisc mq master br0 state UP group default qlen 1000
   link/ether 00:10:18:72:16:66 brd ff:ff:ff:ff:ff:ff
7: br0: <BROADCAST,MULTICAST,UP,LOWER_UP> mtu 1500 qdisc noqueue state UP group default qlen 1000
   link/ether 00:10:18:72:16:66 brd ff:ff:ff:ff:ff:ff
   inet 10.92.10.122/24 brd 10.92.10.255 scope global noprefixroute br0
     valid_lft forever preferred_lft forever
   inet6 fe80::6b72:c989:1193:d011/64 scope link noprefixroute
     valid_lft forever preferred_lft forever
```

3. Route模式

Route模式属于一种特殊模式，使用Route模式，虚拟交换机连接到KVM主机物理网卡，在不使用NAT模式的情况下来回传输流量。使用Route模式时，所有虚拟机都位于自己的子网中，通过虚拟交换机进行路由。外部网络不配置的路由信息是没法发现这些虚拟机的，并且不能访问虚拟机。

4. Isolate模式

Isolate模式也属于一种特殊模式，使用Isolate模式，连接到虚拟交换机的虚拟机可以相互通信，也可以与主机物理机通信，但其通信不会传到KVM主机外，也不能从KVM主机机外部接收通信。

3.6.2 配置KVM桥接网络

了解KVM环境基本的网络架构后，需要学习对其进行配置，下文介绍KVM桥接网络配置。

第1步，使用命令“yum install bridge-utils”安装桥接网络组件。

```
[root@kvm-centos7 ~]# yum install bridge-utils
已加载插件：fastestmirror, langpacks
Loading mirror speeds from cached hostfile
 * base: mirrors.aliyun.com
 * extras: mirrors.aliyun.com
 * updates: mirrors.aliyun.com
......
软件包 bridge-utils-1.5-9.el7.x86_64 已安装并且是最新版本
```

```
无须任何处理
```

第 2 步，选择 KVM 主机物理网卡作为虚拟机对外通桥接网卡，推荐选择非 KVM 主机管理网络，案例中选择使用 KVM 主机 enp2s0f1 作为桥接网卡，查看配置文件。

```
[root@kvm-centos7 ~]# cat /etc/sysconfig/network-scripts/ifcfg-enp2s0f1
TYPE=Ethernet
PROXY_METHOD=none
BROWSER_ONLY=no
BOOTPROTO=dhcp
DEFROUTE=yes
IPV4_FAILURE_FATAL=no
IPV6INIT=yes
NAME=enp2s0f1
UUID=00421d64-dbef-47e2-a793-0305d05ecf79
DEVICE=enp2s0f1
ONBOOT=yes
```

第 3 步，通过复制网卡配置文件创建桥接网卡 br1。

```
[root@kvm-centos7 network-scripts]# cp ifcfg-enp2s0f1 ifcfg-br1
[root@kvm-centos7 network-scripts]# ls ifcfg-br1
ifcfg-br1
[root@kvm-centos7 network-scripts]# cat ifcfg-br1
TYPE=Ethernet
PROXY_METHOD=none
BROWSER_ONLY=no
BOOTPROTO=dhcp
DEFROUTE=yes
IPV4_FAILURE_FATAL=no
NAME=enp2s0f1
UUID=00421d64-dbef-47e2-a793-0305d05ecf79
DEVICE=enp2s0f1
ONBOOT=yes
```

第 4 步，修改 KVM 主机配置 enp2s0f1 文件，注意添加 BRIDGE 和 NM_CONTROLLED 两个参数。

```
[root@kvm-centos7 network-scripts]# vi ifcfg-enp2s0f1
TYPE=Ethernet
PROXY_METHOD=none
BROWSER_ONLY=no
BOOTPROTO=dhcp  #如果网络中存在 DHCP 服务器可以选择，也可手动指定
DEFROUTE=yes
IPV4_FAILURE_FATAL=no
NAME=enp2s0f1
UUID=00421d64-dbef-47e2-a793-0305d05ecf79
DEVICE=enp2s0f1
ONBOOT=yes
BRIDGE=br1  #将物理网卡桥接到 br1 网卡，必须配置
NM_CONTROLLED=no  #配置是否 NetworkManger 服务控制该网络接口
```

第 5 步，修改桥接网卡 br1 配置文件，注意修改 TYPE 类型。

```
[root@kvm-centos7 network-scripts]# vi ifcfg-br1
TYPE=Bridge  #将 TYPE 类型由 Ethernet 修改为 Bridge，必须配置
PROXY_METHOD=none
BROWSER_ONLY=no
BOOTPROTO=dhcp  #如果网络中存在 DHCP 服务器，可以选择，也可手动指定
DEFROUTE=yes
IPV4_FAILURE_FATAL=no
NAME=br1
```

```
UUID=00421d64-dbef-47e2-a793-0305d05ecf79
DEVICE=br1
ONBOOT=yes
```

第6步，重启网络服务，使用命令“brctl show”查看KVM主机物理网卡与桥接网卡的关系是否正确，配置正确则br1与enp2s0f1之间建立桥接关系。

```
[root@kvm-centos7 ~]# brctl show
bridge name     bridge id               STP enabled     interfaces
br0             8000.001018721666       no              ens2f1
br1             8000.d8d385678b61       no              enp2s0f1
virbr0          8000.525400895ebd       yes             virbr0-nic
```

第7步，使用命令“virsh edit centos7-03”修改虚拟机配置文件。

```
[root@kvm-centos7 ~]# virsh edit centos7-03
<interface type='bridge'>
    <mac address='52:54:00:bd:15:7a'/>
    <source bridge='br1'/>   #修改为新的桥接网卡br1
    <model type='virtio'/>
    <address type='pci' domain='0x0000' bus='0x00' slot='0x03' function='0x0'/>
```

第8步，使用命令“brctl show”查看虚拟机与桥接网卡是否建立映射。

```
[root@kvm-centos7 ~]# brctl show
bridge name     bridge id               STP enabled     interfaces
br0             8000.001018721666       no              ens2f1
br1             8000.d8d385678b61       no              enp2s0f1
                                                        vnet0
virbr0          8000.525400895ebd       yes             virbr0-nic
```

第9步，开启虚拟机电源，使用VNC Viewer工具查看虚拟机获取的IP地址和对外通信情况，虚拟机新获取的IP地址为10.92.10.127/24，与KVM主机10.92.10.7/24处于同一网段，说明虚拟机不处于NAT模式，如图3-6-1所示。

```
QEMU (centos7-03) - VNC Viewer
[root@localhost ~]# ip addr
1: lo: <LOOPBACK,UP,LOWER_UP> mtu 65536 qdisc noqueue state UNKNOWN qlen 1
    link/loopback 00:00:00:00:00:00 brd 00:00:00:00:00:00
    inet 127.0.0.1/8 scope host lo
       valid_lft forever preferred_lft forever
    inet6 ::1/128 scope host
       valid_lft forever preferred_lft forever
2: eth0: <BROADCAST,MULTICAST,UP,LOWER_UP> mtu 1500 qdisc pfifo_fast state UP qlen 1000
    link/ether 52:54:00:bd:15:7a brd ff:ff:ff:ff:ff:ff
    inet 10.92.10.127/24 brd 10.92.10.255 scope global dynamic eth0
       valid_lft 86196sec preferred_lft 86196sec
    inet6 fe80::969f:deb2:306e:95c4/64 scope link
       valid_lft forever preferred_lft forever
[root@localhost ~]# ping 119.6.6.6
PING 119.6.6.6 (119.6.6.6) 56(84) bytes of data.
64 bytes from 119.6.6.6: icmp_seq=1 ttl=251 time=5.52 ms
64 bytes from 119.6.6.6: icmp_seq=2 ttl=251 time=5.51 ms
64 bytes from 119.6.6.6: icmp_seq=3 ttl=251 time=9.03 ms
64 bytes from 119.6.6.6: icmp_seq=4 ttl=251 time=4.09 ms
^C
--- 119.6.6.6 ping statistics ---
4 packets transmitted, 4 received, 0% packet loss, time 3004ms
rtt min/avg/max/mdev = 4.097/6.042/9.034/1.824 ms
[root@localhost ~]# ping 51cto.com
PING 51cto.com (47.95.170.110) 56(84) bytes of data.
64 bytes from 47.95.170.110 (47.95.170.110): icmp_seq=1 ttl=91 time=39.2 ms
From gateway (10.92.10.254) icmp_seq=2 Redirect Network(New nexthop: 10.92.10.197 (10.92.10.197))
From gateway (10.92.10.254): icmp_seq=2 Redirect Network(New nexthop: 10.92.10.197 (10.92.10.197))
64 bytes from 47.95.170.110 (47.95.170.110): icmp_seq=2 ttl=91 time=38.4 ms
64 bytes from 47.95.170.110 (47.95.170.110): icmp_seq=3 ttl=91 time=39.6 ms
64 bytes from 47.95.170.110 (47.95.170.110): icmp_seq=4 ttl=91 time=39.0 ms
^C
--- 51cto.com ping statistics ---
4 packets transmitted, 4 received, +1 errors, 0% packet loss, time 3004ms
rtt min/avg/max/mdev = 38.433/39.099/39.670/0.465 ms
[root@localhost ~]# ping ovirt.org
PING ovirt.org (8.43.85.224) 56(84) bytes of data.
64 bytes from www.ovirt.org (8.43.85.224): icmp_seq=1 ttl=49 time=374 ms
64 bytes from www.ovirt.org (8.43.85.224): icmp_seq=2 ttl=49 time=395 ms
64 bytes from www.ovirt.org (8.43.85.224): icmp_seq=3 ttl=49 time=382 ms
From gateway (10.92.10.254) icmp_seq=4 Redirect Network(New nexthop: 10.92.10.197 (10.92.10.197))
From gateway (10.92.10.254): icmp_seq=4 Redirect Network(New nexthop: 10.92.10.197 (10.92.10.197))
^C
--- ovirt.org ping statistics ---
4 packets transmitted, 3 received, +1 errors, 25% packet loss, time 3754ms
rtt min/avg/max/mdev = 374.361/383.955/395.296/8.635 ms
[root@localhost ~]#
```

图3-6-1

第 10 步，配置使用 SecureCRT 从外部连接到虚拟机，访问正常，能够通过 SSH 访问虚拟机说明外部用户也可以访问虚拟机提供的服务，如图 3-6-2 所示。

```
kvm-centos7 - SecureCRT
[admin@localhost ~]$ ip addr
1: lo: <LOOPBACK,UP,LOWER_UP> mtu 65536 qdisc noqueue state UNKNOWN qlen 1
    link/loopback 00:00:00:00:00:00 brd 00:00:00:00:00:00
    inet 127.0.0.1/8 scope host lo
       valid_lft forever preferred_lft forever
    inet6 ::1/128 scope host
       valid_lft forever preferred_lft forever
2: eth0: <BROADCAST,MULTICAST,UP,LOWER_UP> mtu 1500 qdisc pfifo_fast state UP qlen 1000
    link/ether 52:54:00:bd:15:7a brd ff:ff:ff:ff:ff:ff
    inet 10.92.10.127/24 brd 10.92.10.255 scope global dynamic eth0
       valid_lft 86088sec preferred_lft 86088sec
    inet6 fe80::969f:deb2:306e:95c4/64 scope link
       valid_lft forever preferred_lft forever
[admin@localhost ~]$ ping ovirt.org
PING ovirt.org (8.43.85.224) 56(84) bytes of data.
64 bytes from www.ovirt.org (8.43.85.224): icmp_seq=1 ttl=49 time=399 ms
64 bytes from www.ovirt.org (8.43.85.224): icmp_seq=2 ttl=49 time=366 ms
64 bytes from www.ovirt.org (8.43.85.224): icmp_seq=3 ttl=49 time=400 ms
64 bytes from www.ovirt.org (8.43.85.224): icmp_seq=4 ttl=49 time=402 ms
^C
--- ovirt.org ping statistics ---
4 packets transmitted, 4 received, 0% packet loss, time 3000ms
rtt min/avg/max/mdev = 366.590/392.247/402.458/14.866 ms
[admin@localhost ~]$ ping 51cto.com
PING 51cto.com (47.95.170.110) 56(84) bytes of data.
64 bytes from 47.95.170.110 (47.95.170.110): icmp_seq=1 ttl=91 time=38.0 ms
64 bytes from 47.95.170.110 (47.95.170.110): icmp_seq=2 ttl=91 time=38.0 ms
64 bytes from 47.95.170.110 (47.95.170.110): icmp_seq=3 ttl=91 time=37.6 ms
64 bytes from 47.95.170.110 (47.95.170.110): icmp_seq=4 ttl=91 time=38.5 ms
64 bytes from 47.95.170.110 (47.95.170.110): icmp_seq=5 ttl=91 time=38.6 ms
^C
--- 51cto.com ping statistics ---
5 packets transmitted, 5 received, 0% packet loss, time 4006ms
rtt min/avg/max/mdev = 37.647/38.193/38.693/0.413 ms
[admin@localhost ~]$
```

图 3-6-2

第 11 步，使用命令“ip addr”查看 IP 地址，通过图 3-6-3 可以看到，虚拟机目前使用的是 DHCP 分配的地址，生产环境可以根据实际情况配置静态 IP 地址（静态 IP 地址的配置方法可以参考前文）。

```
QEMU (centos7-03) - VNC Viewer
[root@localhost ~]# ip addr
1: lo: <LOOPBACK,UP,LOWER_UP> mtu 65536 qdisc noqueue state UNKNOWN qlen 1
    link/loopback 00:00:00:00:00:00 brd 00:00:00:00:00:00
    inet 127.0.0.1/8 scope host lo
       valid_lft forever preferred_lft forever
    inet6 ::1/128 scope host
       valid_lft forever preferred_lft forever
2: eth0: <BROADCAST,MULTICAST,UP,LOWER_UP> mtu 1500 qdisc pfifo_fast state UP qlen 1000
    link/ether 52:54:00:bd:15:7a brd ff:ff:ff:ff:ff:ff
    inet 10.92.10.127/24 brd 10.92.10.255 scope global dynamic eth0
       valid_lft 86232sec preferred_lft 86232sec
    inet6 fe80::969f:deb2:306e:95c4/64 scope link
       valid_lft forever preferred_lft forever
[root@localhost ~]# cat /etc/sysconfig/network-scripts/ifcfg-eth0
TYPE=Ethernet
PROXY_METHOD=none
BROWSER_ONLY=no
BOOTPROTO=dhcp
DEFROUTE=yes
IPV4_FAILURE_FATAL=no
IPV6INIT=yes
IPV6_AUTOCONF=yes
IPV6_DEFROUTE=yes
IPV6_FAILURE_FATAL=no
IPV6_ADDR_GEN_MODE=stable-privacy
NAME=eth0
UUID=2ae08e8b-529c-4799-b5c9-43008fb742c9
DEVICE=eth0
ONBOOT=yes
[root@localhost ~]# _
```

图 3-6-3

通过上述配置操作，外部用户可以通过 SecureCRT 管理虚拟机，同时虚拟机如果安装有服务程序也可以对外提供服务。至此，完成 KVM 主机的基本网络配置。

3.7 虚拟机日常操作

通过对前文的学习，相信读者已经掌握了Linux虚拟机和Windows虚拟机的创建，以及通过调整KVM主机网络让虚拟机可以从内、外部访问。在生产环境会根据需要对相关参数进行调整。整体来说KVM对Linux虚拟机硬件的支持都非常不错，对Windows虚拟机硬件的支持略差一些，本节将重点介绍基于Windows虚拟机常用的参数调整和性能优化。

3.7.1 调整虚拟机硬件

生产环境中比较常见的虚拟机硬件有CPU、内存、硬盘以及网卡。对于虚拟机硬盘容量调整在前文已经进行了介绍，本节介绍如何调整虚拟机CPU、内存等。

1. 虚拟机CPU调整

虚拟机CPU调整分为Linux虚拟机和Windows虚拟机两种情况，需特别注意Windows 7虚拟机可能存在一些特殊的调整方式。

（1）Linux虚拟机CPU调整

对于Linux虚拟机，调整CPU数量实际就是修改虚拟机配置文件。

第1步，使用命令“virsh dominfo centos7-03”查看虚拟机硬件信息，虚拟机centos7-03目前CPU数量为1。

```
[root@kvm-centos7 ~]# virsh dominfo centos7-03
Id:             2
名称:           centos7-03
UUID:           6cf38c45-be49-4612-afe9-114aa203b3f0
OS 类型:        hvm
状态:           running
CPU:            1   #CPU 数量为 1
CPU 时间:       20.3s
最大内存:       2097152 KiB
使用的内存:     2097152 KiB
持久:           是
自动启动:       禁用
管理的保存:     否
安全性模式:     selinux
安全性 DOI:     0
安全性标签:     system_u:system_r:svirt_t:s0:c404,c721 (enforcing)
```

第2步，调整虚拟机CPU数量前需要关闭虚拟机电源，使用命令“virsh shutdown centos7-03”关闭虚拟机电源。

```
[root@kvm-centos7 ~]# virsh shutdown centos7-03
域 centos7-03 被关闭
```

第3步，使用命令“virsh edit centos7-03”修改虚拟机配置文件，将CPU数量调整为4。

```
[root@kvm-centos7 ~]# virsh edit centos7-03
<domain type='kvm'>
  <name>centos7-03</name>
  <uuid>6cf38c45-be49-4612-afe9-114aa203b3f0</uuid>
  <memory unit='KiB'>2097152</memory>
  <currentMemory unit='KiB'>2097152</currentMemory>
  <vcpu placement='static'>4</vcpu>  #调整 CPU 数量为 4
```

```
  <os>
```

第 4 步，使用命令“virsh start centos7-03”打开虚拟机电源。

```
[root@kvm-centos7 ~]# virsh start centos7-03
域 centos7-03 已开始
```

第 5 步，使用命令“virsh dominfo centos7-03”查看虚拟机硬件信息，调整后虚拟机 centos7-03 CPU 数量为 4。

```
[root@kvm-centos7 ~]# virsh dominfo centos7-03
Id:             2
名称:           centos7-03
UUID:           6cf38c45-be49-4612-afe9-114aa203b3f0
OS 类型:        hvm
状态:           running
CPU:            4   #调整后 CPU 数量为 4
CPU 时间:       18.2s
最大内存:       2097152 KiB
使用的内存:     2097152 KiB
持久:           是
自动启动:       禁用
管理的保存:     否
安全性模式:     selinux
安全性 DOI:     0
安全性标签:     system_u:system_r:svirt_t:s0:c684,c911 (enforcing)
```

（2）Windows Server 2012 虚拟机 CPU 调整

Windows Server 2012 虚拟机 CPU 调整与 Linux 虚拟机类似，同样是修改虚拟机配置文件。

第 1 步，使用命令“virsh dominfo win2012”查看虚拟机硬件信息，虚拟机 win2012 目前 CPU 数量为 2。

```
root@kvm-server:/home/admin# virsh dominfo win2012
Id:             8
Name:           win2012
UUID:           1aaea0ca-d562-43dd-8c6b-6e67f4e3b806
OS Type:        hvm
State:          running
CPU(s):         2   #CPU 数量为 2
CPU time:       43402.9s
Max memory:     4194304 KiB
Used memory:    4194304 KiB
Persistent:     yes
Autostart:      disable
Managed save:   no
Security model: apparmor
Security DOI:   0
Security label: libvirt-1aaea0ca-d562-43dd-8c6b-6e67f4e3b806 (enforcing)
```

第 2 步，使用命令“virsh edit win2012”修改虚拟机配置文件，将 CPU 数量调整为 4。

```
root@kvm-server:/home/admin# virsh edit win2012
<domain type='kvm'>
  <name>win2012</name>
  <uuid>1aaea0ca-d562-43dd-8c6b-6e67f4e3b806</uuid>
  <memory unit='KiB'>4194304</memory>
  <currentMemory unit='KiB'>4194304</currentMemory>
  <vcpu placement='static'>4</vcpu>  #调整 CPU 数量为 4
  <os>
```

第 3 步，使用命令“virsh start win2012”打开虚拟机电源。

```
root@kvm-server:/home/admin# virsh start win2012
Domain win2012 started
```

第 4 步，使用命令“virsh dominfo win2012”查看虚拟机硬件信息，调整后虚拟机 win2012 CPU 数量为 4。

```
root@kvm-server:/home/admin# virsh dominfo win2012
Id:             8
Name:           win2012
UUID:           1aaea0ca-d562-43dd-8c6b-6e67f4e3b806
OS Type:        hvm
State:          running
CPU(s):         4   #调整后 CPU 数量为 4
CPU time:       2.8s
Max memory:     4194304 KiB
Used memory:    4194304 KiB
Persistent:     yes
Autostart:      disable
Managed save:   no
Security model: apparmor
Security DOI:   0
Security label: libvirt-1aaea0ca-d562-43dd-8c6b-6e67f4e3b806 (enforcing)
```

第 5 步，使用 VNC Viewer 工具连接到虚拟机 win2012，查看 CPU 数量为 4，如图 3-7-1 所示。

图 3-7-1

（3）Windows 7 虚拟机 CPU 调整

Windows 7 虚拟机 CPU 调整同样是修改虚拟机配置文件，但需要注意 Windows 7 操作系统 2 个以上 CPU 调整后不能识别的问题。

第 1 步，使用命令“virsh dominfo win7-02”查看虚拟机硬件信息，虚拟机 win7-02 目前 CPU 数量为 2。

```
root@kvm-server:/home/admin# virsh dominfo win7-02
Id:             49
Name:           win7-02
UUID:           27e0e8ed-2c9c-4c3f-8b18-9a88790d8737
OS Type:        hvm
State:          running
CPU(s):         2   #CPU 数量为 2
CPU time:       3.2s
Max memory:     4194304 KiB
Used memory:    4194304 KiB
Persistent:     yes
Autostart:      disable
Managed save:   no
Security model: apparmor
Security DOI:   0
Security label: libvirt-27e0e8ed-2c9c-4c3f-8b18-9a88790d8737 (enforcing)
```

第 2 步，使用命令“virsh edit win7-02”修改虚拟机配置文件，将 CPU 数量调整为 4。

```
root@kvm-server:/home/admin# virsh edit win7-02
<domain type='kvm'>
  <name>win7-02</name>
  <uuid>27e0e8ed-2c9c-4c3f-8b18-9a88790d8737</uuid>
  <memory unit='KiB'>4194304</memory>
  <currentMemory unit='KiB'>4194304</currentMemory>
  <vcpu placement='static'>4</vcpu>   #调整 CPU 数量为 4
  <os>
```

第 3 步，使用命令“virsh start win7-02”打开虚拟机电源。

```
root@kvm-server:/home/admin# virsh start win7-02
Domain win7-02 started
```

第 4 步，使用命令“virsh dominfo win7-02”查看虚拟机硬件信息，调整后虚拟机 win7-02 CPU 数量为 4。

```
root@kvm-server:/home/admin# virsh dominfo win7-02
Id:             50
Name:           win7-02
UUID:           27e0e8ed-2c9c-4c3f-8b18-9a88790d8737
OS Type:        hvm
State:          running
CPU(s):         4   #调整后 CPU 数量为 4
CPU time:       3.1s
Max memory:     4194304 KiB
Used memory:    4194304 KiB
Persistent:     yes
Autostart:      disable
Managed save:   no
Security model: apparmor
Security DOI:   0
Security label: libvirt-27e0e8ed-2c9c-4c3f-8b18-9a88790d8737 (enforcing)
```

第 5 步，使用 VNC Viewer 工具连接到虚拟机 win7-02，查看 CPU 数量，发现 CPU 数量没有变化，依然为 2，如图 3-7-2 所示。其原因是 KVM 默认每个 CPU 模拟一个 socket，必须修改虚拟机 CPU 的 topology sockets 参数，才能使用多个 CPU。

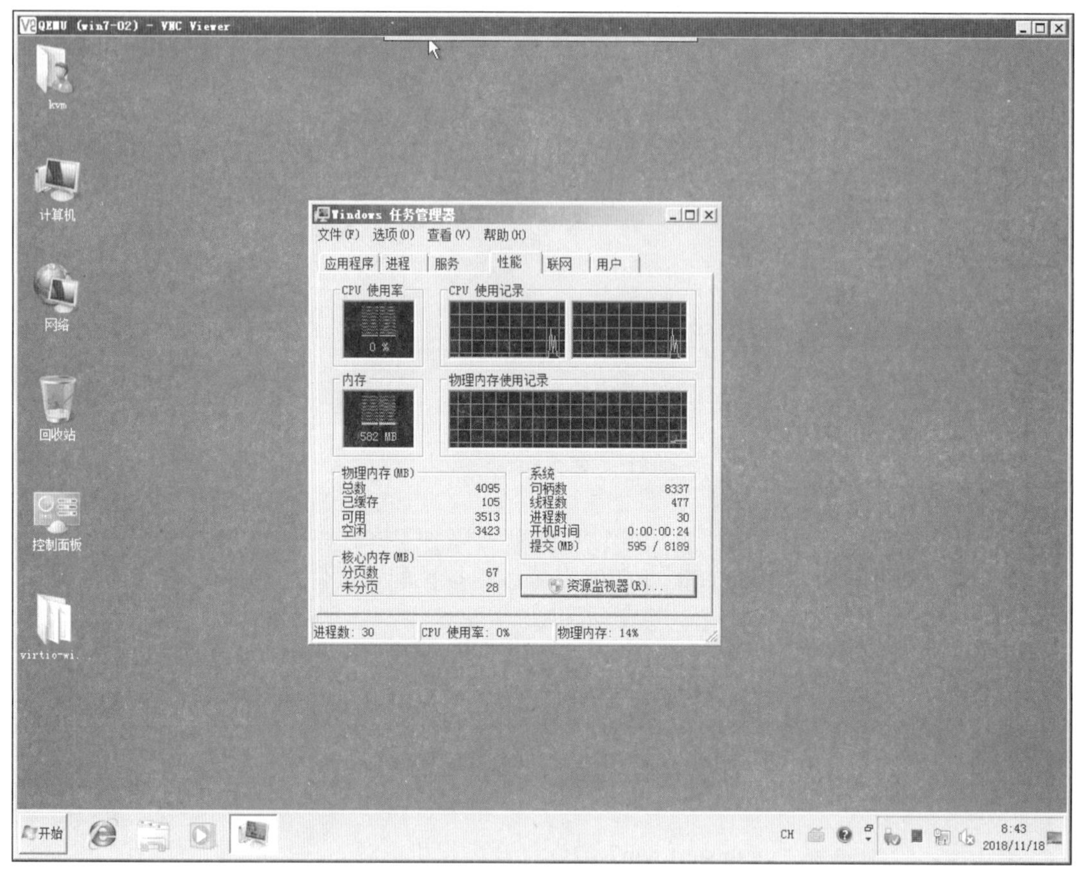

图 3-7-2

第 6 步，使用命令“virsh edit win7-02”修改虚拟机配置文件，注意数量的匹配。

```
root@kvm-server:/home/admin# virsh dominfo win7-02
<domain type='kvm'>
  <name>win7-02</name>
  <uuid>27e0e8ed-2c9c-4c3f-8b18-9a88790d8737</uuid>
  <memory unit='KiB'>4194304</memory>
  <currentMemory unit='KiB'>4194304</currentMemory>
  <vcpu placement='static'>4</vcpu>   #调整 CPU 数量为 4
  <os>
  <cpu mode='custom' match='exact' check='partial'>
    <model fallback='allow'>Westmere</model>
    <topology sockets='1' cores='4' threads='1'/>  #修改 topology sockets 参数
  </cpu>
```

第 7 步，关闭虚拟机电源后再使用 VNC Viewer 工具连接到虚拟机 win7-02，查看 CPU 数量为 4，如图 3-7-3 所示。

至此，Linux 虚拟机和 Windows 虚拟机 CPU 调整完成。需要注意的是，硬件调整需要在虚拟机电源关闭的情况下进行。

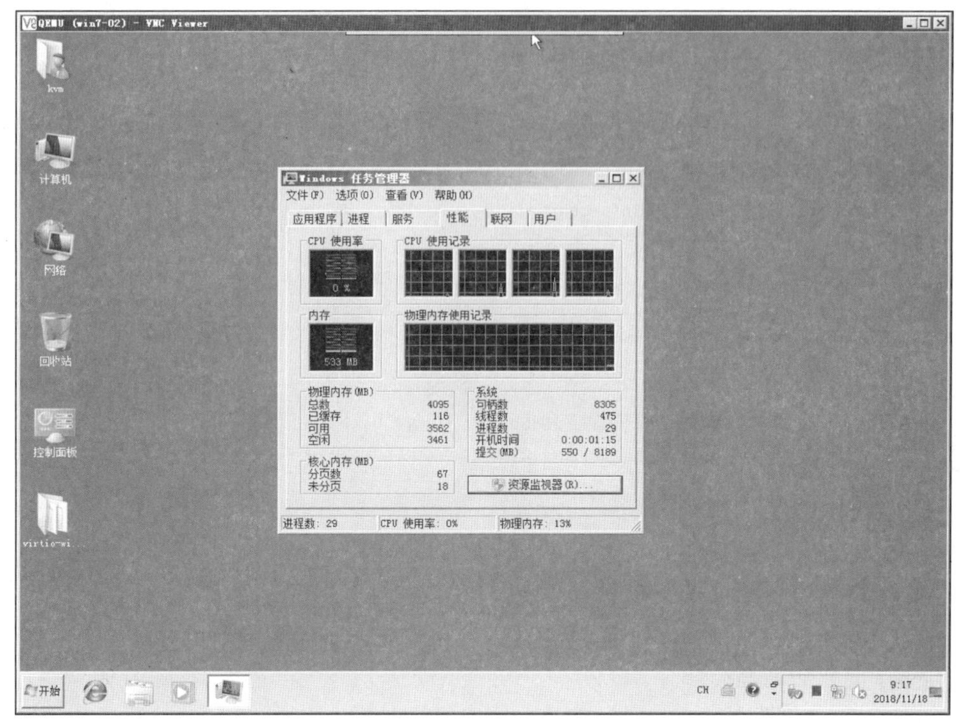

图 3-7-3

2. 虚拟机内存调整

虚拟机内存调整与 CPU 调整类似，也是修改虚拟机配置文件，注意配置文件中内存的单位为 KB。

第 1 步，使用命令“virsh dominfo centos7-03”查看虚拟机硬件信息，虚拟机 centos7-03 目前内存为“2097152 KiB”，也是就是 2GB。

```
[root@kvm-centos7 ~]# virsh dominfo centos7-03
Id:             2
名称:           centos7-03
UUID:           6cf38c45-be49-4612-afe9-114aa203b3f0
OS 类型:        hvm
状态:           running
CPU:            4
CPU 时间:       20.3s
最大内存:       2097152 KiB
使用的内存:     2097152 KiB
持久:           是
自动启动:       禁用
管理的保存:     否
安全性模式:     selinux
安全性 DOI:     0
安全性标签:     system_u:system_r:svirt_t:s0:c404,c721 (enforcing)
```

第 2 步，使用命令“virsh edit centos7-03”修改虚拟机配置文件，调整内存为 4GB。

```
[root@kvm-centos7 ~]# virsh edit centos7-03
<domain type='kvm'>
  <name>centos7-03</name>
  <uuid>6cf38c45-be49-4612-afe9-114aa203b3f0</uuid>
  <memory unit='KiB'>4194304</memory>  #调整内存为 4GB，单位 KB
```

```
<currentMemory unit='KiB'>4194304</currentMemory>
<vcpu placement='static'>4</vcpu>
<os>
```

第 3 步，使用命令“virsh dominfo centos7-03”查看虚拟机硬件信息，调整后虚拟机 centos7-03 内存为 4GB。

```
[root@kvm-centos7 ~]# virsh dominfo centos7-03
Id:             4
名称:           centos7-03
UUID:           6cf38c45-be49-4612-afe9-114aa203b3f0
OS 类型:        hvm
状态:           running
CPU:            4
CPU 时间:       18.4s
最大内存:       4194304 KiB  #调整后内存为 4GB
使用的内存:     4194304 KiB
持久:           是
自动启动:       禁用
管理的保存:     否
安全性模式:     selinux
安全性 DOI:     0
安全性标签:     system_u:system_r:svirt_t:s0:c533,c593 (enforcing)
```

至此，虚拟机内存调整完成。Windows 虚拟机调整与 Linux 虚拟机调整方式一样，不存在类似 CPU 调整的差异问题。

3. 虚拟机网卡调整

生产环境虚拟机除了 CPU、内存调整外，另外比较常见的就是网卡调整。

第 1 步，使用 VNC Viewer 工具连接到虚拟机 win2012，虚拟机配置 1 个以太网卡，如图 3-7-4 所示。

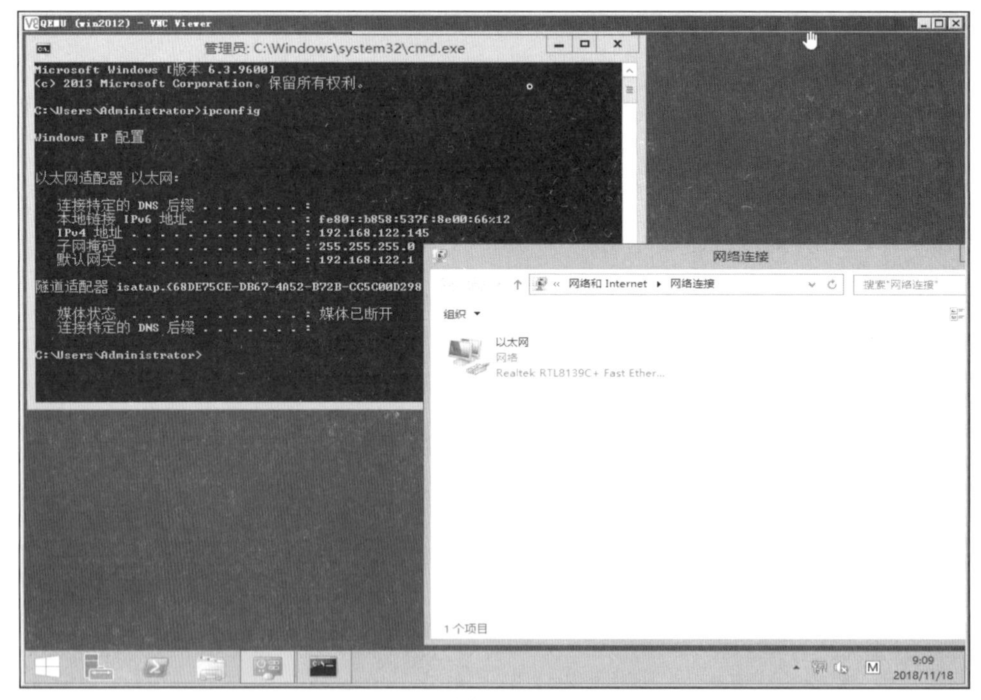

图 3-7-4

第 2 步，使用命令“virsh edit win2012”修改虚拟机配置文件，新增网卡配置文件，重新生成 mac address 和 address type。

```
root@kvm-server:/home/admin# virsh dominfo win2012
    <interface type='bridge'>
      <mac address='52:54:00:73:2e:d3'/>
      <source bridge='virbr0'/>
      <model type='rtl8139'/>
      <address type='pci' domain='0x0000' bus='0x00' slot='0x03' function='0x0'/>
    </interface>
    <interface type='bridge'>
      <source bridge='virbr0'/>
      <model type='rtl8139'/>
    </interface>
```

第 3 步，使用 VNC Viewer 工具连接到虚拟机 win2012，调整后虚拟机配置 2 个以太网卡，如图 3-7-5 所示。

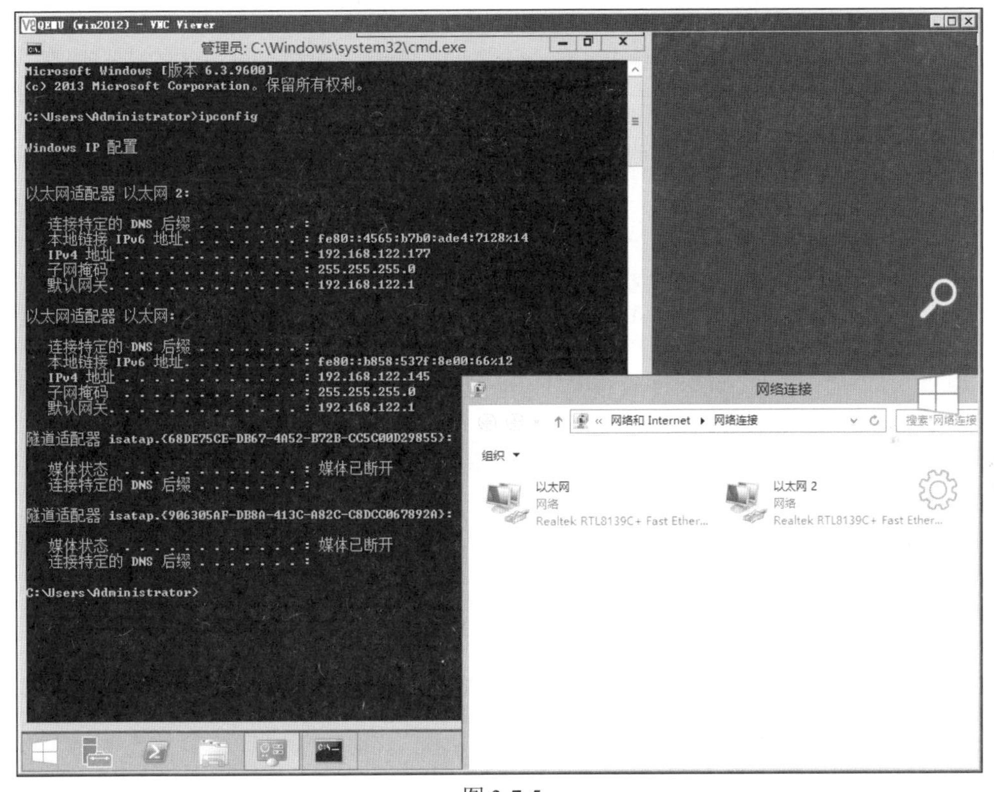

图 3-7-5

第 4 步，使用命令“virsh edit win2012”修改虚拟机配置文件，新增网卡配置文件，mac address 和 address type 已经重新生成。

```
root@kvm-server:/home/admin# virsh dominfo win2012
    <interface type='bridge'>
      <mac address='52:54:00:73:2e:d3'/>
      <source bridge='virbr0'/>
      <model type='rtl8139'/>
      <address type='pci' domain='0x0000' bus='0x00' slot='0x03' function='0x0'/>
    </interface>
```

```
<interface type='bridge'>
  <mac address='52:54:00:43:e1:2d'/>
  <source bridge='virbr0'/>
  <model type='rtl8139'/>
  <address type='pci' domain='0x0000' bus='0x00' slot='0x06' function='0x0'/>
</interface>
```

3.7.2 使用虚拟机快照

在生产环境中，虚拟机快照使用非常广泛。如需要对虚拟机更新应用或升级，但不确定操作是否会对虚拟机造成很大的影响的时候，可以创建快照记录虚拟机操作前的状态，如果更新或升级出现问题，可以快速回退到操作前的状态。特别注意：快照不是备份工具。下文介绍如何创建使用虚拟机快照。

第1步，使用命令“virsh start win7-02”为虚拟机创建第1个快照。

```
root@kvm-server:/home/admin# virsh snapshot-create win7-02
Domain snapshot 1542544562 created
```

第2步，使用命令“virsh snapshot-list win7-02”查看虚拟机win7-02的快照情况，注意Name是快照唯一的编号。

```
root@kvm-server:/home/admin# virsh snapshot-list win7-02
 Name                 Creation Time             State
--------------------------------------------------------------------------------------------------
 1542544562           2018-11-18 12:36:02     +0000 running
```

第3步，使用命令“virsh snapshot-current win7-02”查看快照文件信息。

```
root@kvm-server:/home/admin# virsh snapshot-current win7-02
<domainsnapshot>
  <name>1542544562</name>
  <state>running</state>
  <creationTime>1542544562</creationTime>
  <memory snapshot='internal'/>
  <disks>
    <disk name='hda' snapshot='internal'/>
    <disk name='hdb' snapshot='no'/>
  </disks>
  <domain type='kvm'>
    <name>win7-02</name>
    <uuid>27e0e8ed-2c9c-4c3f-8b18-9a88790d8737</uuid>
    <memory unit='KiB'>4194304</memory>
    <currentMemory unit='KiB'>4194304</currentMemory>
    <vcpu placement='static'>4</vcpu>
    <resource>
      <partition>/machine</partition>
    </resource>
   ……（省略）
  </cookie>
</domainsnapshot>
```

第4步，使用VNC Viewer工具连接到虚拟机win7-02复制桌面文件夹，如图3-7-6所示。

第5步，使用命令“virsh snapshot-revert win7-02 1542544562”恢复虚拟机快照，注意编号的唯一性，执行恢复快照操作虚拟机会重启。

```
root@kvm-server:/home/admin# virsh snapshot-revert win7-02 1542544562
```

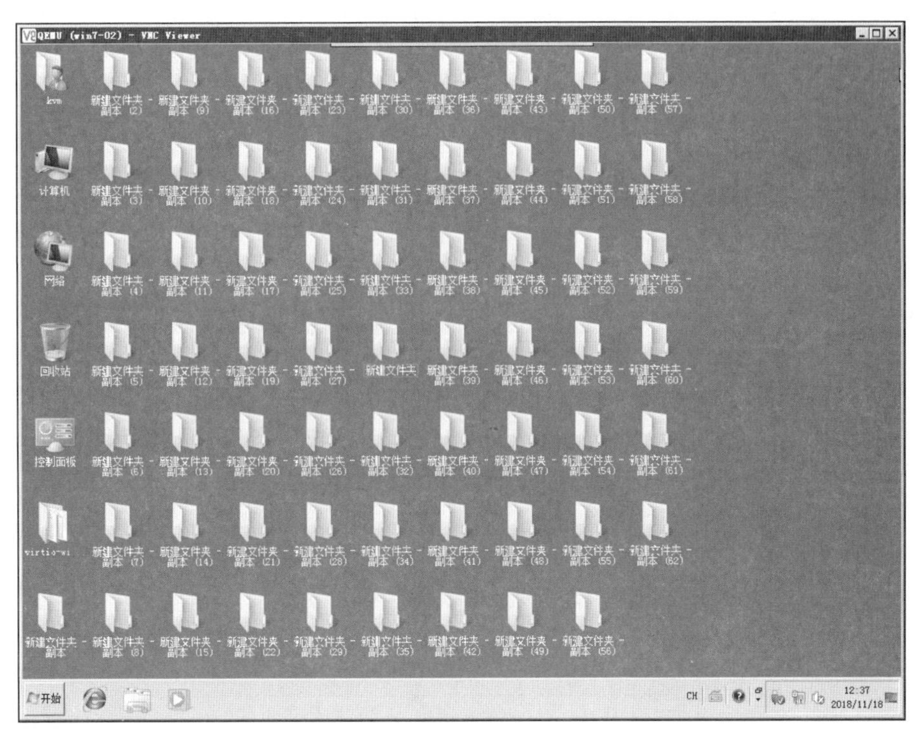

图 3-7-6

第 6 步，使用 VNC Viewer 工具连接到虚拟机 win7-02，虚拟机已经恢复到操作之前的状态，如图 3-7-7 所示。

图 3-7-7

第7步，一台虚拟机可以创建多个快照，使用命令“virsh snapshot-create win7-02”再次创建快照。

```
root@kvm-server:/home/admin# virsh snapshot-create win7-02
Domain snapshot 1542544831 created
```

第8步，使用命令“virsh snapshot-list win7-02”查看虚拟机 win7-02 的快照情况，目前虚拟机有2个快照，注意 Name 是快照唯一的编号。

```
root@kvm-server:/home/admin# virsh snapshot-list win7-02
 Name                 Creation Time             State
------------------------------------------------------------------------------------------------
 1542544562          2018-11-18 12:36:02     +0000 running
 1542544831          2018-11-18 12:40:31     +0000 running
```

第9步，使用命令“virsh snapshot-delete win7-02 1542544562”删除快照。

```
root@kvm-server:/home/admin# virsh snapshot-delete win7-02 1542544562
Domain snapshot 1542544562 deleted
```

第10步，使用命令“virsh snapshot-list win7-02”查看虚拟机 win7-02 删除快照后的情况，虚拟机只存在编号为 1542544831 的快照。

```
root@kvm-server:/home/admin# virsh snapshot-list win7-02
 Name                 Creation Time              State
------------------------------------------------------------------------------------------------
 1542544831          2018-11-18 12:40:31     +0000 running
```

使用 virsh 命令对虚拟机进行快照操作相对简单，可以在开机状态下进行。需要注意的是，恢复快照虚拟机会重启。最后再提醒读者：快照不是备份工具，过多的快照可能导致虚拟机运行缓慢甚至无法启动。

3.7.3 备份恢复虚拟机

在生产环境中，最常用的备份方式一般有两种：一是复制虚拟机镜像和配置文件备份；二是使用快照进行备份。读者肯定会问，前文说过快照不是备份工具，那么为什么在此处又会介绍使用快照对虚拟机进行备份恢复呢？原因很简单，KVM 平台备份工具相对较少，只要虚拟机不保留过多的快照在 KVM 平台，也可作为一种备份方式。下文介绍如何通过复制和快照备份恢复虚拟机。

1. 通过复制方式备份恢复虚拟机

第1步，使用命令“virsh list --all”查看虚拟机运行情况，选择虚拟机 centos7-04 进行备份恢复。

```
root@kvm-server:/# virsh list --all
 Id    Name                           State
------------------------------------------------------------------------------------------------
 43    centos7-04                     running
 46    centos7-03                     running
 68    win2012                        running
 71    win7-02                        running
 84    win2012-virtio                 running
 -     centos7                        shut off
 -     centos7-02                     shut off
 -     win7                           shut off
 -     win7-kvm                       shut off
```

第 2 步，使用命令“cp /vm-linux/centos7-04.qcow2 /vm-backup”复制虚拟机镜像文件，虚拟机处于运行状态也可以复制，推荐使用单独的目录存放虚拟机备份文件。

```
root@kvm-server:/vm-backup# cp /vm-linux/centos7-04.qcow2 /vm-backup
root@kvm-server:/vm-backup# ls
centos7-04.qcow2
```

第 3 步，通过复制文件的方式（不能忘记复制虚拟机配置文件），使用命令“cp /etc/libvirt/qemu/centos7-04.xml /vm-backup/”复制虚拟机配置文件。

```
root@kvm-server:/vm-backup# cp /etc/libvirt/qemu/centos7-04.xml /vm-backup/
root@kvm-server:/vm-backup# ls
centos7-04.qcow2  centos7-04.xml
```

第 4 步，模拟虚拟机 centos7-04 出现故障，已经不在运行列表。

```
root@kvm-server:/vm-backup# virsh list --all
 Id    Name                           State
----------------------------------------------------------------------------------------------------
 46    centos7-03                     running
 68    win2012                        running
 71    win7-02                        running
 84    win2012-virtio                 running
 -     centos7                        shut off
 -     centos7-02                     shut off
 -     win7                           shut off
 -     win7-kvm                       shut off
```

第 5 步，查看原虚拟机目录和配置文件目录，均无 centos7-04 相关信息。

```
root@kvm-server:/vm-backup# ls /vm-linux
centos7-02.qcow2  centos7-03.qcow2  centos7.raw
root@kvm-server:/vm-backup# ls /etc/libvirt/qemu/
centos7-02.xml  centos7.xml     win2012-virtio.xml        win7-02.xml  win7.xml
centos7-03.xml  networks        win2012.xml               win7-kvm.xml
```

第 6 步，通过复制或移动方式还原备份文件到原目录。

```
root@kvm-server:/vm-backup# cp centos7-04.qcow2 /vm-linux
root@kvm-server:/vm-backup# ls /vm-linux
centos7-02.qcow2  centos7-03.qcow2  centos7-04.qcow2  centos7.raw
root@kvm-server:/vm-backup# cp centos7-04.xml /etc/libvirt/qemu
root@kvm-server:/vm-backup# ls /etc/libvirt/qemu/
centos7-02.xml  centos7-04.xml    networks                  win2012.xml  win7-kvm.xml
centos7-03.xml  centos7.xml       win2012-virtio.xml        win7-02.xml  win7.xml
```

第 7 步，还原后需要重新注册虚拟机，使用命令“virsh define /etc/libvirt/qemu/centos7-04.xml”注册虚拟机。

```
root@kvm-server:/vm-backup# virsh define /etc/libvirt/qemu/centos7-04.xml
Domain centos7-04 defined from /etc/libvirt/qemu/centos7-04.xml
```

第 8 步，使用命令“virsh list --all”查看虚拟机运行情况，虚拟机 centos7-04 已重新注册，但处于关闭状态。

```
root@kvm-server:/vm-backup# virsh list --all
 Id    Name                           State
----------------------------------------------------------------------------------------------------
 46    centos7-03                     running
 68    win2012                        running
 71    win7-02                        running
 84    win2012-virtio                 running
```

```
 -    centos7                      shut off
 -    centos7-02                   shut off
 -    centos7-04                   shut off
 -    win7                         shut off
 -    win7-kvm                     shut off
```

第 9 步，使用命令“virsh start centos7-04”启动虚拟机。

```
root@kvm-server:/vm-linux# virsh start centos7-04
Domain centos7-04 started
```

第 10 步，使用命令“virsh console centos7-04”连接到虚拟机控制台，虚拟机已经启动成功，恢复正常运行。

```
root@kvm-server:/vm-backup# virsh console centos7-04
Connected to domain centos7-04
Escape character is ^]
CentOS Linux 7 (Core)
Kernel 3.10.0-693.el7.x86_64 on an x86_64
kvm-centos login: root
Password:
Last login: Tue Nov 13 00:07:25 on ttyS0
[root@kvm-centos ~]#
```

第 11 步，使用命令“virsh list --all”查看虚拟机运行情况，虚拟机 centos7-04 处于运行状态。

```
root@kvm-server:/vm-backup# virsh list --all
 Id   Name                         State
----------------------------------------------------------------------------------------------
 46   centos7-03                   running
 68   win2012                      running
 71   win7-02                      running
 84   win2012-virtio               running
 85   centos7-04                   running
 -    centos7                      shut off
 -    centos7-02                   shut off
 -    win7                         shut off
 -    win7-kvm                     shut off
```

至此，通过复制文件的方式备份和恢复虚拟机完成，这种方式操作起来比较麻烦，适用于小规模环境，复制文件备份时若虚拟机处于运行状态，推荐在虚拟机访问量小的时候进行，当然也可以通过创建脚本的方式实现自动备份。

2. 通过快照备份恢复虚拟机

通过快照备份恢复虚拟机是作者不太建议的方式，但在 KVM 虚拟化中依旧在使用。对于使用快照备份来说，作者不建议创建过多的快照备份，以避免快照过多导致的各种问题。另外，作者看到网上不少的文档使用命令“qemu-img snapshot”创建快照，但在实际使用中，这个命令创建的快照的字节数都是 0，也就是说什么也没保存下来。在生产环境中，强烈推荐使用“virsh snapshot-create”命令创建快照。

第 1 步，使用命令“virsh snapshot-list centos7-03”查看虚拟机 centos7-03 是否已创建快照。

```
root@kvm-server:/vm-backup# virsh snapshot-list centos7-03
 Name               Creation Time          State
----------------------------------------------------------------------------------------------
```

第 2 步，使用命令“virsh snapshot-create centos7-03”创建快照。

```
root@kvm-server:/vm-backup# virsh snapshot-create centos7-03
Domain snapshot 1544863104 created
```

第 3 步，使用命令“virsh snapshot-list centos7-03”查看虚拟机的快照情况，注意 Name 是快照唯一的编号。

```
root@kvm-server:/vm-backup# virsh snapshot-list     centos7-03
 Name                 Creation Time                  State
--------------------------------------------------------------------------------------------
 1544863104           2018-12-15 08:38:24 +0000      running
```

第 4 步，使用命令“virsh snapshot-create centos7-03”再次创建快照。

```
root@kvm-server:/vm-backup# virsh snapshot-create centos7-03
Domain snapshot 1544863205 created
```

第 5 步，使用命令“virsh snapshot-list centos7-03”查看虚拟机的快照情况，注意虚拟机目前已经有 2 个不同时间的快照，可以理解为 2 个备份。

```
root@kvm-server:/vm-backup# virsh snapshot-list     centos7-03
 Name                 Creation Time                  State
--------------------------------------------------------------------------------------------
 1544863104           2018-12-15 08:38:24 +0000      running
 1544863205           2018-12-15 08:40:05 +0000      running
```

第 6 步，如果虚拟机或应用程序出现故障，可以根据时间还原到故障前的状态，如使用使命“virsh snapshot-revert centos7-03 1544863104”恢复虚拟机到当时的状态。需特别注意，恢复后当前的数据就被还原到快照时的状态，要谨慎操作。

```
root@kvm-server:/vm-backup# virsh snapshot-revert centos7-03 1544863104
```

第 7 步，使用命令“virsh console centos7-03”查看恢复快照后虚拟机的运行状态，一定要确认虚拟机的运行是否存在问题。

```
root@kvm-server:/vm-backup# virsh console centos7-03
Connected to domain centos7-03
Escape character is ^]
CentOS Linux 7 (Core)
Kernel 3.10.0-693.el7.x86_64 on an x86_64

kvm-centos login: root
Password:
Last login: Sat Dec 15 16:37:39 on ttyS0
```

第 8 步，过多的虚拟机快照可能会影响虚拟机的运行速度或导致虚拟机故障，定期删除虚拟机快照是必须的，使用命令“virsh snapshot-delete centos7-03 1544863104”即可删除对应的快照。

```
root@kvm-server:/vm-backup# virsh snapshot-delete centos7-03 1544863104
Domain snapshot 1544863104 deleted
```

第 9 步，使用命令“virsh snapshot-list centos7-03”查看虚拟机的快照情况，虚拟机目前只保留 1 个快照，相当于 1 个备份。

```
root@kvm-server:/vm-backup# virsh snapshot-list centos7-03
 Name              Creation Time                State
------------------------------------------------------------------------------------------
 1544863205        2018-12-15 08:40:05 +0000    running
```

至此，使用虚拟机快照备份和恢复操作完成。与复制文件相比，快照方式简单明了，在KVM虚拟化备份工具不太多的情况下，生产环境中这种方式的确在被大量使用。再强调一次，过多的虚拟机快照可能会影响虚拟机的运行速度或导致虚拟机故障，定期删除虚拟机快照是必须的。

3.7.4 虚拟机常见的性能优化

前文在介绍创建虚拟机时，对于虚拟机的很多参数都使用默认参数。而在生产环境中，特别是对于Windows虚拟机，需要对一些参数进行调整以便优化虚拟机性能。

1. Windows虚拟机驱动程序安装

KVM新的版本对Windows虚拟机支持已经有了很大的提升，但还是存在一些驱动程序不能识别的安装问题。

第1步，使用VNC Viewer工具连接到虚拟机win2012，在设备管理器中可以看到PCI设备图标上存在感叹号，如图3-7-8所示。

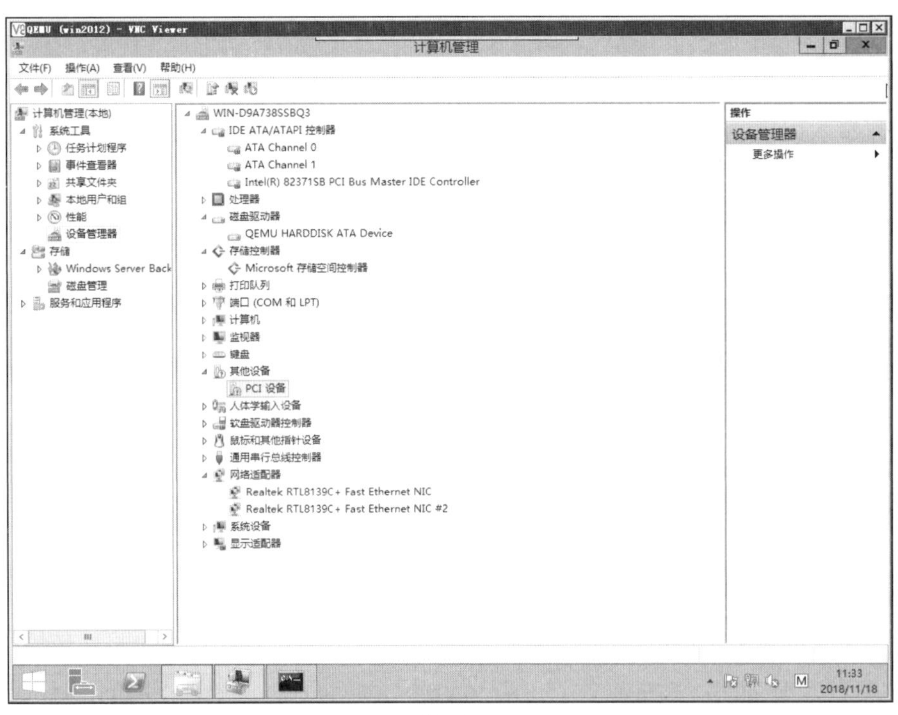

图3-7-8

第2步，访问https://fedorapeople.org/groups/virt/virtio-win/direct-downloads/archive-virtio/virtio-win-0.1.160-1/下载virtio驱动程序，解压后如图3-7-9所示，其中Balloon是内存气球驱动程序，NetKVM是网络驱动程序，vioserial是控制台驱动程序，viostor是存储相关驱动程序。

第3步，更新PCI设备驱动程序，如图3-7-10所示，单击“下一步”按钮。

第4步，系统提示找到驱动程序，VirtIO Balloon Driver为内存气球驱动程序，如图3-7-11所示，单击“安装”按钮。

第5步，成功安装virtio驱动程序，如图3-7-12所示，单击“关闭”按钮。

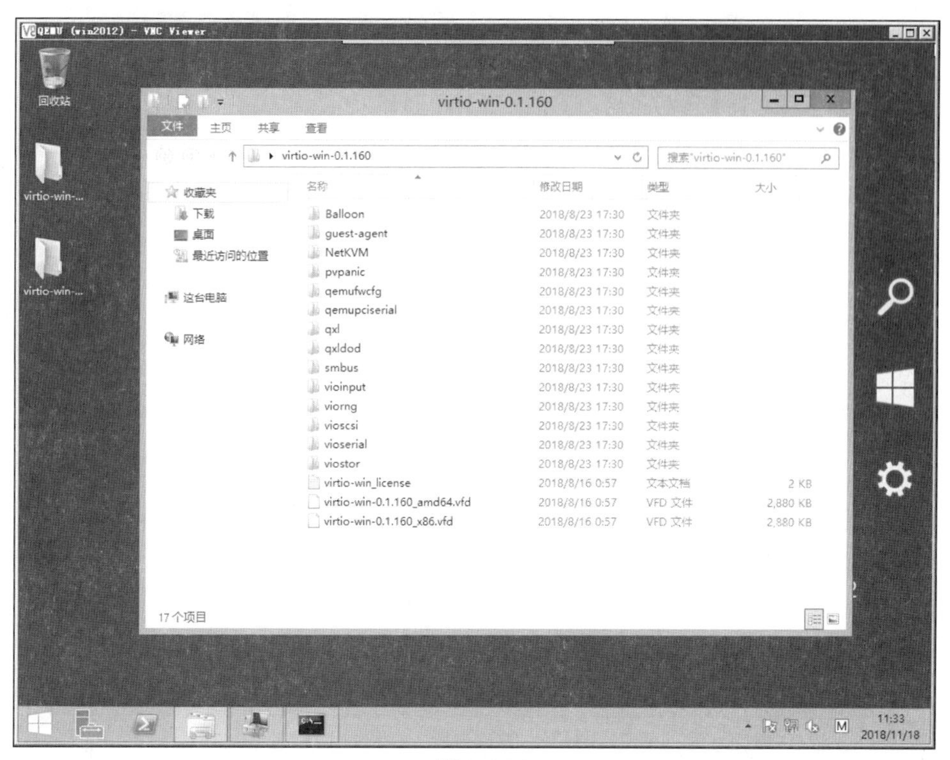
QEMU (win2012) - VNC Viewer
回收站
virtio-win-...
virtio-win-...
virtio-win-0.1.160
文件 主页 共享 查看
virtio-win-0.1.160
搜索"virtio-win-0.1.160"
收藏夹
下载
桌面
最近访问的位置
这台电脑
网络
名称 修改日期 类型 大小
Balloon 2018/8/23 17:30 文件夹
guest-agent 2018/8/23 17:30 文件夹
NetKVM 2018/8/23 17:30 文件夹
pvpanic 2018/8/23 17:30 文件夹
qemufwcfg 2018/8/23 17:30 文件夹
qemupciserial 2018/8/23 17:30 文件夹
qxl 2018/8/23 17:30 文件夹
qxldod 2018/8/23 17:30 文件夹
smbus 2018/8/23 17:30 文件夹
vioinput 2018/8/23 17:30 文件夹
viorng 2018/8/23 17:30 文件夹
vioscsi 2018/8/23 17:30 文件夹
vioserial 2018/8/23 17:30 文件夹
viostor 2018/8/23 17:30 文件夹
virtio-win_license 2018/8/16 0:57 文本文档 2 KB
virtio-win-0.1.160_amd64.vfd 2018/8/16 0:57 VFD 文件 2,880 KB
virtio-win-0.1.160_x86.vfd 2018/8/16 0:57 VFD 文件 2,880 KB
17 个项目
11:33
2018/11/18

图 3-7-9

QEMU (win2012) - VNC Viewer
计算机管理
文件(F) 操作(A) 查看(V) 帮助(H)
计算机管理(本地)
系统工具
任务计划程序
事件查看器
共享文件夹
本地用户和组
性能
设备管理器
存储
Windows Server Back
磁盘管理
服务和应用程序
WIN-D9A738SSBQ3
IDE ATA/ATAPI 控制器
ATA Channel 0
操作
设备管理器
更多操作
更新驱动程序软件 - PCI 设备
浏览计算机上的驱动程序文件
在以下位置搜索驱动程序软件:
C:\Users\Administrator\Desktop\virtio-win-0.1.160
浏览(R)...
包括子文件夹(I)
从计算机的设备驱动程序列表中选取(L)
此列表将显示与该设备兼容的已安装的驱动程序软件，以及与该设备处于同一类别下的所有驱动程序软件。
下一步(N) 取消
11:33
2018/11/18

图 3-7-10

图 3-7-11

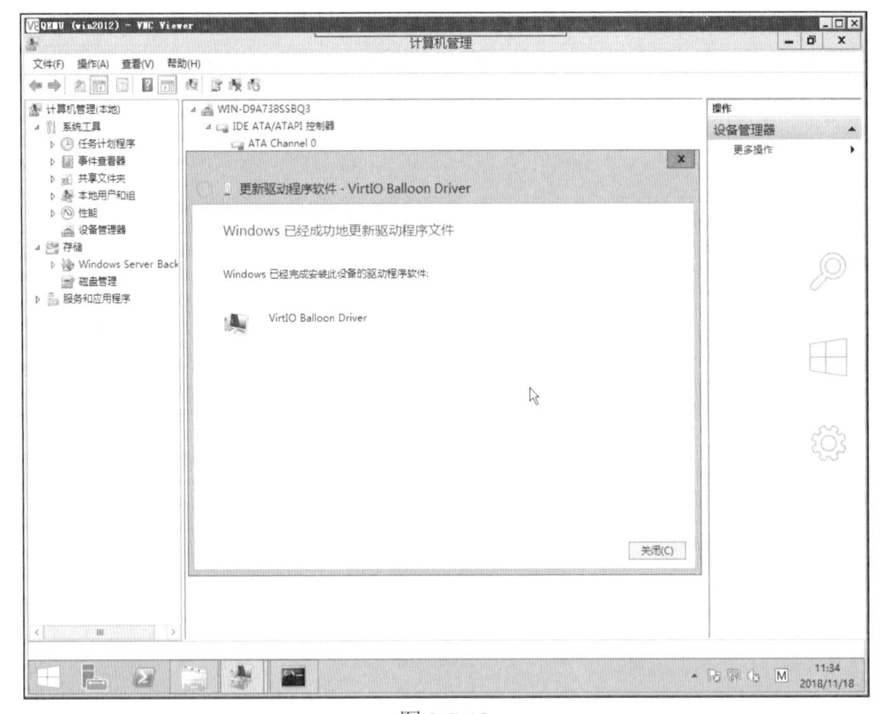

图 3-7-12

第 6 步，查看虚拟机 win2012 设备管理器，已经没有未安装驱动程序的硬件，如图 3-7-13 所示。

图 3-7-13

2. 调整虚拟机默认网卡

在 KVM 环境下安装虚拟机，默认使用的是 RTL8139 网卡，该网卡速度为 100 Mbit/s，在生产环境中需要进行调整。

第 1 步，使用 VNC Viewer 工具连接到虚拟机 win2012，查看默认安装的 RTL8139 网卡信息，其速度为 100 Mbit/s，如图 3-7-14 所示。

图 3-7-14

第 2 步，KVM 内核可以直接支持 E1000 类型，因此修改虚拟机配置文件即可。使用命令“virsh edit win2012”修改配置文件，将原 model type 修改为 e1000。

```
root@kvm-server:/home/admin# virsh edit win2012
<interface type='bridge'>
     <mac address='52:54:00:73:2e:d3'/>
     <source bridge='virbr0'/>
     <model type='e1000'/>  #修改网卡类型为e1000
     <address type='pci' domain='0x0000' bus='0x00' slot='0x03' function='0x0'/>
   </interface>
```

第 3 步，使用 VNC Viewer 工具连接到虚拟机 win2012 虚拟机，网络适配器已从 RTL8139 变为 Intel PRO 100/1000MT，其速度为 1 Gbit/s，如图 3-7-15 所示。

图 3-7-15

3. 调整虚拟机使用 virtio 网卡

将虚拟机默认网卡从 RTL8139 调整为 E1000，可以实现 100 Mbit/s 到 1 Gbit/s 的升级。但在生产环境中，10 Gbit/s 网络已经被大量使用，在 KVM 环境下也已经可以支持，使用 virtio 网卡可以支持 10 Gbit/s 网络。

第 1 步，使用命令“virsh edit win7-02”修改配置文件，将原 model type 修改为 virtio。

```
root@kvm-server:/home/admin# virsh edit win7-02
   <interface type='bridge'>
     <mac address='52:54:00:3c:37:19'/>
     <source bridge='virbr0'/>
     <model type='virtio'/>
     <address type='pci' domain='0x0000' bus='0x00' slot='0x03' function='0x0'/>
   </interface>
```

第 2 步，关闭电源重启虚拟机，系统会找到未安装驱动程序的以太网控制器，如图 3-7-16 所示。

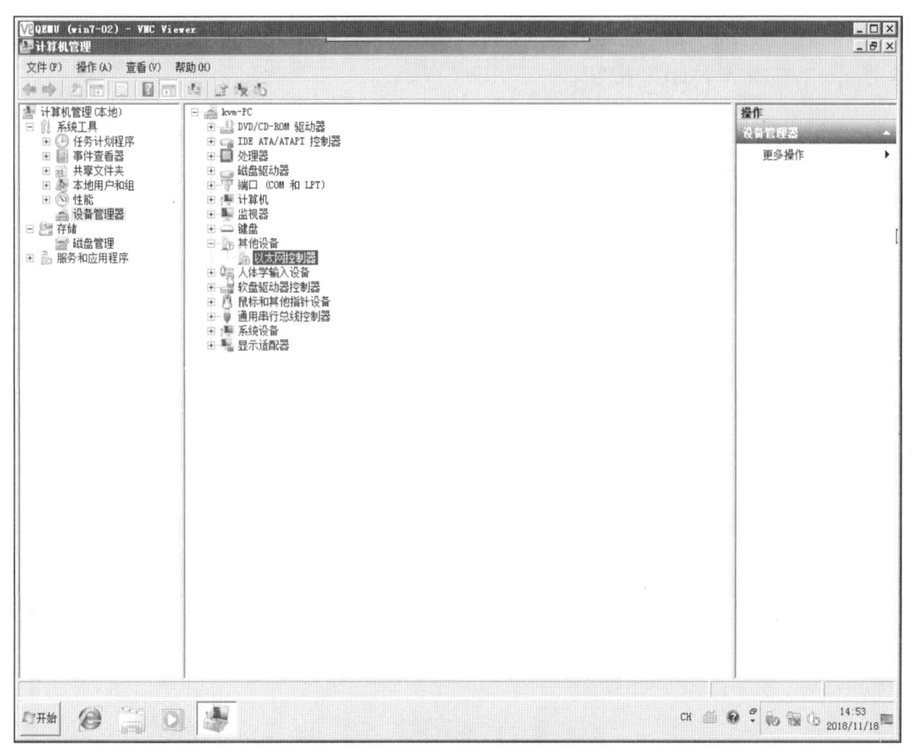

图 3-7-16

第 3 步，更新以太网控制器驱动程序，如图 3-7-17 所示，单击“下一步”按钮。

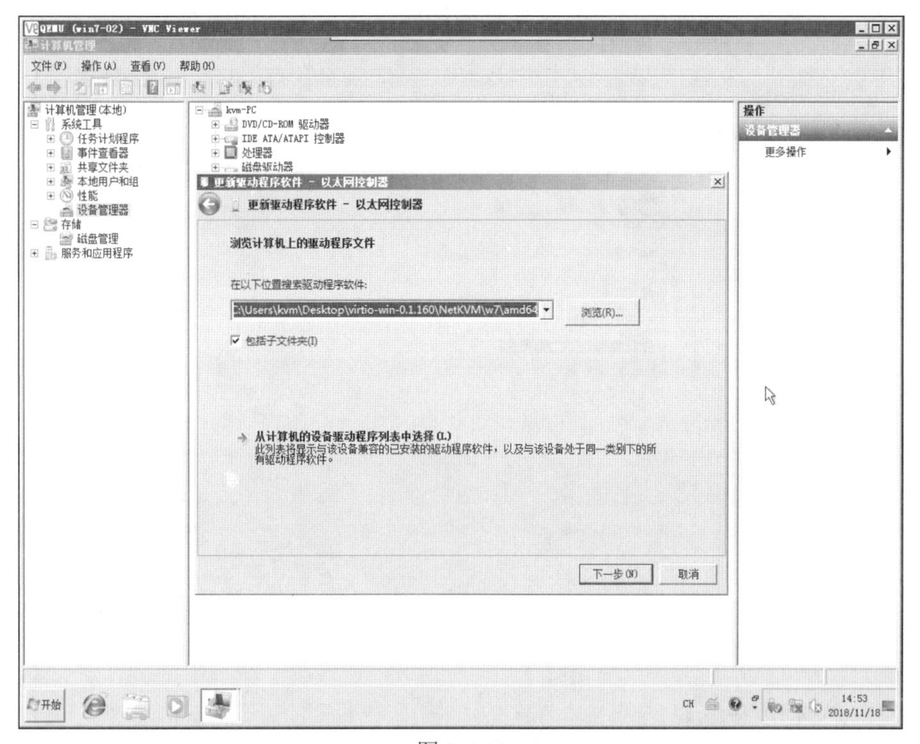

图 3-7-17

第 4 步，以太网控制器驱动程序使用的是 Red Hat VirtIO Ethernet Adapter 驱动程序，如图 3-7-18 所示，单击“关闭”按钮。

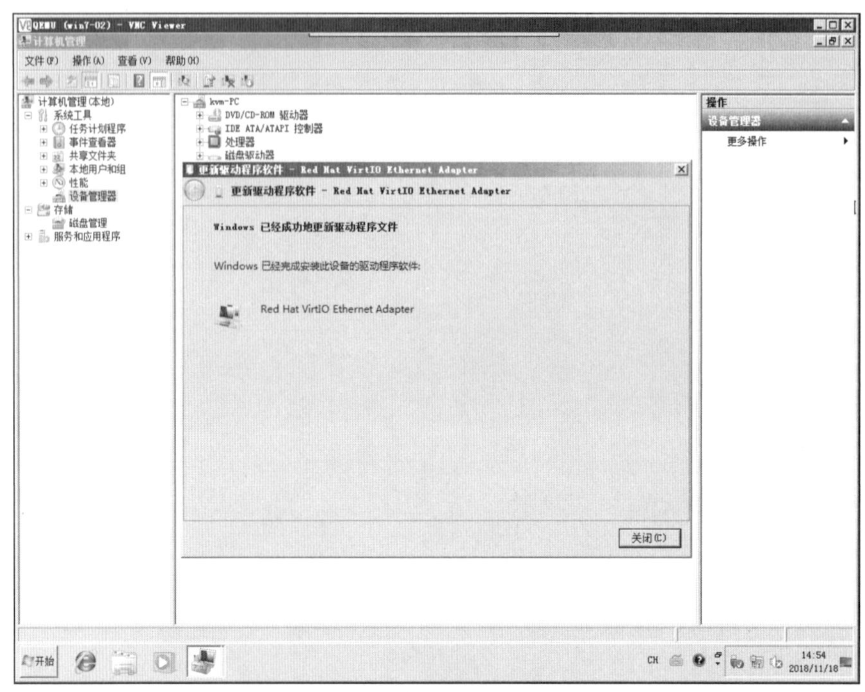

图 3-7-18

第 5 步，在设备管理器中可以看到新安装的网络适配器，如图 3-7-19 所示。

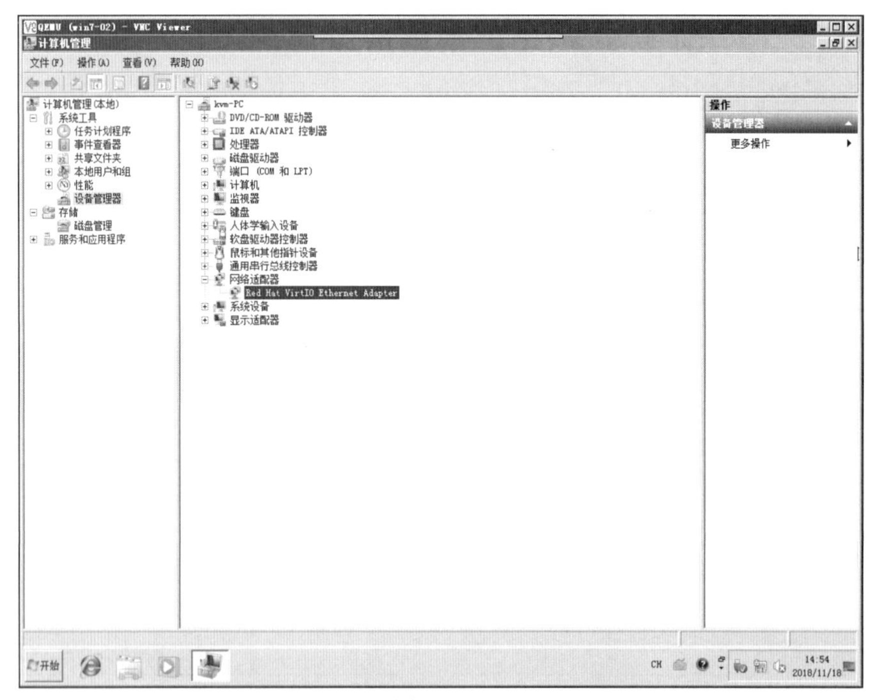

图 3-7-19

第 6 步，查看新的网络适配器属性，速度为 10 Gbit/s，如图 3-7-20 所示。

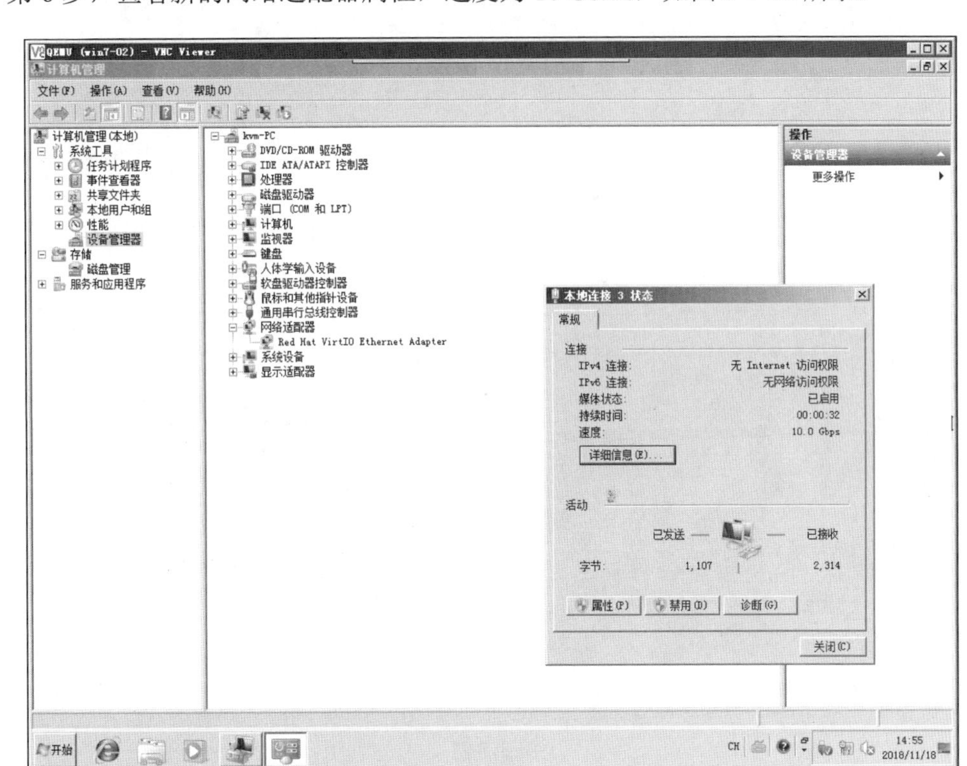

图 3-7-20

4. 调整 Windows 虚拟机存储控制器

对于 Windows 虚拟机来说，在初始安装时如果没有指定存储控制器参数，默认会使用 IDE 设备。IDE 设备属于比较旧的设备，其传输效率相对较低，推荐在安装 Windows 7 以后的操作系统时使用 virtio 设备。

第 1 步，使用命令“virt-install”新部署虚拟机 win2012-viroio，注意 bus 参数使用 virtio。

```
root@kvm-server:/vm-iso# virt-install  --name win2012-virtio --vcpus 2 --ram 4096 --boot cdrom --cdrom=/vm-iso/cn_windows_server_2012_r2.iso --disk path=/vm-win/win2012-virtio.qcow2,size=20,format=qcow2,bus=virtio  --network bridge=virbr0 --os-type=windows  --vnc --vncport=5906 --vnclisten=0.0.0.0 --noautoconsole --force
Starting install...
Allocating 'win2012-virtio.qcow2'
|  20 GB  00:00:00
Domain installation still in progress. You can reconnect to
the console to complete the installation process.
```

第 2 步，使用 VNC Viewer 工具连接到虚拟机 win2012-virtio，安装操作系统，如图 3-7-21 所示，单击“下一步”按钮。

第 3 步，使用 virtio 存储控制器 Windows Server 2012 R2 系统不能直接识别硬盘信息，如图 3-7-22 所示，单击“加载驱动程序”。

第 4 步，通过浏览的方式添加驱动程序，如图 3-7-23 所示，单击“浏览”按钮。

第 5 步，操作系统安装镜像中已经集成 virtio 驱动程序，直接选择使用，如图 3-7-24 所示，单击“确定”按钮。

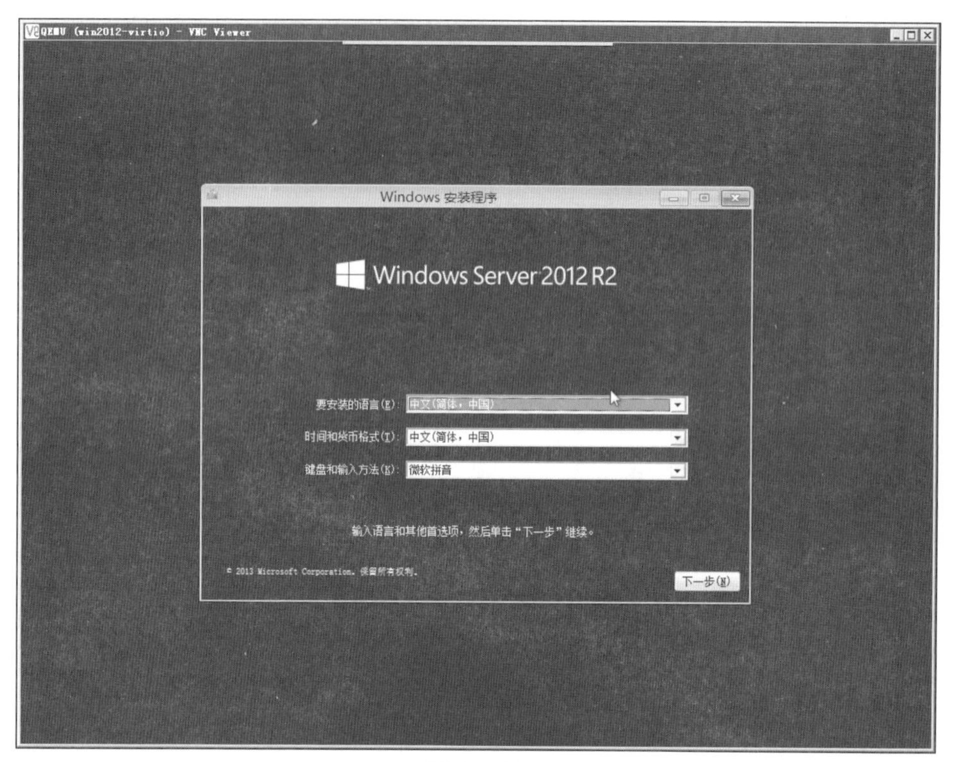
QEMU (win2012-virtio) - VNC Viewer
Windows 安装程序
Windows Server 2012 R2
要安装的语言(E):
中文(简体，中国)
时间和货币格式(T):
中文(简体，中国)
键盘和输入方法(K):
微软拼音
输入语言和其他首选项，然后单击“下一步”继续。
下一步(N)

图 3-7-21

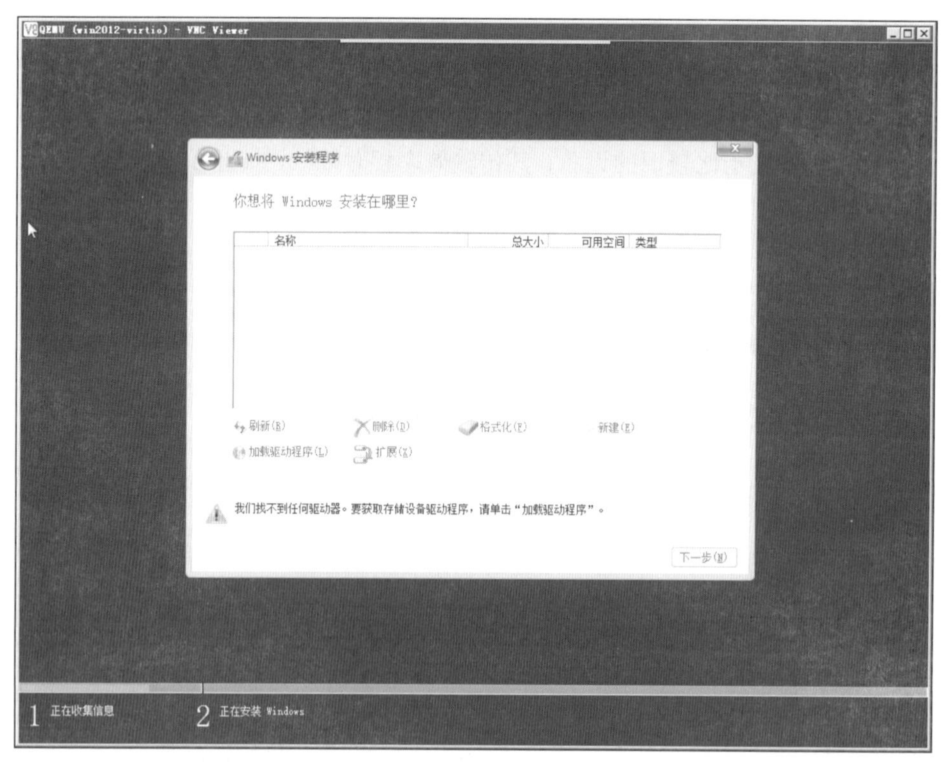
QEMU (win2012-virtio) - VNC Viewer
Windows 安装程序
你想将 Windows 安装在哪里?
名称
总大小
可用空间
类型
刷新(R)
删除(D)
格式化(F)
新建(E)
加载驱动程序(L)
扩展(X)
我们找不到任何驱动器。要获取存储设备驱动程序，请单击“加载驱动程序”。
下一步(N)
1 正在收集信息
2 正在安装 Windows

图 3-7-22

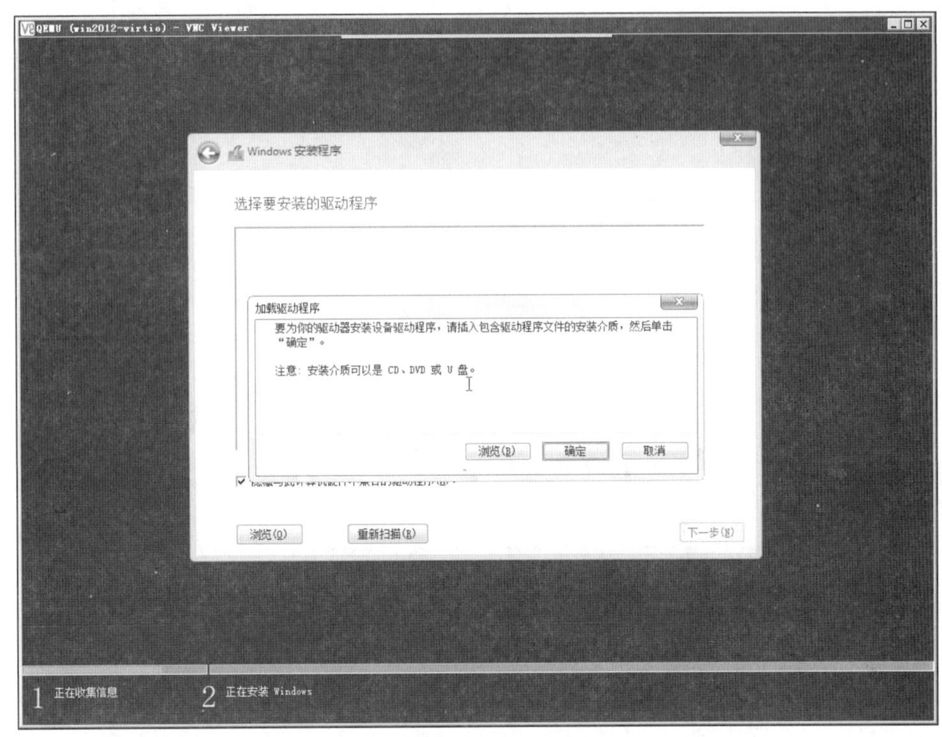

图 3-7-23

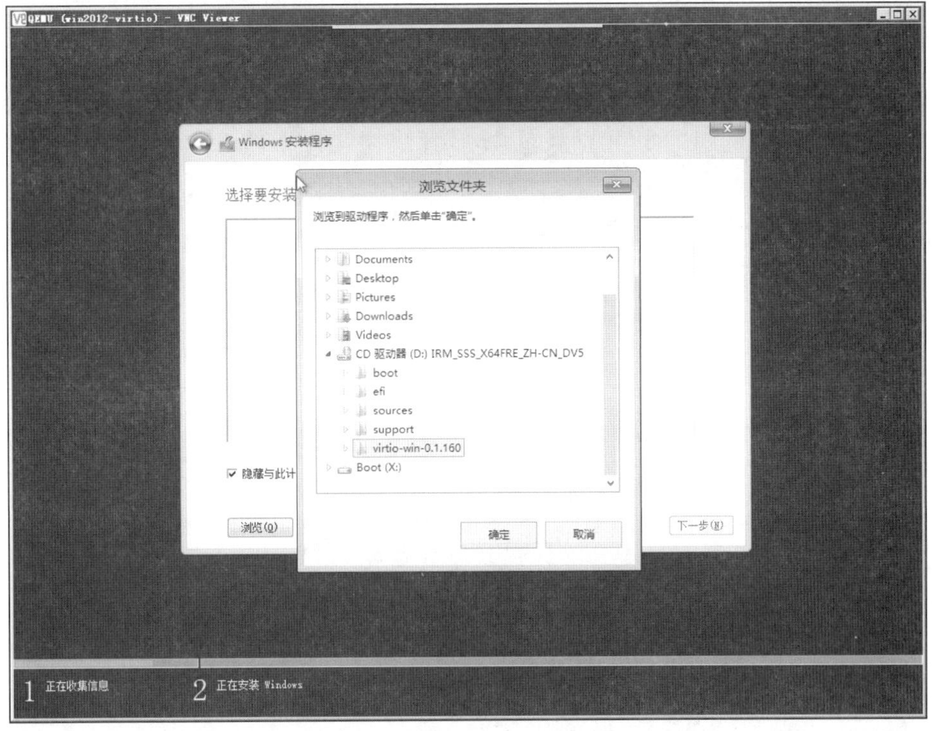

图 3-7-24

第 6 步，成功加载 virtio 驱动程序，如图 3-7-25 所示，单击“下一步”按钮。

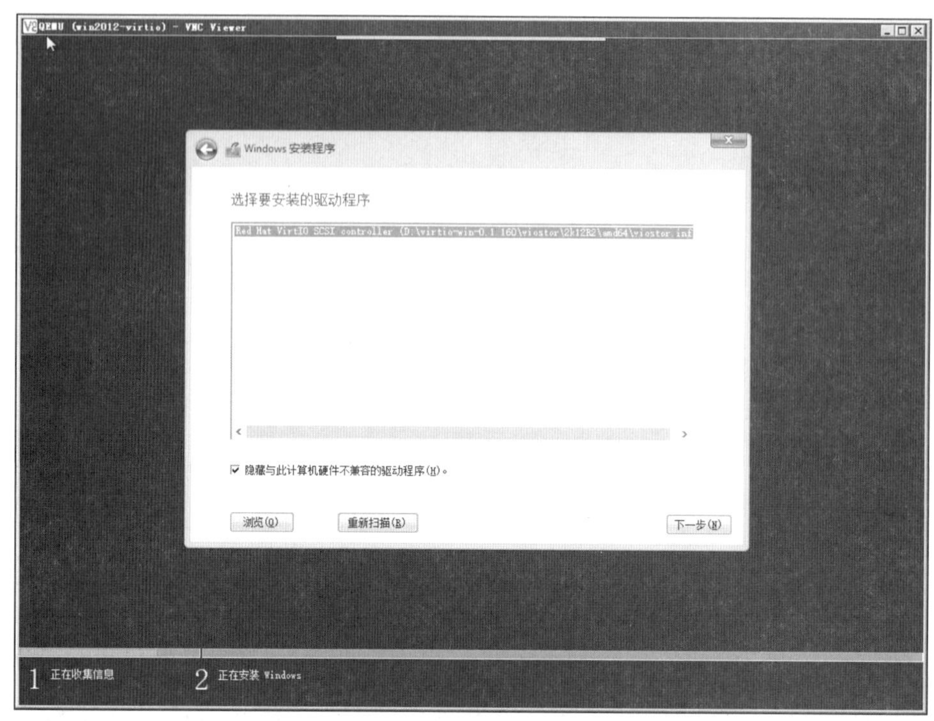

图 3-7-25

第 7 步，加载 virtio 驱动程序后操作系统识别到硬盘，如图 3-7-26 所示，单击“下一步”按钮进行安装。

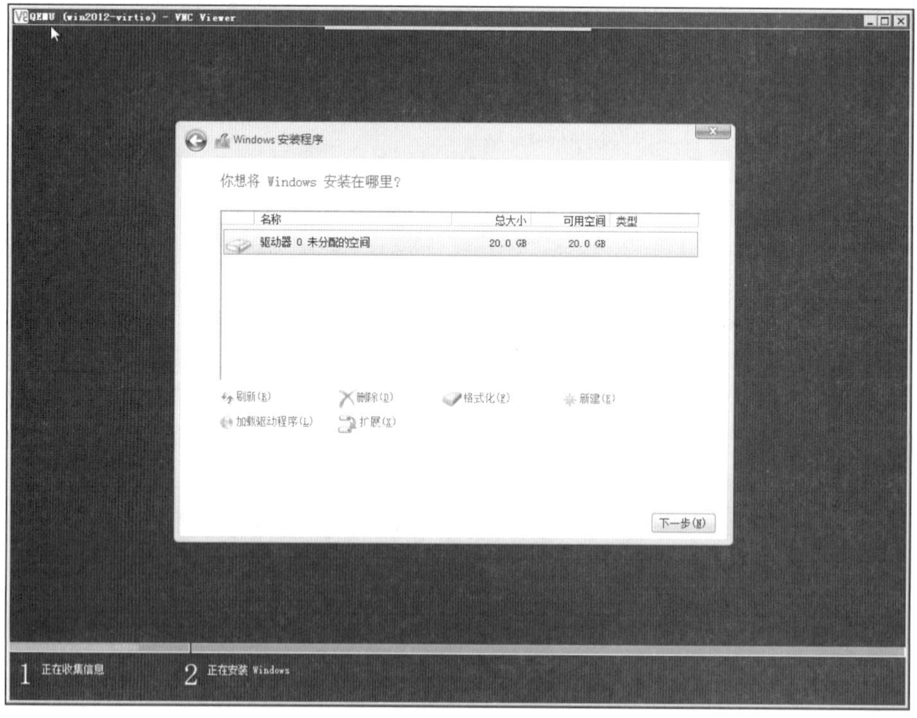

图 3-7-26

第 8 步，开始安装 Windows Server 2012 R2 操作系统，如图 3-7-27 所示。

图 3-7-27

第 9 步，安装完成后，查看磁盘驱动器类型，类型为 Red Hat VirtIO SCSI Disk Device，如图 3-7-28 所示。

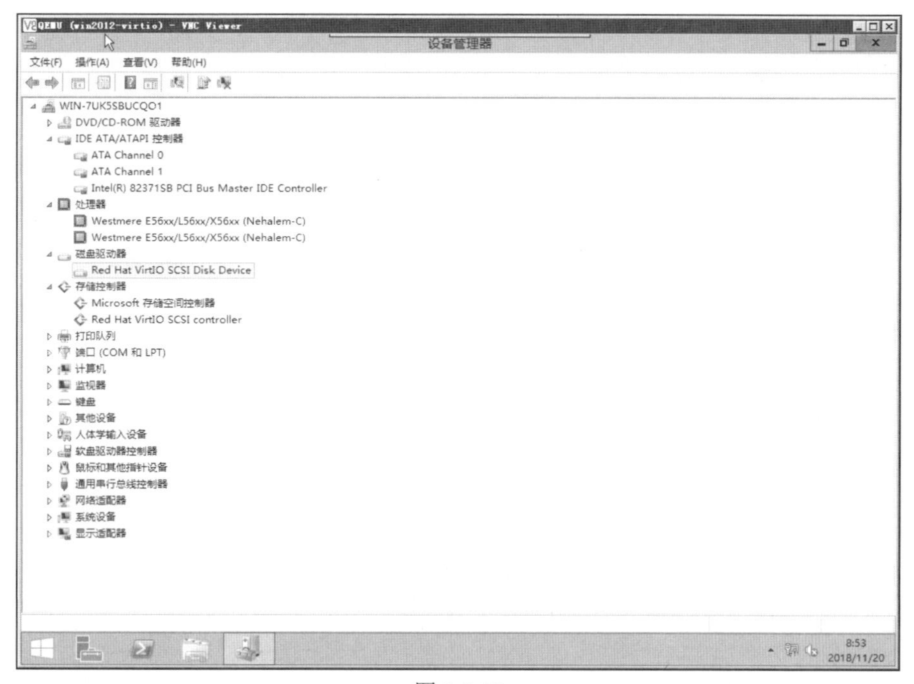

图 3-7-28

至此，Windows 虚拟机安装 Red Hat VirtIO SCSI Disk Device 驱动程序完成，使用该驱动程序存储设备的传输速度相对使用 IDE 类型的驱动程序得到提升。对于已安装操作系统的 Windows 虚拟机，如果已经使用 IDE 类型的驱动程序，不太建议进行替换，在生产环境测试替换可能造成 Windows 操作系统不稳定或运行过程中出现蓝屏的情况。

3.8 本章小结

本章介绍了如何在 Linux 系统上部署使用 KVM 虚拟化平台，包括 Linux 虚拟机、Windows 虚拟机的创建和日常使用。需要注意的是，在生产环境中使用 KVM 虚拟化平台，前期的规划设计非常重要，从硬盘格式的选择、网络的设计以及日常的备份等，如果没有做好规划设计，后期调整相对比较麻烦。

第 4 章　部署使用 oVirt 平台

通过前文的学习，相信读者已能使虚拟机在 KVM 环境下运行起来，并且能够根据生产环境的实际情况对虚拟机进行调整。细心的读者会发现，前文大量使用了命令行，并且都是基于单机 KVM 环境运行的，如果大量使用，其管理界面并不友好。本章介绍企业级虚拟化平台 oVirt 的部署和使用。

本章要点

- 为什么使用 oVirt 平台。
- 部署 oVirt 平台。
- 将主机加入 oVirt 平台管理。
- 配置使用存储。
- 创建使用虚拟机。
- 配置 oVirt 平台高可用。
- 备份和恢复虚拟机。
- 将物理服务器迁移到 oVirt 平台。
- 跨平台迁移虚拟机到 oVirt 平台。

4.1　为什么使用 oVirt 平台

整体来说，在开源 KVM 虚拟化环境中，管理平台的使用是一大难题。一些企业在使用开源 KVM 虚拟化架构的时候选择根据企业的实际情况自行开发的管理端，但对于中小企业来说，这样强大的开发能力几乎是不具备的。当然，开源社区也有不少的管理平台，但从功能和稳定程度来说与商业软件还是存在差距。Red Hat 公司针对这样的情况，推出了 Red Hat 企业级虚拟化管理平台 RHEV，但是使用 RHEV 需要订阅 RHN 服务。对于不愿意订阅 RHN 服务的企业，推荐使用 RHEV 的社区开源版本 oVirt。

4.1.1　oVirt 平台概述

oVirt 是 Red Hat 虚拟化管理平台 RHEV 的开源版本。该项目起源于 Qumranet 公司，该公司在 2008 年被 Red Hat 收购之后，其原有的用 C#开发的虚拟化管理软件被用 Java 改写，并在 2011 年开源为 oVirt 项目。2011 年 11 月，Red Hat 举办了第一次 oVirt 研讨会（oVirt Workshop），与 IBM、英特尔、思科、Canonical、NetApp 以及 SUSE 等一同宣布 oVirt 社区的成立。利用 oVirt 管理 KVM 虚拟机和网络，企业可以快速地搭建起一个私有云环境。从这一点看来，oVirt 的定位和另一个知名云计算项目 OpenStack 的定位是有些类似的。不过有意思的是，oVirt 实际上是 Red Hat 的企业级虚拟化解决方案 RHEV 的上游项目，而这些支持 oVirt 项目的厂商，也同时是 OpenStack 项目的参与者。

那么为什么要使用 oVirt 平台？主要有以下几点原因。

1）KVM 使用命令行或 virt-manager 管理工具，更多是基于单机管理模式。

2）单机 KVM 不具备数据中心。集群等高级管理特性。

3）oVirt 平台属于社区开源版本，免费使用，不需要订阅服务。

4）oVirt 平台源于 Red Hat RHEV 平台，其功能、稳定性得到肯定。

5）oVirt 平台通过 WEB 方式进行管理，管理界面非常友好。

4.1.2 oVirt 平台的特点

oVirt 平台由两部分组成：oVirt Engine 管理端和 oVirt Node 节点主机。

oVirt Engine 属于 oVirt 管理端，提供完整的企业级虚拟化平台管理能力，对应 RHEV 虚拟化架构的 RHEV-M 管理平台，可在 RHEL 操作系统或 CentOS 操作系统运行，基于 Web 浏览器进行操作，其功能类似于 VMware vSphere 虚拟化架构中的 vCenter Server。

oVirt Node 属于节点主机，用于运行虚拟机，其内核基于 KVM，对应 RHEV 虚拟化架构的 RHEV-H 主机，是由 fedaro 16 订制而成的，也可以通过在 Linux 操作系统上安装 VDSM 服务得到一个 oVirt 节点主机。

oVirt 平台主要功能如下。

1）对节点主机的管理，支持基于数据中心和集群的管理。

2）对虚拟机的管理，可以完成虚拟机的创建、快照、删除以及基于模板的复制等常见操作。

3）迁移，在线或离线迁移虚拟机。

4）高可用，当一台节点主机宕机，在另一台节点主机上自动开启虚拟机。

5）主机和虚拟机性能的查看与统计。

6）支持主流的 NFS、iSCSI、FC 等存储方式。

7）支持多端口的网络接口绑定。

8）提供命令行界面，可以完成 GUI 的大部分功能。

9）活动目录集成。

10）提供 Python API，可以通过 API 进行二次开发。

11）没有授权限制。

关于 oVirt 的详细介绍，可以访问官方网站查阅。

4.2 部署 oVirt 平台

了解 oVirt 平台概念后就可以开始部署，本节将分别介绍 oVirt Engine 管理端和 oVirt Node 节点主机的部署。

4.2.1 部署 oVirt Engine 管理端

开始部署 oVirt Engine 管理端之前，先了解一下需要的条件，主要分为服务器、CentOS 版本以及 YUM 源 3 个方面。

服务器：oVirt Engine 管理端支持在虚拟机和物理服务器上安装，oVirt Engine 管理端安装最低需要 4GB 内存和 20GB 硬盘空间。

CentOS 版本：推荐使用 CentOS 7.5 以及之后的版本，也支持其他 Linux 版本，但可能需要额外依赖包。

YUM 源：oVirt Engine 管理端在安装时需要使用官方提供的 YUM 源或其他源，需要注意的是，oVirt Engine 管理端 YUM 源站点位于国外，可能存在延时问题，亚洲镜像站点访问速度也一般。

准备好安装 oVirt Engine 管理端的服务器后，就可以开始部署 oVirt Engine 管理端，本节实战操作使用 KVM 主机上安装好的 1 台虚拟机进行。

第 1 步，使用命令“yum install ovirt-engine”部署会提示没有可用的软件包，其原因是 Linux 操作系统自带的 YUM 源不提供 ovirt-engine 相关组件，需要重新配置 YUM 源。

```
[root@centos7-ovirt ~]# yum install ovirt-engine
Loaded plugins: fastestmirror
Loading mirror speeds from cached hostfile
 * base: mirrors.163.com
 * extras: mirrors.nwsuaf.edu.cn
 * updates: mirrors.nwsuaf.edu.cn
No package oVirt-Engine available.
```

第 2 步，使用命令“yum install https://resources.ovirt.org/pub/yum-repo/ovirt-release42.rpm”在线下载安装配置 YUM 源，写作本书的时候 oVirt 最新版本为 4.2.7。

```
[root@centos7-ovirt ~]# yum install https://resources.ovirt.org/pub/yum-repo/ovirt-release42.rpm
Loaded plugins: fastestmirror
ovirt-release42.rpm
| 12 kB  00:00:00
Examining /var/tmp/yum-root-BYkiJ7/ovirt-release42.rpm: ovirt-release42-4.2.7-1.el7.noarch
Marking /var/tmp/yum-root-BYkiJ7/ovirt-release42.rpm to be installed
Resolving Dependencies
--> Running transaction check
---> Package ovirt-release42.noarch 0:4.2.7-1.el7 will be installed
--> Finished Dependency Resolution
Dependencies Resolved
=============================================================================
 Package              Arch      Version         Repository              Size
=============================================================================
Installing:
 ovirt-release42      noarch    4.2.7-1.el7     /ovirt-release42        11 k
Transaction Summary
=============================================================================
Install  1 Package

Total size: 11 k
Installed size: 11 k
Is this ok [y/d/N]: y
Downloading packages:
Running transaction check
Running transaction test
Transaction test succeeded
Running transaction
  Installing:ovirt-release42-4.2.7-1.el7.noarch  1/1
  Verifying: ovirt-release42-4.2.7-1.el7.noarch  1/1
Installed:
  ovirt-release42.noarch 0:4.2.7-1.el7
Complete!
```

第 3 步，更新 YUM 源后重新使用命令“yum install ovirt-engine”部署，由于使用在线下载安装方式，且软件包较大，下载安装时间可能较长，实战操作使用的 CentOS 7 操作系统 mini 安装模式，需要下载的软件包大小为 515MB，注意安装完成后一定要出现“Complete!”才代表安装成功。

```
[root@centos7-ovirt ~]# yum install ovirt-engine
Loaded plugins: fastestmirror
Loading mirror speeds from cached hostfile
 * base: mirrors.163.com
 * extras: mirrors.163.com
 * ovirt-4.2: mirror.linux.duke.edu
 * ovirt-4.2-epel: mirrors.aliyun.com
 * updates: mirrors.163.com
……（省略）
Install  1 Package (+362 Dependent packages)
Upgrade  1 Package (+ 21 Dependent packages)
Total download size: 515 M
Is this ok [y/d/N]:y
……（省略）
Dependency Updated:
  audit.x86_64 0:2.8.1-3.el7_4.1              audit-libs.x86_64 0:2.8.1-3.el7_4.1
  bind-libs-lite.x86_64 32:9.9.4-61.el7_4.1   bind-license.noarch 32:9.9.4-61.el7_4.1
  libselinux.x86_64 0:2.5-12.el7              libselinux-python.x86_64 0:2.5-12.el7
……（省略）
Complete!
```

第 4 步，安装完成后使用命令“engine-setup”配置 oVirt Engine 管理端，注意查看注释。安装结束一定要出现“Execution of setup completed successfully”提示才代表 oVirt Engine 管理端安装成功。

```
[root@ovirt etc]# engine-setup
[ INFO ] Stage: Initializing
[ INFO ] Stage: Environment setup
         Configuration files: ['/etc/oVirt-Engine-setup.conf.d/10-packaging-jboss.conf', '/etc/oVirt-Engine-setup.conf.d/10-packaging.conf']
         Log file: /var/log/oVirt-Engine/setup/oVirt-Engine-setup-20180112003218ax8gkk.log
         Version: otopi-1.7.8 (otopi-1.7.8-1.el7)
[ INFO ] Stage: Environment packages setup
[ INFO ] Stage: Programs detection
[ INFO ] Stage: Environment setup
[ INFO ] Stage: Environment customization
               --== PRODUCT OPTIONS ==--
         Configure Engine on this host (Yes, No) [Yes]:
         Configure ovirt-provider-ovn (Yes, No) [Yes]:#配置虚拟网络
         Configure Image I/O Proxy on this host (Yes, No) [Yes]:#Image I/O Proxy 可以支持向 oVirt 平台上传虚拟机的磁盘镜像
         Configure WebSocket Proxy on this host (Yes, No) [Yes]:#WebSocket Proxy 支持使用 noVNC 远程登录虚拟机图形管理界面
         * Please note * : Data Warehouse is required for the engine.
         If you choose to not configure it on this host, you have to configure
         it on a remote host, and then configure the engine on this host so
         that it can access the database of the remote Data Warehouse host.
         Configure Data Warehouse on this host (Yes, No) [Yes]:#在本机配置数据库
         Configure VM Console Proxy on this host (Yes, No) [Yes]:#支持访问虚拟机的控制台
               --== PACKAGES ==--
[ INFO ] Checking for product updates...
[ INFO ] No product updates found
[ INFO ] Checking for product updates...
```

```
[ INFO ] No product updates found
          --== NETWORK CONFIGURATION ==--
Host fully qualified DNS name of this server [centos7-ovirt.bdnetlab.com]:
[WARNING] Failed to resolve centos7-ovirt.bdnetlab.com using DNS, it can be resolved only locally
        Setup can automatically configure the firewall on this system.
        Note: automatic configuration of the firewall may overwrite current settings.
        NOTICE: iptables is deprecated and will be removed in future releases
        Do you want Setup to configure the firewall? (Yes, No) [Yes]:
[ INFO ] firewalld will be configured as firewall manager.
          --== DATABASE CONFIGURATION ==--
     Where is the DWH database located? (Local, Remote) [Local]: Data WareHouse #使用本地数据库
     Setup can configure the local postgresql server automatically for the DWH to run. This may conflict
with existing applications.
     Would you like Setup to automatically configure postgresql and create DWH database, or prefer
to perform that manually? (Automatic, Manual) [Automatic]: #自动配置 PostgreSQL 数据库
     Where is the Engine database located? (Local, Remote) [Local]: #数据库位于本机
     Setup can configure the local postgresql server automatically for the engine to run. This may
conflict with existing applications.
     Would you like Setup to automatically configure postgresql and create Engine database, or prefer
to perform that manually? (Automatic, Manual) [Automatic]: #使用安装程序自动创建数据库
          --== OVIRT-ENGINE CONFIGURATION ==--
        Engine admin password:
        Confirm engine admin password: #管理员 admin 的密码
        Application mode (Virt, Gluster, Both) [Both]:
        Use default credentials (admin@internal) for ovirt-provider-ovn (Yes, No) [Yes]: #使用默认
的 admin@internal 内部管理员账号
          --== STORAGE CONFIGURATION ==--
        Default SAN wipe after delete (Yes, No) [No]: #删除虚拟机的虚拟磁盘后会擦除存储设备上的对应块
          --== PKI CONFIGURATION ==--
        Organization name for certificate [bdnetlab.com]:
          --== APACHE CONFIGURATION ==--
        Setup can configure the default page of the web server to present the application home page.
This may conflict with existing applications.
        Do you wish to set the application as the default page of the web server? (Yes, No) [Yes]:
#使用 Apache 作为 Web 服务器端软件
        Setup can configure apache to use SSL using a certificate issued from the internal CA.
        Do you wish Setup to configure that, or prefer to perform that manually? (Automatic, Manual)
[Automatic]: #自动配置 CA 证书
          --== SYSTEM CONFIGURATION ==--
          --== MISC CONFIGURATION ==--
        Please choose Data Warehouse sampling scale:
        (1) Basic
        (2) Full
        (1, 2)[1]: #使用基本的数据库示例初始化数据
          --== END OF CONFIGURATION ==--
[ INFO ] Stage: Setup validation
[WARNING] Less than 16384MB of memory is available
          --== CONFIGURATION PREVIEW ==--
        Application mode                        : both
        Default SAN wipe after delete           : False
        Firewall manager                        : firewalld
        Update Firewall                         : True
        Host FQDN                               : centos7-ovirt.bdnetlab.com
        Configure local Engine database         : True
        Set application as default page         : True
        Configure Apache SSL                    : True
        Engine database secured connection      : False
        Engine database user name               : engine
        Engine database name                    : engine
        Engine database host                    : localhost
```

```
        Engine database port                    : 5432
        Engine database host name validation    : False
        Engine installation                     : True
        PKI organization                        : bdnetlab.com
        Set up ovirt-provider-ovn               : True
        Configure WebSocket Proxy               : True
        DWH installation                        : True
        DWH database host                       : localhost
        DWH database port                       : 5432
        Configure local DWH database            : True
        Configure Image I/O Proxy               : True
        Configure VMConsole Proxy               : True

        Please confirm installation settings (OK, Cancel) [OK]:
[ INFO ] Stage: Transaction setup
[ INFO ] Stopping engine service
[ INFO ] Stage: Misc configuration
[ INFO ] Stage: Package installation
[ INFO ] Stage: Misc configuration
[ INFO ] Upgrading CA
[ INFO ] Initializing PostgreSQL
[ INFO ] Creating PostgreSQL 'engine' database
[ INFO ] Configuring PostgreSQL
[ INFO ] Creating PostgreSQL 'ovirt_engine_history' database
[ INFO ] Configuring PostgreSQL
[ INFO ] Creating CA
[ INFO ] Creating/refreshing DWH database schema
[ INFO ] Configuring Image I/O Proxy
[ INFO ] Setting up ovirt-vmconsole proxy helper PKI artifacts
[ INFO ] Setting up ovirt-vmconsole SSH PKI artifacts
[ INFO ] Configuring WebSocket Proxy
[ INFO ] Creating/refreshing Engine database schema
[ INFO ] Creating/refreshing Engine 'internal' domain database schema
[ INFO ] Adding default OVN provider to database
[ INFO ] Adding OVN provider secret to database
[ INFO ] Setting a password for internal user admin
[ INFO ] Stage: Transaction commit
[ INFO ] Stage: Closing up
[ INFO ] Starting engine service
[ INFO ] Starting dwh service
[ INFO ] Restarting ovirt-vmconsole proxy service
                --== SUMMARY ==--
       [ INFO ] Restarting httpd
        Please use the user 'admin@internal' and password specified in order to login
        Web access is enabled at:
           http://centos7-ovirt.bdnetlab.com:80/oVirt-Engine
           https://centos7-ovirt.bdnetlab.com:443/oVirt-Engine
        Internal CA ED:63:41:96:A6:71:C8:CF:84:E6:1D:05:AA:88:54:F1:8E:A5:63:01
        SSH fingerprint: SHA256:8hnatBUF88TZ8bMIaEYEapovnlDfsyJ15wp+gm3sqSw
[WARNING] Less than 16384MB of memory is available
                --== END OF SUMMARY ==--
 [ INFO ] Stage: Clean up
        Log file is located at /var/log/oVirt-Engine/setup/oVirt-Engine-setup-20181124034717-un8qgs.log
[ INFO ] Generating answer file '/var/lib/oVirt-Engine/setup/answers/20181124035549-setup.conf'
[ INFO ] Stage: Pre-termination
[ INFO ] Stage: Termination
[ INFO ] Execution of setup completed successfully
```

第 5 步，完成后使用浏览器连接 oVirt Engine 管理端，如图 4-2-1 所示。

第 6 步，登录 oVirt Engine 管理端后的页面如图 4-2-2 所示。

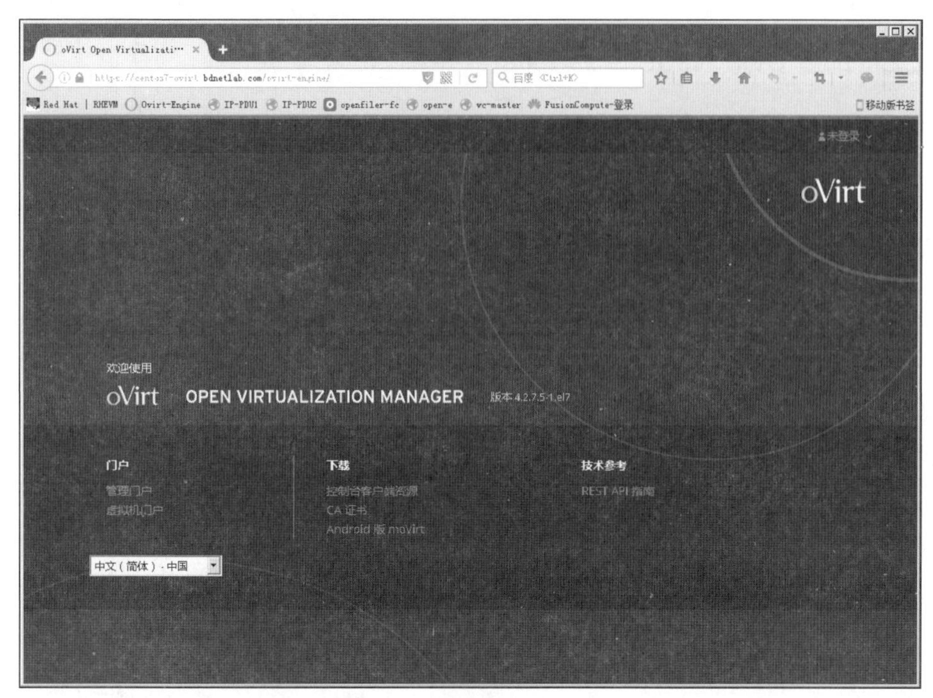

图 4-2-1

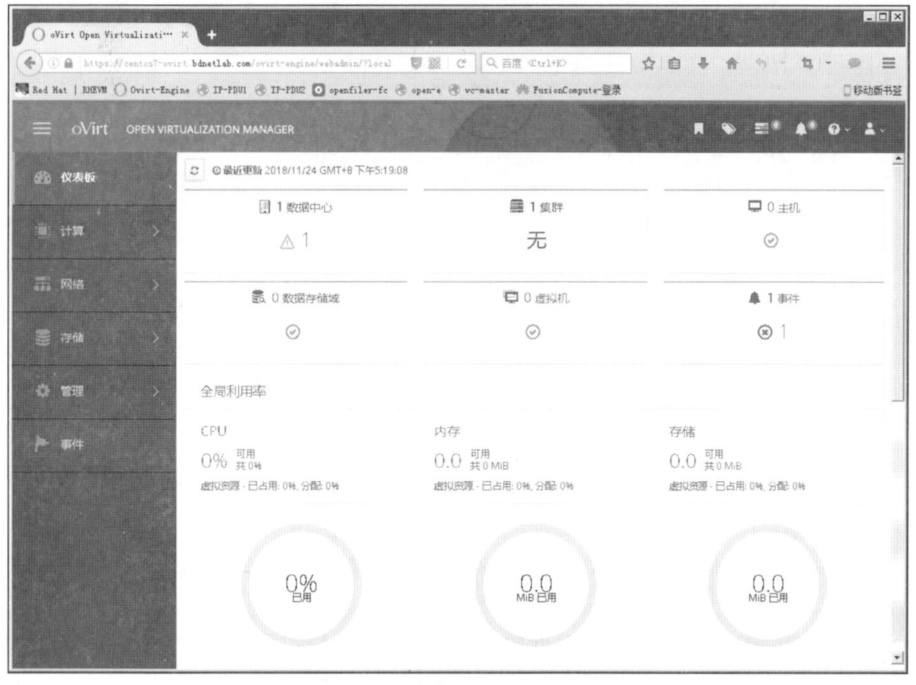

图 4-2-2

整体来说，oVirt Engine 管理端的安装不存在难度，主要在于配置参数。配置参数时一定要仔细，结合生产环境的实际情况进行。如常见的全限定域名（Fully Qualified Domain Name，FQDN）解析问题，一些生产环境可能未配置 DNS 服务器，可以在/etc/hosts 中添加本地域名解析即可。

4.2.2 部署 oVirt Node 节点主机

部署完成 oVirt Engine 管理端后就可以部署 oVirt Node 节点主机。oVirt Node 节点主机分为专用的节点和通过安装 VDSM 将 Linux 主机变为节点主机的节点，专用节点主机需要使用官方提供的 ISO 文件进行安装，其实质就是精简版的 KVM，为虚拟机的运行提供环境。本节先介绍专用节点端的部署。通过安装 VDSM 将 Linux 主机变为节点主机的相关部署后文会介绍。

在开始部署前，先了解一下需要的条件。oVirt Node 节点主机由于其特殊性，不保证在虚拟机上能够安装成功，即使通过修改文件参数成功安装，在后续使用过程中也可能存在问题，所以推荐 oVirt Node 节点主机使用物理服务器进行安装。oVirt Node 节点主机最低需要 4GB 内存，生产环境根据实际情况进行配置。

访问官网下载 oVirt Node 节点主机安装 ISO 文件后，即可在物理服务器上安装。本节实战操作使用 2 台物理服务器安装 oVirt Node 节点主机。

第 1 步，访问官方下载 oVirt Node 镜像文件，oVirt Node 目前提供多个版本下载，特别注意版本匹配建议，如图 4-2-3 所示。

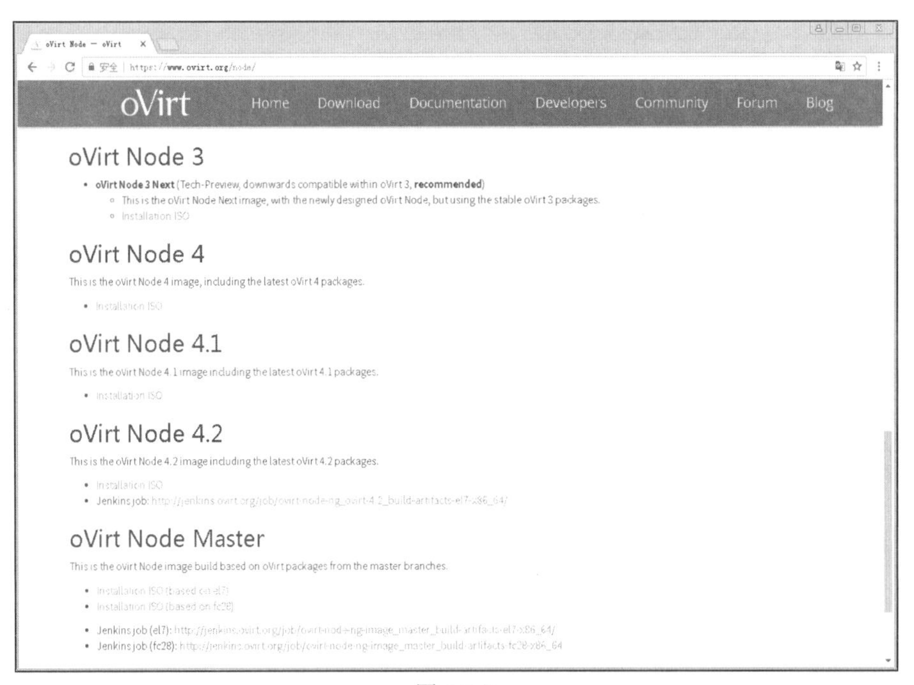

图 4-2-3

第 2 步，下载 oVirt Node 镜像文件进行安装，其引导过程类似于 Linux 操作系统的安装，如图 4-2-4 所示。

第 3 步，进入 oVirt Node 节点主机安装主界面，与 Linux 操作系统的安装界面基本相同，如图 4-2-5 所示，单击“Continue”按钮。

第 4 步，完成 oVirt Node 节点主机基本参数配置，如图 4-2-6 所示，单击“Begin Installation”开始部署。

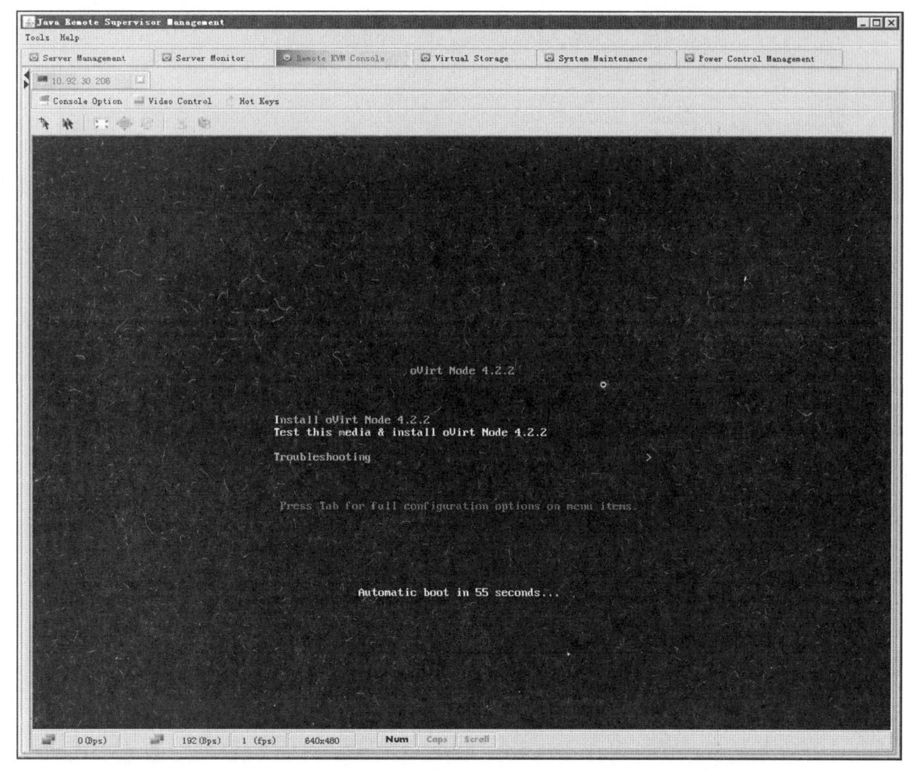
Java Remote Supervisor Management
Tools Help
Server Management
Server Monitor
Remote KVM Console
Virtual Storage
System Maintenance
Power Control Management
10.92.30.208
Console Option
Video Control
Hot Keys
oVirt Node 4.2.2
Install oVirt Node 4.2.2
Test this media & install oVirt Node 4.2.2
Troubleshooting
Press Tab for full configuration options on menu items.
Automatic boot in 55 seconds...
Num
Caps
Scroll

图 4-2-4

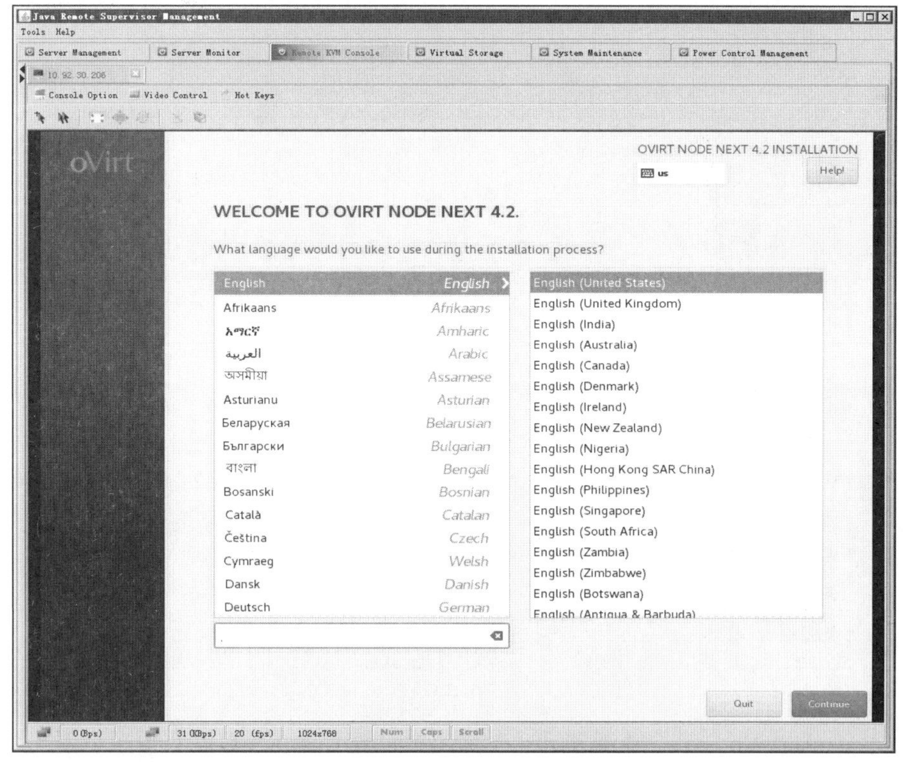
Java Remote Supervisor Management
Tools Help
Server Management
Server Monitor
Remote KVM Console
Virtual Storage
System Maintenance
Power Control Management
10.92.30.208
Console Option
Video Control
Hot Keys
oVirt
OVIRT NODE NEXT 4.2 INSTALLATION
us
Help!
WELCOME TO OVIRT NODE NEXT 4.2.
What language would you like to use during the installation process?
English English
Afrikaans Afrikaans
አማርኛ Amharic
العربية Arabic
অসমীয়া Assamese
Asturianu Asturian
Беларуская Belarusian
Български Bulgarian
বাংলা Bengali
Bosanski Bosnian
Català Catalan
Čeština Czech
Cymraeg Welsh
Dansk Danish
Deutsch German
English (United States)
English (United Kingdom)
English (India)
English (Australia)
English (Canada)
English (Denmark)
English (Ireland)
English (New Zealand)
English (Nigeria)
English (Hong Kong SAR China)
English (Philippines)
English (Singapore)
English (South Africa)
English (Zambia)
English (Zimbabwe)
English (Botswana)
English (Antigua & Barbuda)
Quit
Continue
Num
Caps
Scroll

图 4-2-5

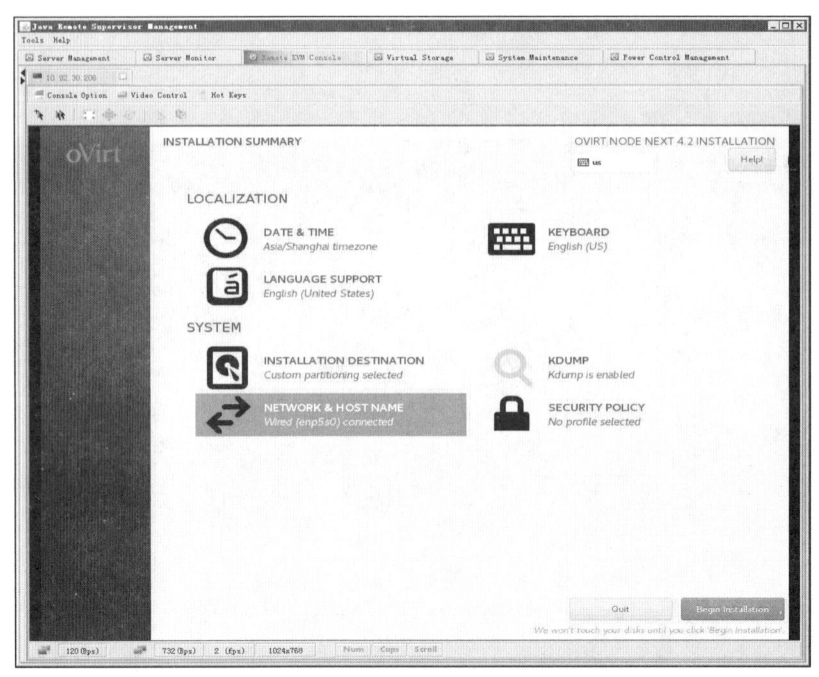

图 4-2-6

第 5 步，完成 oVirt Node 节点主机安装，如图 4-2-7 所示，单击“Reboot”按钮重启 oVirt Node 节点主机。

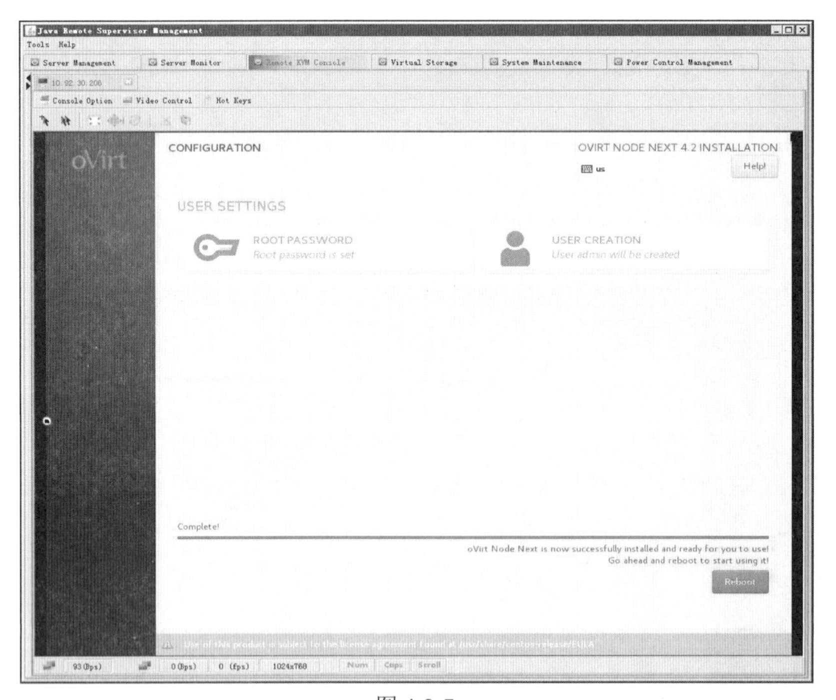

图 4-2-7

第 6 步，oVirt Node 节点主机重启，如图 4-2-8 所示，通过启动菜单可以看到其实质就是 CentOS 操作系统。

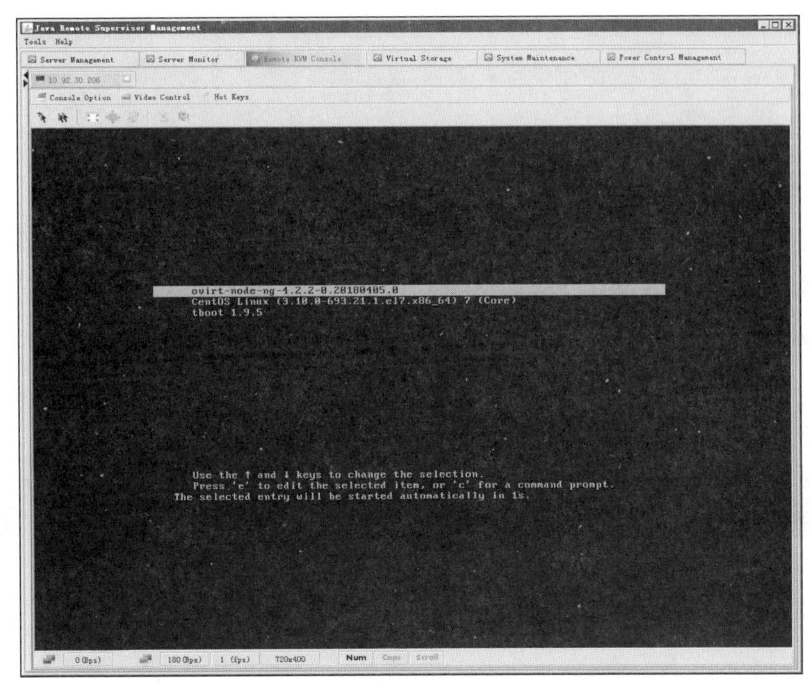

图 4-2-8

第 7 步，使用 root 用户登录 oVirt Node 节点主机，查看 IP 地址并测试网络的连通性，网络连接正常，如图 4-2-9 所示。

图 4-2-9

第 8 步，如果 oVirt Node 节点主机能够访问外网，可以使用命令“yum update”更新操作系统，如图 4-2-10 所示，可以看到需要下载的软件包为 637M。

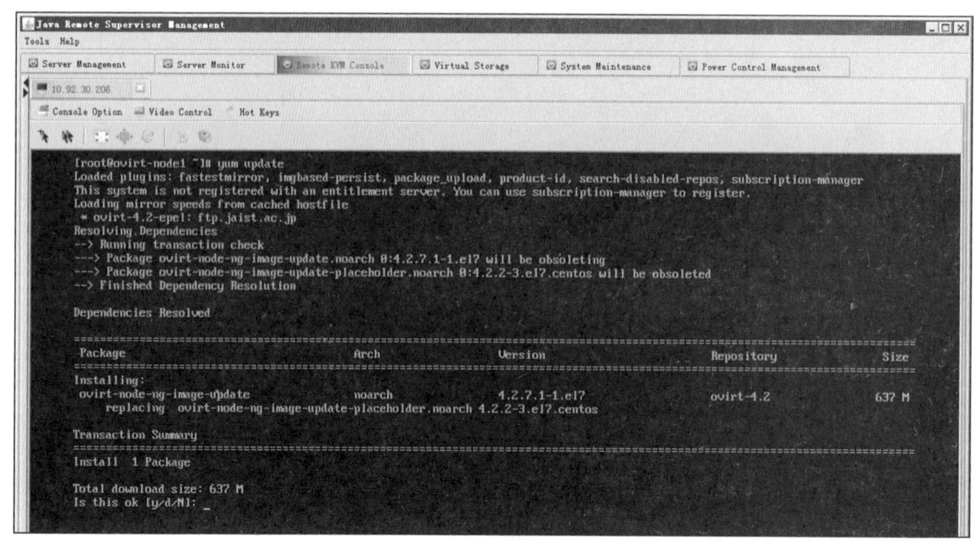

图 4-2-10

至此，oVirt Node 节点主机安装完成。

4.3 将主机加入 oVirt 平台管理

完成 oVirt Engine 管理端和 oVirt Node 节点主机部署后，就可以将节点主机加入管理端进行统一管理，同时，一些高级特性也需要管理端的支持。本节介绍如何将节点主机和普通 Linux 主机安装 VDSM 后加入管理端进行管理。

4.3.1 将 oVirt Node 节点主机加入管理端

实验环境已经部署好 2 台 oVirt Node 节点主机，下文介绍如何将这 2 台节点主机加入 oVirt Engine 管理端。

第 1 步，使用浏览器登录 oVirt Engine 管理端，系统会创建默认（Default）数据中心，如图 4-3-1 所示，本节操作不使用默认数据中心，单击“新建”按钮。

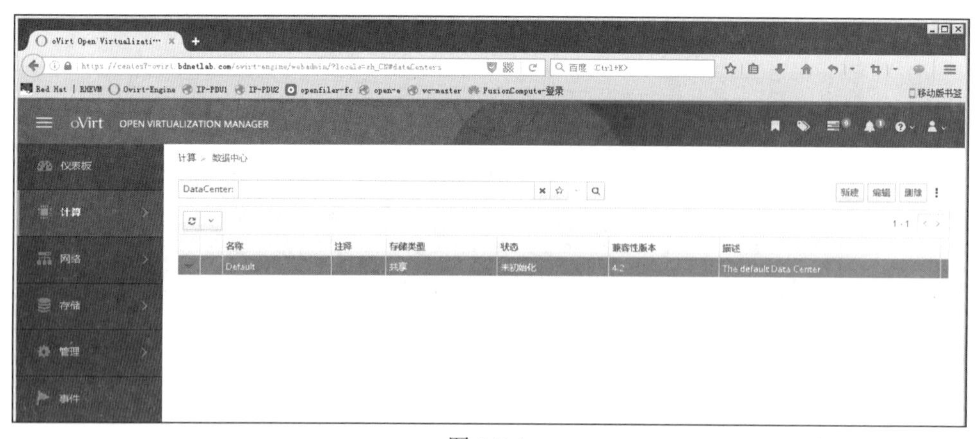

图 4-3-1

第 2 步，输入新建数据中心的相关参数，如图 4-3-2 所示，单击“确定”按钮。

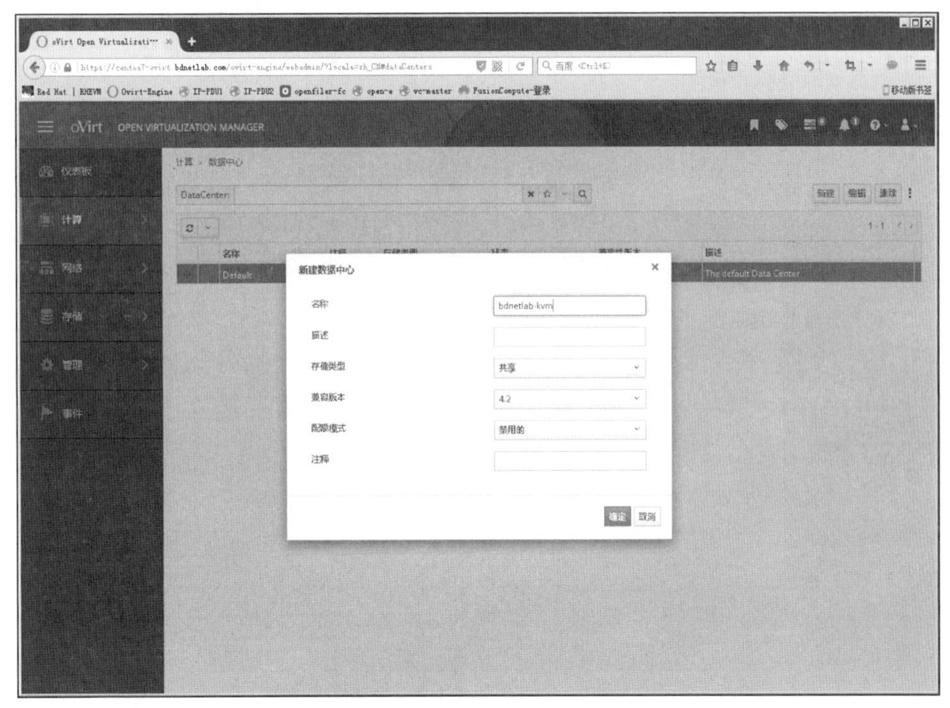

图 4-3-2

第 3 步，创建完数据中心后需要配置集群，如图 4-3-3 所示，单击“配置集群”。

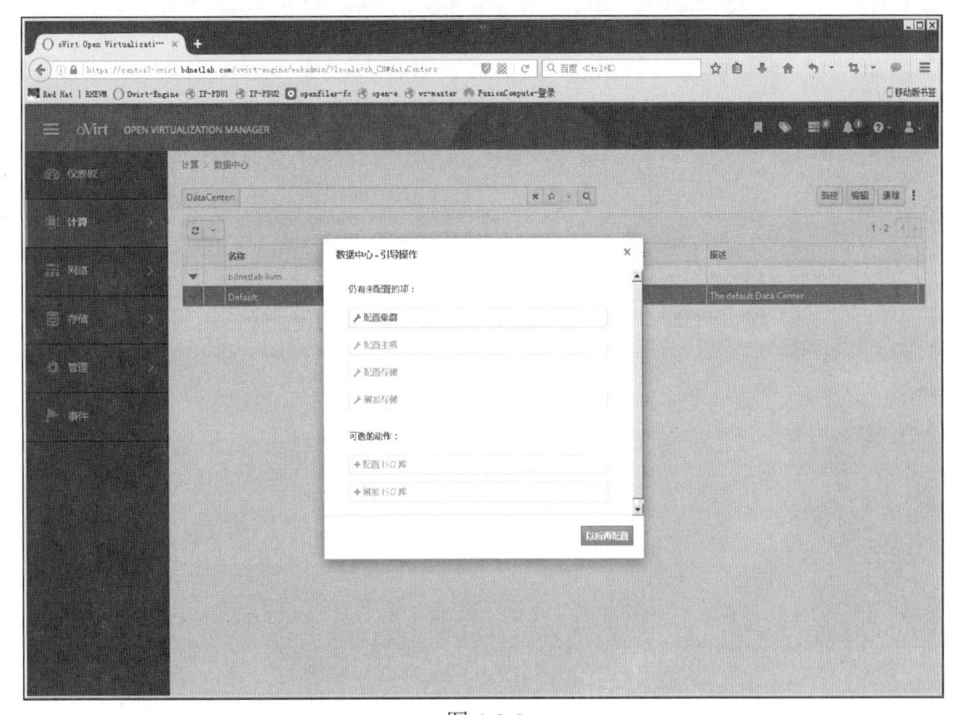

图 4-3-3

第 4 步，进入集群“常规”参数配置界面进行相应配置，如图 4-3-4 所示。

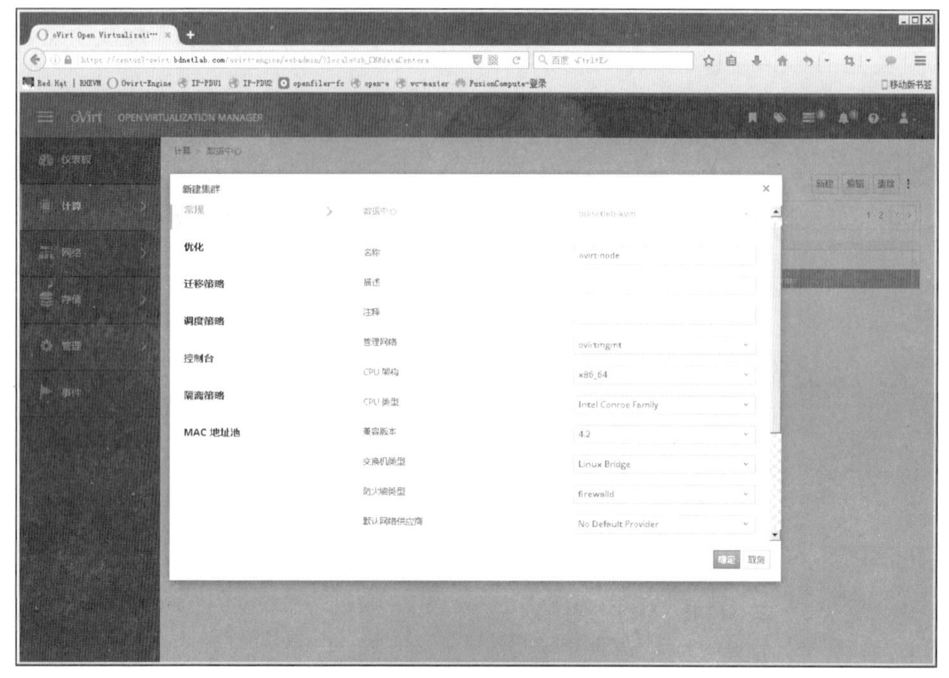

图 4-3-4

第 5 步，进入集群“优化”参数配置界面进行相应配置，如图 4-3-5 所示。

图 4-3-5

第 6 步，进入集群“迁移策略”参数配置界面进行相应配置，如图 4-3-6 所示。

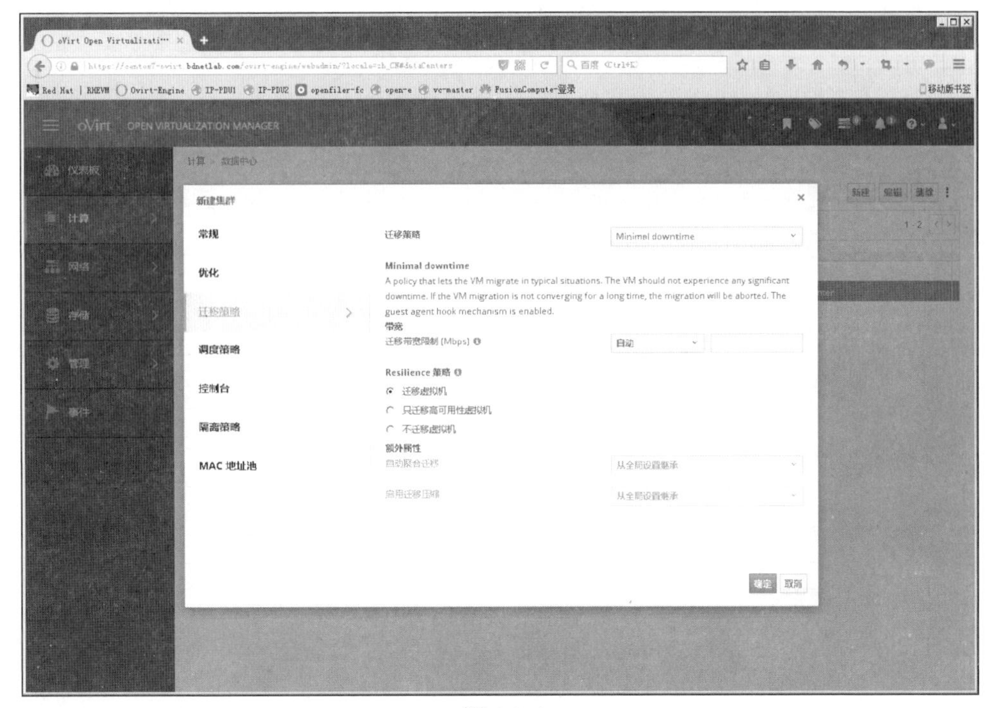

图 4-3-6

第 7 步，进入集群“调整策略”参数配置界面进行相应配置，如图 4-3-7 所示。

图 4-3-7

第 8 步，进入集群“控制台”参数配置界面进行相应配置，如图 4-3-8 所示。

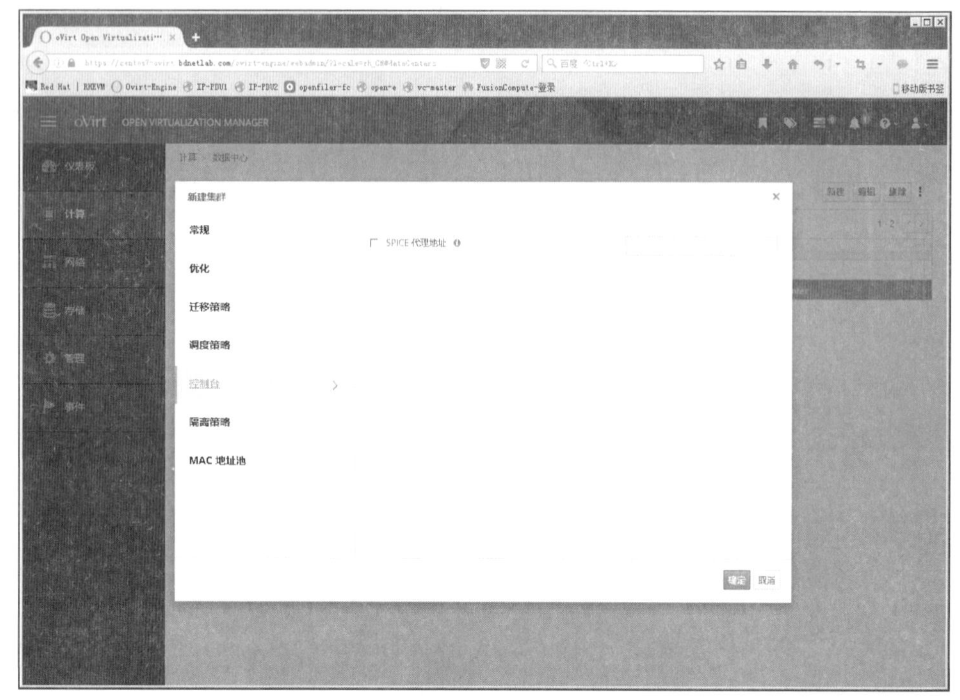

图 4-3-8

第 9 步，进入集群“隔离策略”参数配置界面进行相应配置，如图 4-3-9 所示。

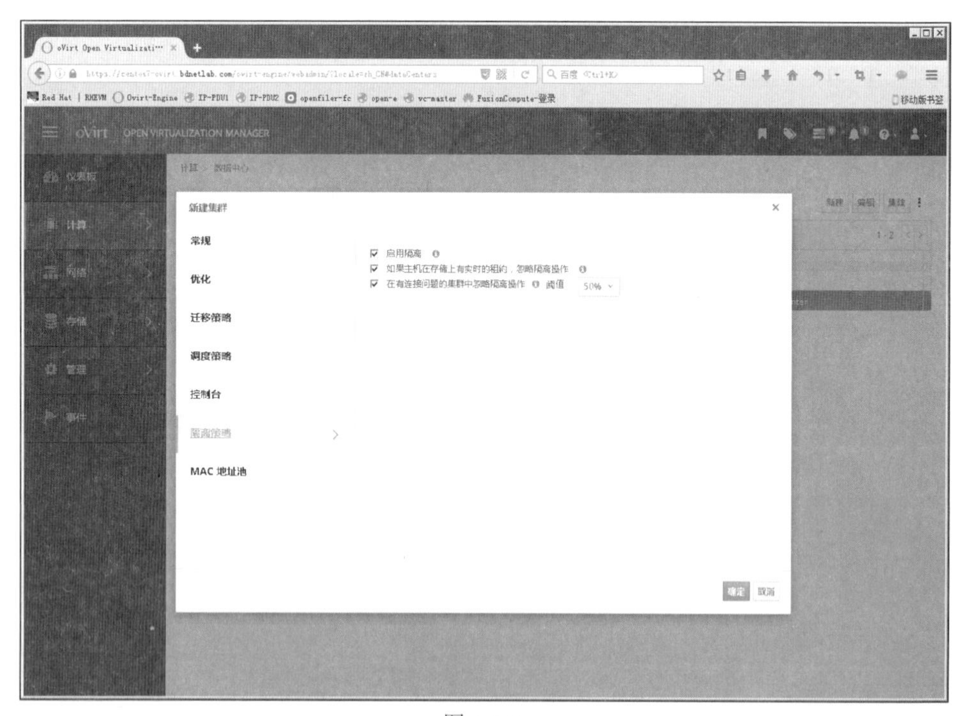

图 4-3-9

第 10 步，进入集群“MAC 地址池”参数配置界面进行相应配置，如图 4-3-10 所示，配置完成后单击“确定”按钮。

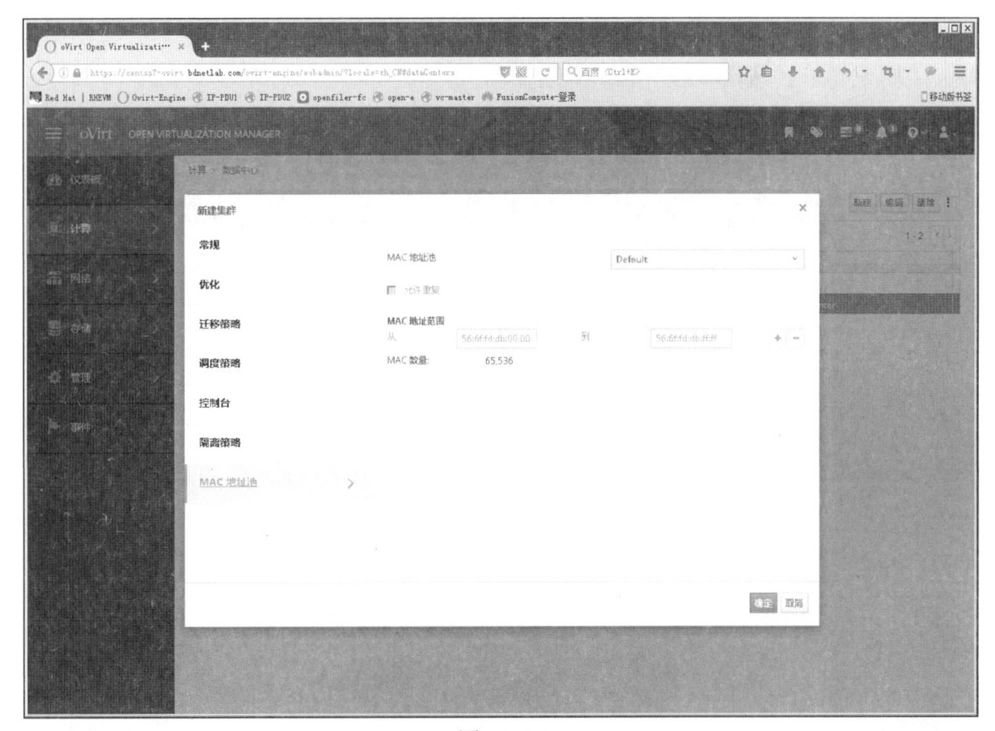

图 4-3-10

第 11 步，初期集群参数使用默认配置即可，后续可以进行调整，集群创建后需要配置主机，如图 4-3-11 所示，单击“配置主机”。

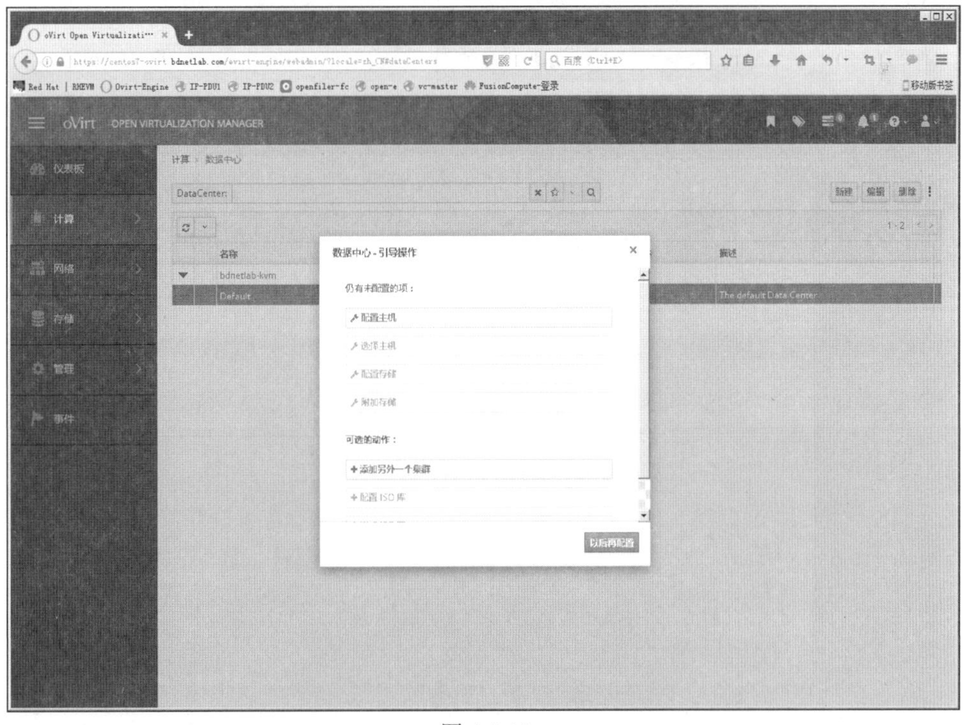

图 4-3-11

第 12 步，进入主机“常规”参数配置界面，如图 4-3-12 所示，选择主机需要加入的集群，输入主机名称、IP 地址以及密码等信息。

图 4-3-12

第 13 步，进入主机“电源管理”参数配置界面进行相关配置，如图 4-3-13 所示。

图 4-3-13

第 14 步，进入主机“SPM”参数配置界面进行相关配置，如图 4-3-14 所示。

图 4-3-14

第 15 步，进入主机“控制台”参数配置界面进行相关配置，如图 4-3-15 所示。

图 4-3-15

第 16 步，进入主机“网络供应商”参数配置界面进行相关配置，如图 4-3-16 所示。

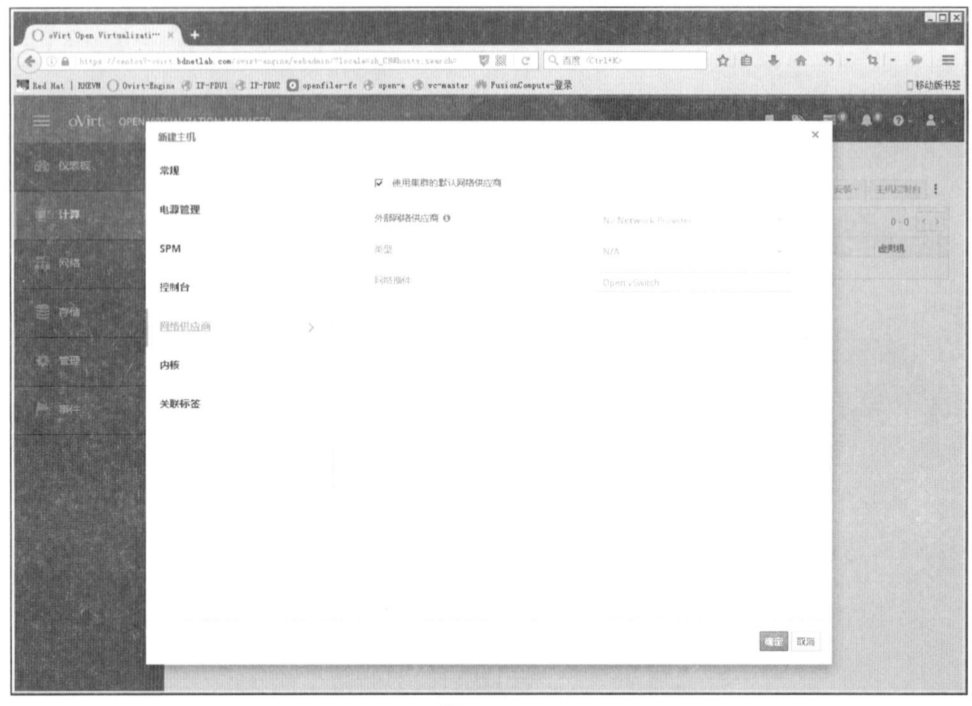

图 4-3-16

第 17 步，进入主机“内核”参数配置界面进行相关配置，如图 4-3-17 所示。

图 4-3-17

第 18 步，进入主机“关联标签”参数配置界面进行相关配置，如图 4-3-18 所示。

图 4-3-18

第 19 步，如果没有配置主机电源管理，系统会出现提示，如图 4-3-19 所示，单击“确定”按钮。

图 4-3-19

第 20 步，开始添加节点主机，主机处于安装状态，如图 4-3-20 所示。

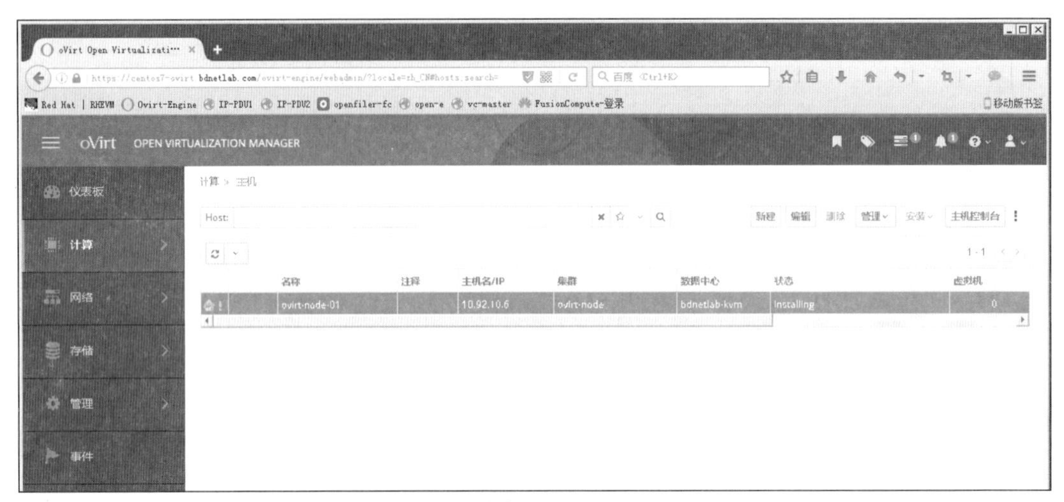

图 4-3-20

第 21 步，查看主机加入 oVirt Engine 管理端状态，如图 4-3-21 所示。

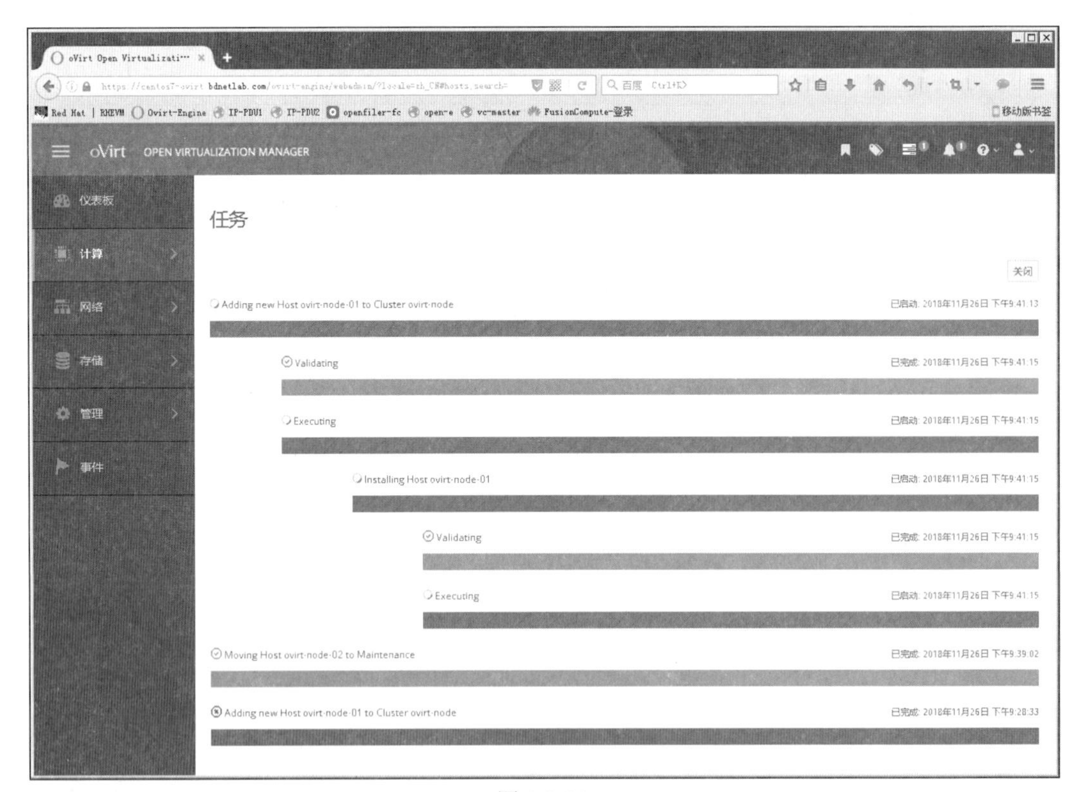

图 4-3-21

第 22 步，完成节点主机的添加，如图 4-3-22 所示。

第 23 步，查看节点主机处于“Up”状态，如图 4-3-23 所示。

第 24 步，查看节点主机的详细信息，如图 4-3-24 所示。

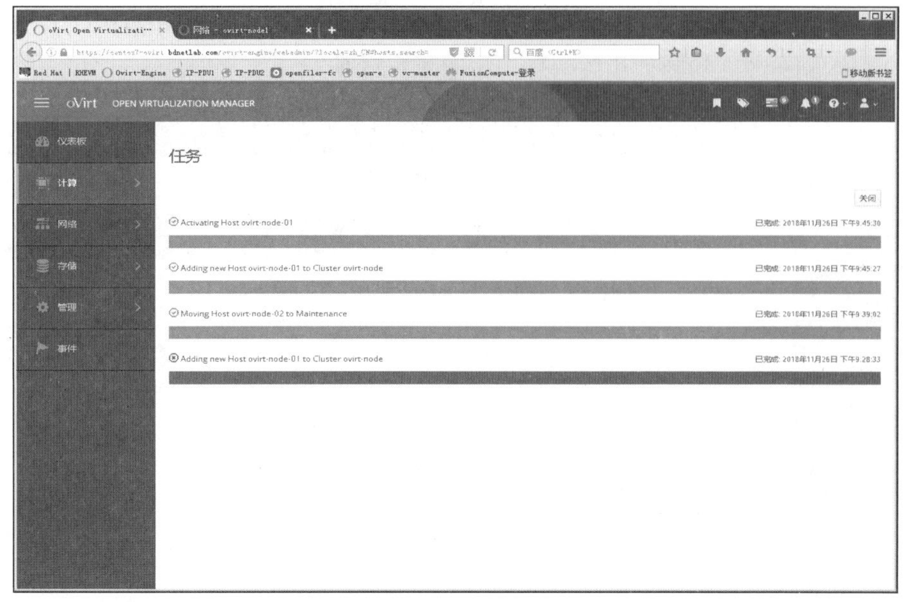

图 4-3-22

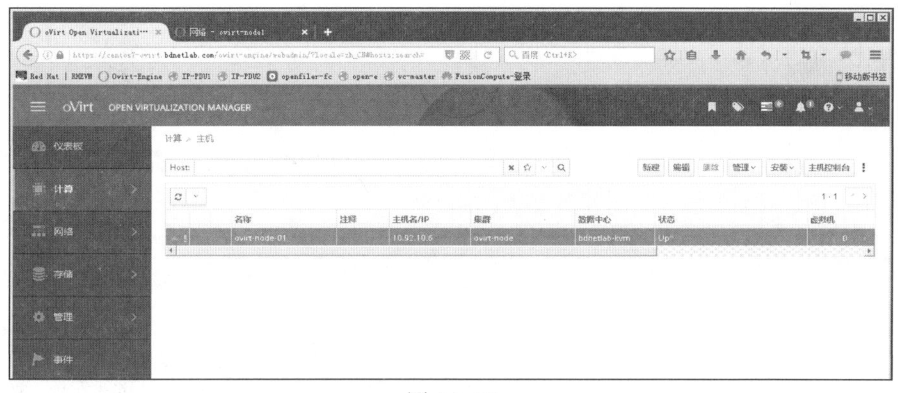

图 4-3-23

图 4-3-24

第 25 步，查看节点主机详细软件信息，可以看到操作系统版本为 RHEL 7，操作系统描述为 oVirt Node 4.2.2 和其他相关信息，如图 4-3-25 所示。

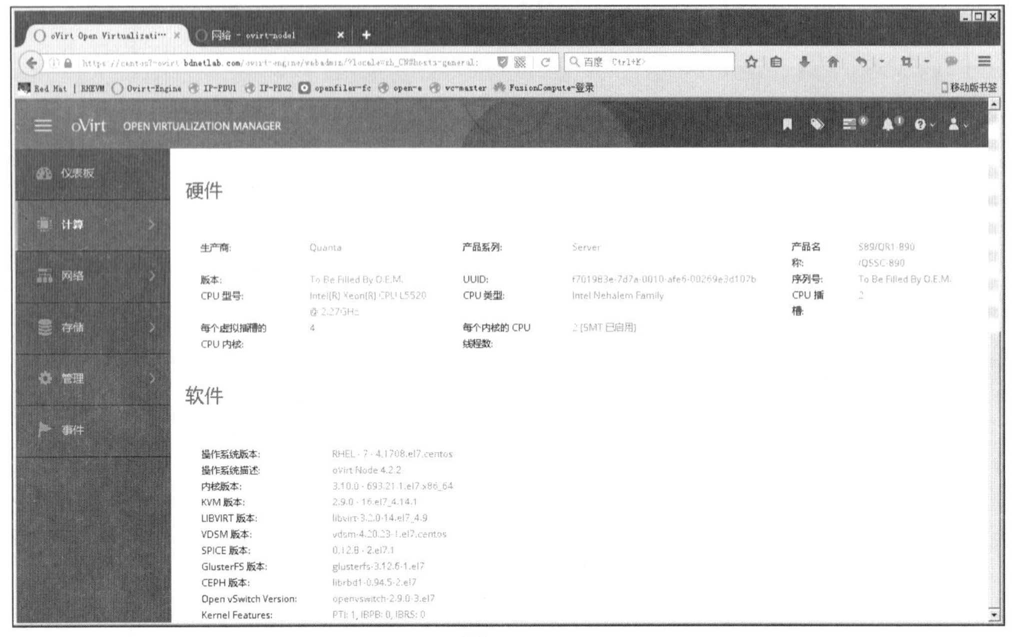

图 4-3-25

第 26 步，使用相同方式将另外一台节点主机加入 oVirt Engine 管理端，结果出现状态为“InstallFailed”的情况，如图 4-3-26 所示。

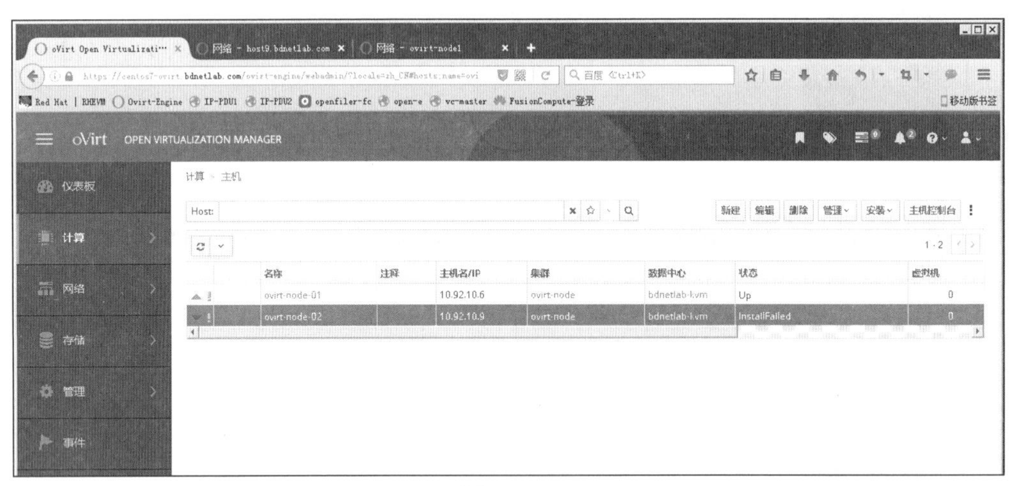

图 4-3-26

第 27 步，如果节点主机安装没有问题，可以打开节点主机控制台查看相关信息，如图 4-3-27 所示。

第 28 步，不能加入 oVirt Engine 管理端比较常见的原因是主机没有添加默认路由，注意节点主机 IP 地址写法，将子网掩码 PREFIX=24 修改为 NETMASK=254.254.255.0 后，重新加入成功，如图 4-3-28 所示。

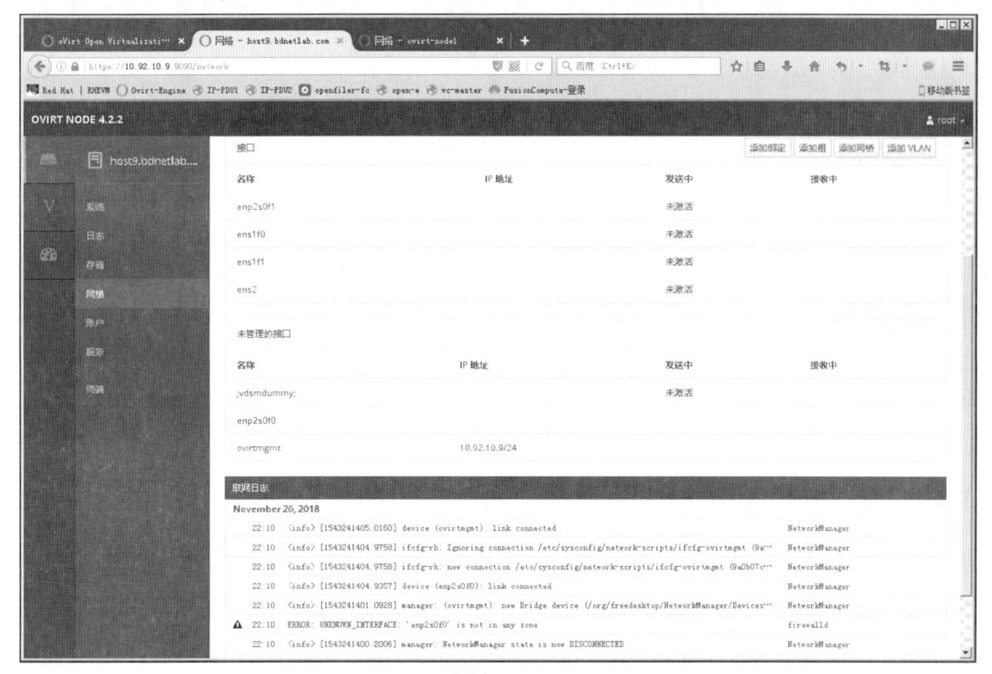

图 4-3-27

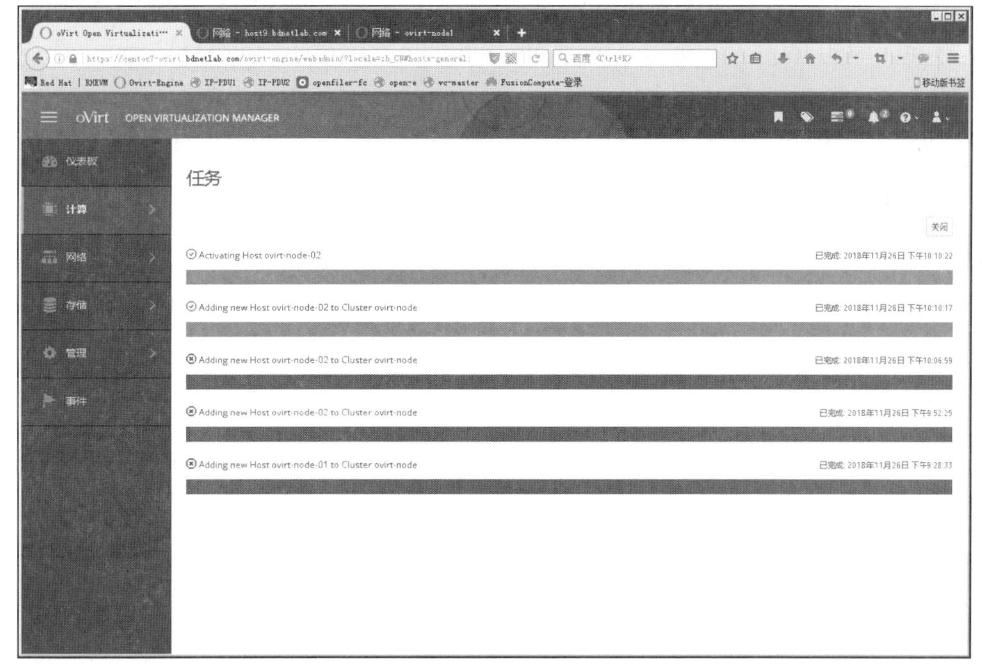

图 4-3-28

第 29 步，节点主机正常加入后，状态变为“Up”，如图 4-3-29 所示。

至此，将 oVirt Node 节点主机加入管理端完成。一般来说，只要节点主机在安装过程中没有出现错误提示，加入管理端就不存在问题，常见的问题就是主机没有添加默认路由，调整子网掩码写法后则加入正常，这是管理端对子网掩码的识别问题。

图 4-3-29

4.3.2 将 KVM 主机加入管理端

将 oVirt Node 节点主机加入管理端后，可以将之前安装的 KVM 主机加入管理端统一进行管理。需要说明的是，oVirt Engine 管理端对于 RHEL、CentOS 操作系统的 KVM 主机管理没有问题，对于 Ubuntu 操作系统的 KVM 主机可能存在运行不稳定的情况。本节介绍将 CentOS 操作系统 KVM 主机加入 oVirt Engine 管理端。

第 1 步，KVM 主机加入 oVirt Engine 管理端前需要安装 oVirt 平台 YUM 源，使用命令“yum install https://resources.ovirt.org/pub/yum-repo/ovirt-release42.rpm”在线配置 YUM 源。

```
[root@host7 ~]# yum install https://resources.ovirt.org/pub/yum-repo/ovirt-release42.rpm
已加载插件：fastestmirror
ovirt-release42.rpm
|  12 kB  00:00:00
Examining /var/tmp/yum-root-BYkiJ7/ovirt-release42.rpm: ovirt-release42-4.2.7-1.el7.noarch
Marking /var/tmp/yum-root-BYkiJ7/ovirt-release42.rpm to be installed
Resolving Dependencies
--> Running transaction check
---> Package ovirt-release42.noarch 0:4.2.7-1.el7 will be installed
--> Finished Dependency Resolution
Dependencies Resolved
=============================================================================
 Package            Arch       Version         Repository               Size
=============================================================================
Installing:
 ovirt-release42    noarch     4.2.7-1.el7     /ovirt-release42         11 k
Transaction Summary
=============================================================================
Install  1 Package

Total size: 11 k
Installed size: 11 k
Is this ok [y/d/N]: y
Downloading packages:
Running transaction check
Running transaction test
Transaction test succeeded
Running transaction
```

```
  Installing:ovirt-release42-4.2.7-1.el7.noarch  1/1
  Verifying: ovirt-release42-4.2.7-1.el7.noarch  1/1
Installed:
  ovirt-release42.noarch 0:4.2.7-1.el7
Complete!
```

第 2 步，使用命令“yum install vdsm”安装插件，已经安装 KVM 的主机需要更新 83MB 左右的插件包，其他 Linux 主机约 200MB。

```
[root@host7 ~]#yum install vdsm
已加载插件：fastestmirror
Loading mirror speeds from cached hostfile
 * base: mirrors.neusoft.edu.cn
 * extras: mirrors.neusoft.edu.cn
 * ovirt-4.2: resources.ovirt.org
……（省略）
总计：83 MB
Is this ok [y/d/N]: y
Downloading packages:
作为依赖被升级：
  augeas-libs.x86_64 0:1.4.0-5.el7_4.1                  glusterfs.x86_64 0:3.12.15-1.el7
  glusterfs-api.x86_64 0:3.12.15-1.el7                  glusterfs-cli.x86_64 0:3.12.15-1.el7
  glusterfs-client-xlators.x86_64 0:3.12.15-1.el7       glusterfs-libs.x86_64 0:3.12.15-1.el7
  net-snmp-libs.x86_64 1:5.7.2-33.el7_4.2
替代：
qemu-img.x86_64 10:1.4.3-156.el7_5.5   qemu-kvm.x86_64 10:1.4.3-156.el7_5.5   qemu-kvm-common.x86_64
10:1.4.3-156.el7_5.5
完毕！
```

第 3 步，插件 VDSM 安装完成后就可以将 KVM 主机加入 oVirt Engine 管理端，选择“主机集群”，输入主机名、IP 地址以及密码等信息，如图 4-3-30 所示，单击“确定”按钮。

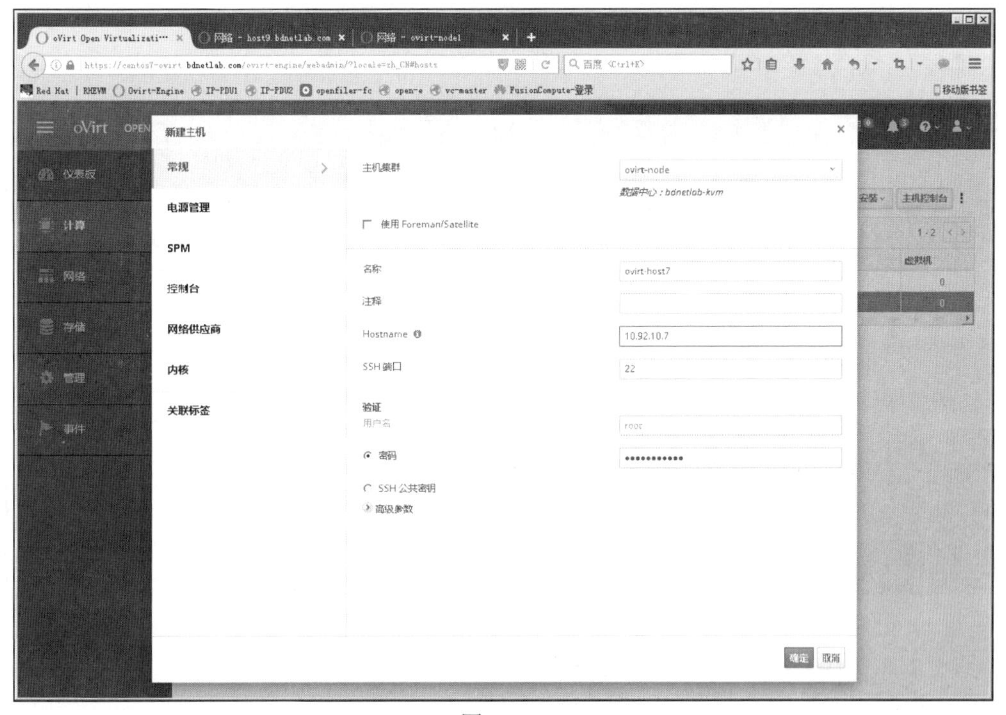

图 4-3-30

第 4 步，如果没有配置主机电源管理，系统会出现提示，如图 4-3-31 所示，单击“确定”按钮。

图 4-3-31

第 5 步，开始添加 KVM 主机，主机处于安装状态，如图 4-3-32 所示。

图 4-3-32

第 6 步，查看 KVM 主机加入 oVirt Engine 管理端状态，如图 4-3-33 所示。

第 7 步，完成 KVM 主机的添加，如图 4-3-34 所示。

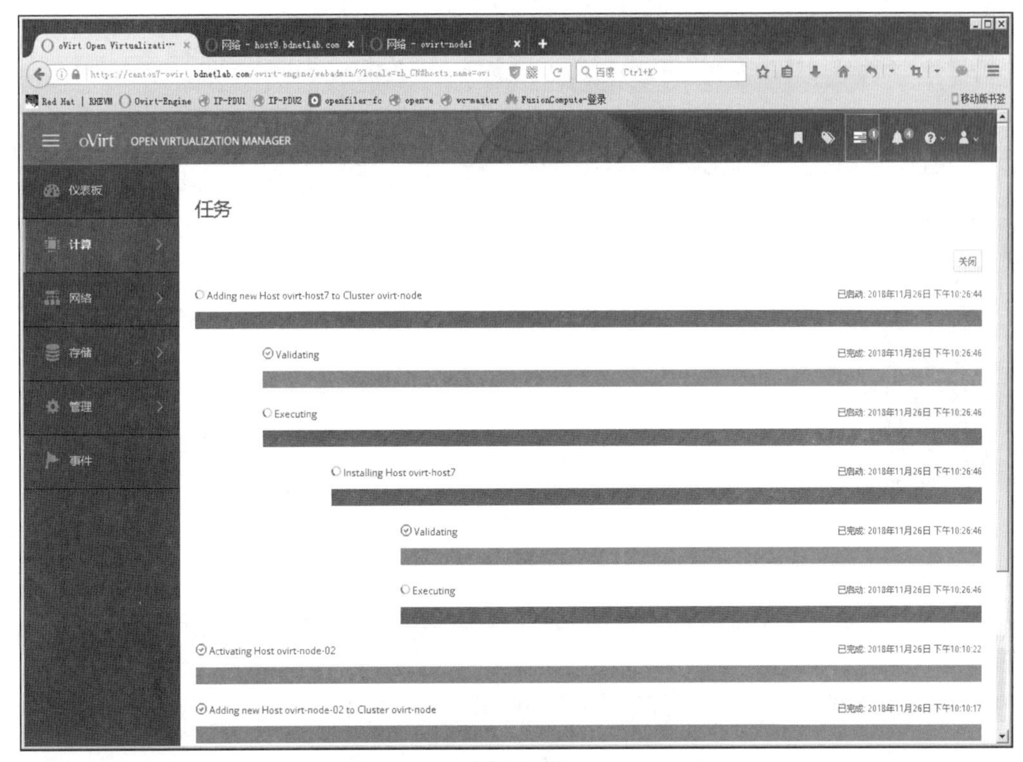
oVirt OPEN VIRTUALIZATION MANAGER
仪表板
计算
网络
存储
管理
事件
任务
关闭
Adding new Host ovirt-host7 to Cluster ovirt-node
Validating
Executing
Installing Host ovirt-host7
Validating
Executing
Activating Host ovirt-node-02
Adding new Host ovirt-node-02 to Cluster ovirt-node

图 4-3-33

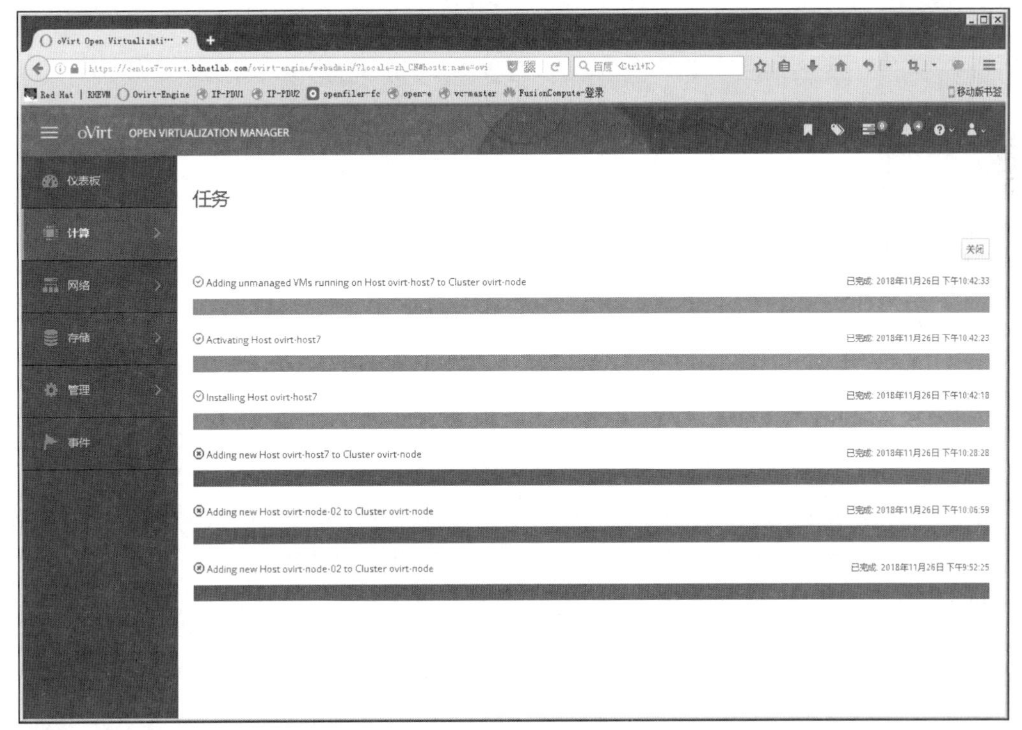
oVirt OPEN VIRTUALIZATION MANAGER
仪表板
计算
网络
存储
管理
事件
任务
关闭
Adding unmanaged VMs running on Host ovirt-host7 to Cluster ovirt-node
Activating Host ovirt-host7
Installing Host ovirt-host7
Adding new Host ovirt-host7 to Cluster ovirt-node
Adding new Host ovirt-node-02 to Cluster ovirt-node
Adding new Host ovirt-node-02 to Cluster ovirt-node

图 4-3-34

第8步，KVM 主机正常加入后，状态变为“Up”，如图 4-3-35 所示。

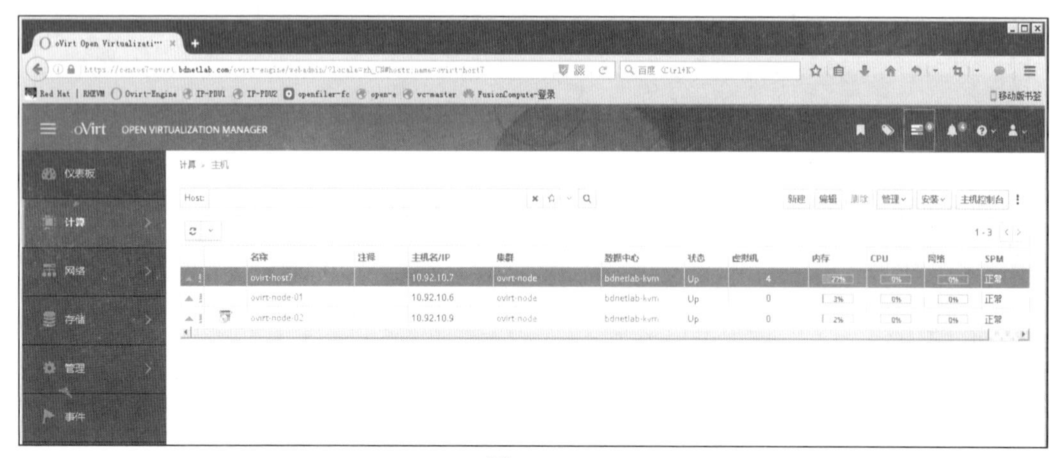

图 4-3-35

第9步，查看 KVM 主机的详细信息，如图 4-3-36 所示。

图 4-3-36

第10步，查看 KVM 主机的详细软件信息，可以看到操作系统版本为 RHEL 7，操作系统描述为 CentOS 7 操作系统，区别于 oVirt Node，如图 4-3-37 所示。

第11步，查看 KVM 主机的详细信息，可以看到 KVM 主机上正在运行的虚拟机的情况，如图 4-3-38 所示。

第12步，查看虚拟机的详细信息，如图 4-3-39 所示。

至此，KVM 主机成功加入 oVirt Engine 管理端，其主要的操作就是在 KVM 主机上安装插件 VDSM，只要正确安装插件 VDSM，加入 oVirt Engine 管理端就基本不存在问题。

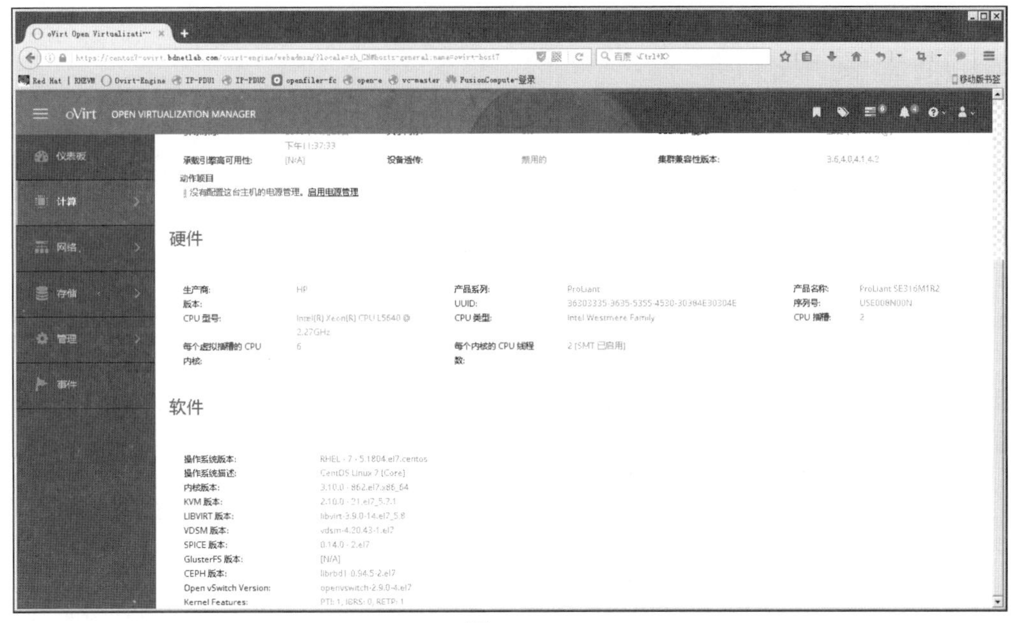

图 4-3-37

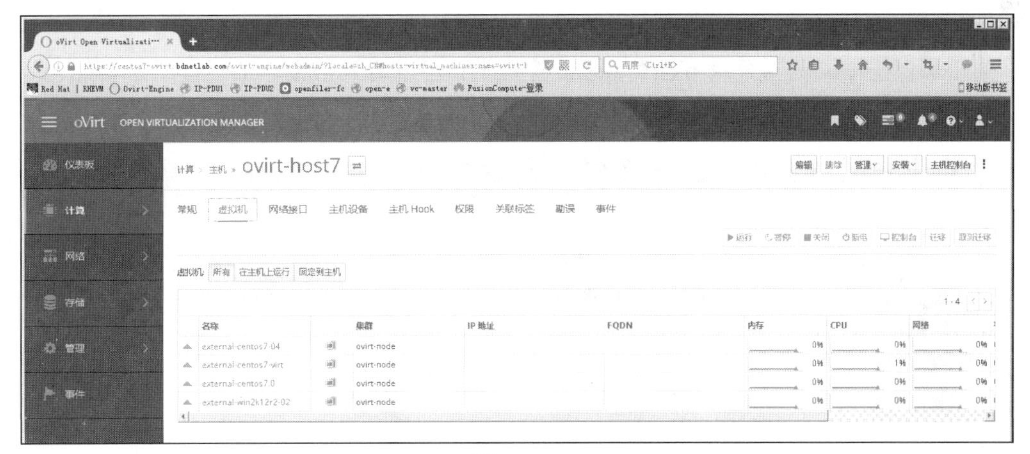

图 4-3-38

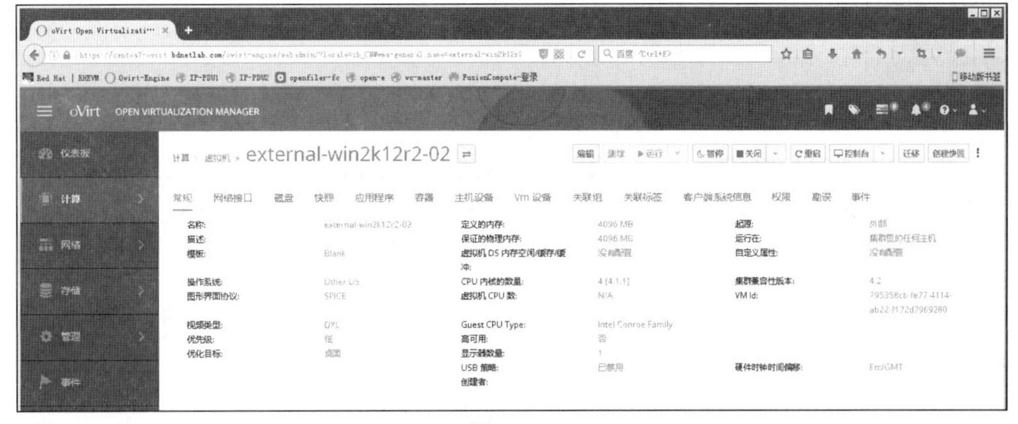

图 4-3-39

4.4 配置使用存储

无论是传统数据中心还是虚拟化数据中心，存储是数据中心正常运行的关键设备。作为企业虚拟化架构实施人员或管理人员，必须考虑如何在企业生产环境构建高可用的存储环境，以保证虚拟化架构的正常运行。IBM、HP、EMC 等专业级存储设备可以提供大容量、高容错以及双机热备等功能，但相对来说价格昂贵。作为性价比较好的 iSCSI 存储在企业生产环境中占有很大的市场份额，特别是使用 10 Gbit/s 网络承载的 iSCSI 存储性能非常不错。

NFS 是在 UNIX 和 Linux 操作系统中流行的网络文件系统，Windows 也将 NFS 作为一个组件，添加配置后可以让其提供 NFS 存储服务。NFS 存储由于配置简单和管理方便，也大量在生产环境中使用。oVirt 环境中对 NFS 的使用相对较多，如镜像域和导出域都会使用到 NFS。本节介绍如何创建使用存储。

4.4.1 配置使用 iSCSI 存储

在实验环境中已经配置了 OPEN-E 作为存储服务器，可以提供 iSCSI/NFS 等多种存储服务，关于 OPEN-E 如何部署可以参考相关文档。本节介绍 oVirt Engine 管理端 iSCSI 的存储配置。

第 1 步，使用浏览器登录 oVirt Engine 管理端，可以看到数据中心处于“未初始化” 状态，如图 4-4-1 所示，其原因是数据中心未配置存储。

图 4-4-1

第 2 步，选择“存储”选项“存储域”配置，在 oVirt 环境下存储是以存储域的方式存在的，默认有 oVirt-image-repository 镜像存储域，如图 4-4-2 所示，在生产环境中推荐新建，单击“新建域”按钮。

第 3 步，出现新建域界面，在“域功能”选择“数据”，用于存储虚拟机文件，存储类型选择“iSCSI”，输入名称，在“发现目标”处输入生产环境中的 iSCSI 存储服务器地址，如果有 CHAP 认证同时需要输入用户名和密码，如图 4-4-3 所示，单击“发现”按钮。

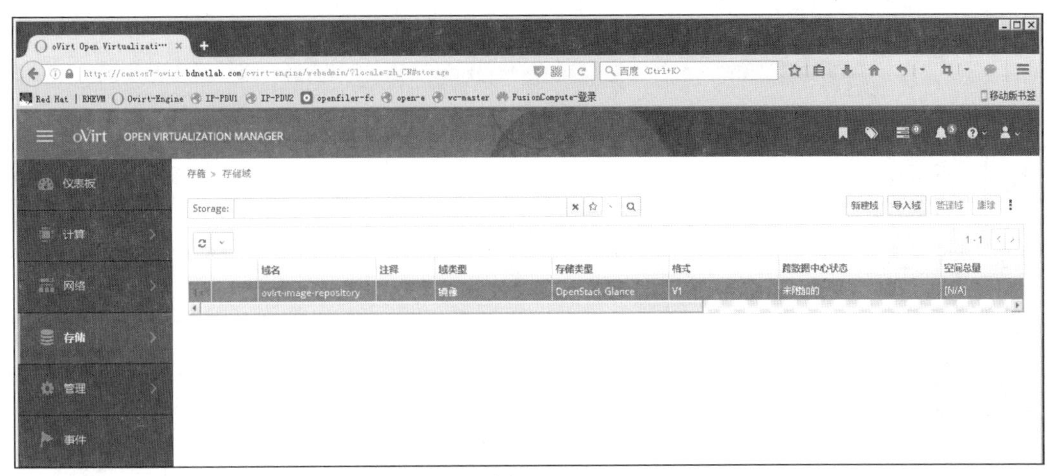

图 4-4-2

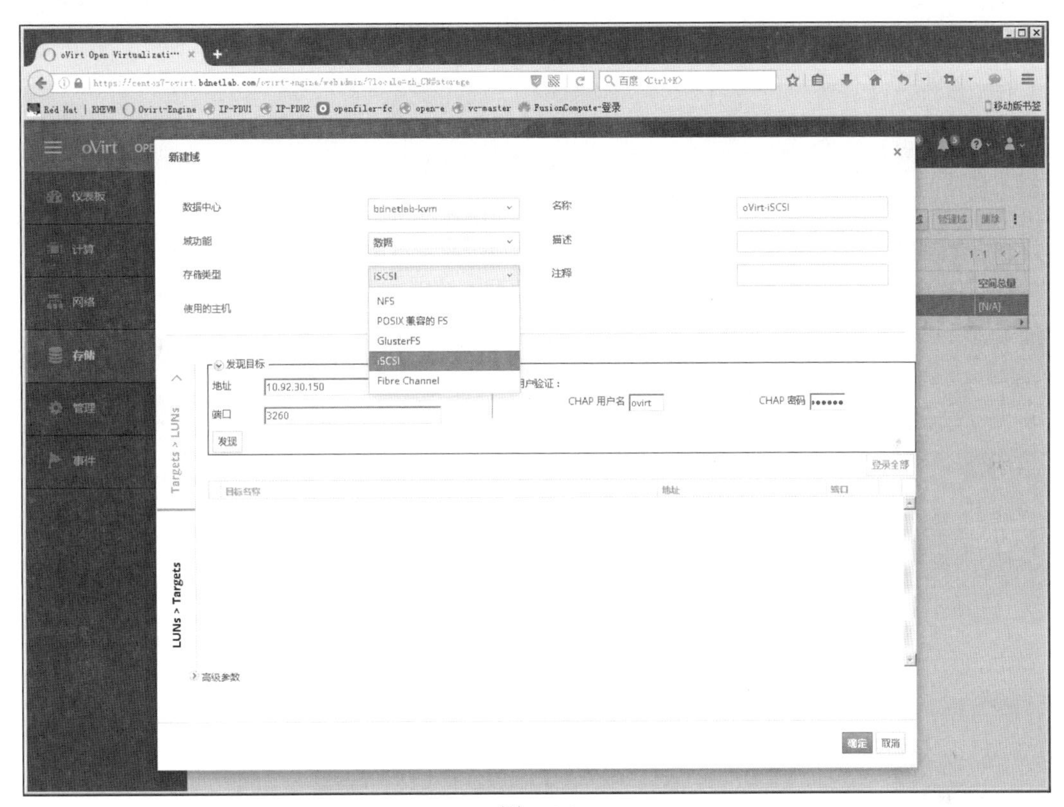

图 4-4-3

第 4 步，系统会查找 iSCSI 存储服务器可以使用的目标，如图 4-4-4 所示。

第 5 步，选择 oVirt Engine 管理端使用的 iSCSI 存储，如图 4-4-5 所示，单击“确定”按钮。

第 6 步，系统将 iSCSI 添加到 oVirt Engine 管理端，添加过程中存储域处于“已锁定”状态，如图 4-4-6 所示。

第 7 步，查看新建 iSCSI 存储的信息，如图 4-4-7 所示。

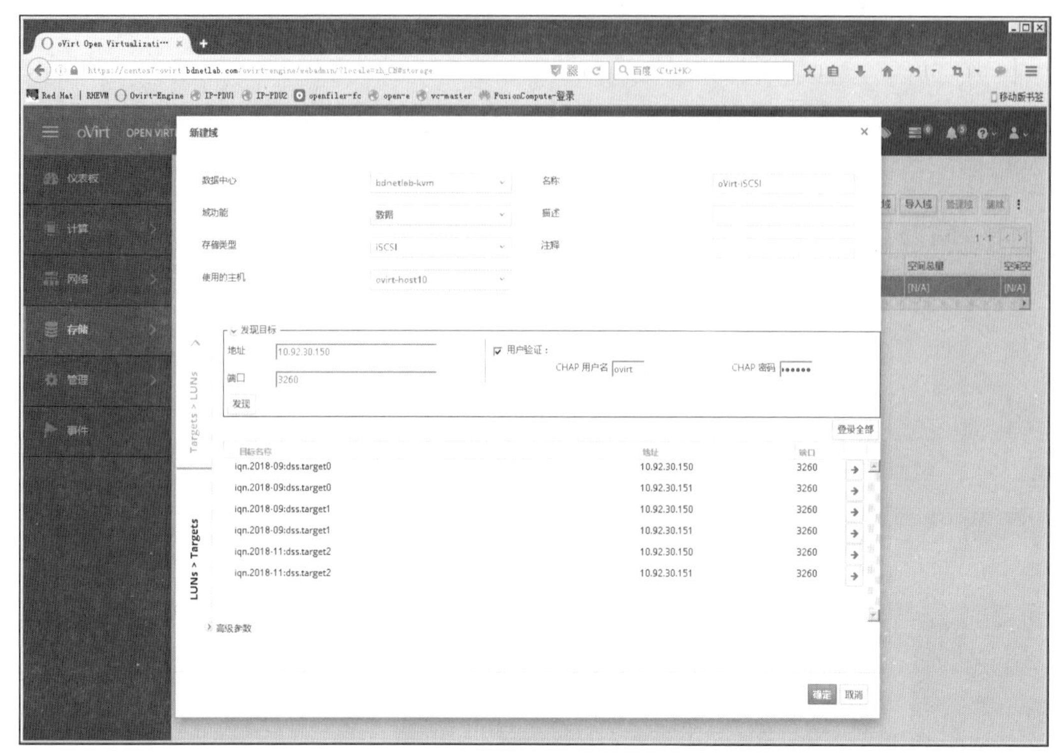
新建域
数据中心
bdnetlab-kvm
名称
oVirt-iSCSI
域功能
数据
描述
存储类型
iSCSI
注释
使用的主机
ovirt-host10
发现目标
地址
10.92.30.150
端口
3260
发现
用户验证：
CHAP 用户名
ovirt
CHAP 密码
登录全部
目标名称
地址
端口
iqn.2018-09:dss.target0 10.92.30.150 3260
iqn.2018-09:dss.target0 10.92.30.151 3260
iqn.2018-09:dss.target1 10.92.30.150 3260
iqn.2018-09:dss.target1 10.92.30.151 3260
iqn.2018-11:dss.target2 10.92.30.150 3260
iqn.2018-11:dss.target2 10.92.30.151 3260
Targets > LUNs
LUNs > Targets
高级参数
确定
取消

图 4-4-4

新建域
数据中心
bdnetlab-kvm
名称
oVirt-iSCSI
域功能
数据
描述
存储类型
iSCSI
注释
使用的主机
ovirt-host10
发现目标
登录全部
目标名称
地址
端口
iqn.2018-09:dss.target0 10.92.30.150 3260
iqn.2018-09:dss.target0 10.92.30.151 3260
iqn.2018-09:dss.target1 10.92.30.150 3260
iqn.2018-09:dss.target1 10.92.30.151 3260
iqn.2018-11:dss.target2 10.92.30.150 3260
LUN ID
239576a4257717279 256 GiB 1 SCST_BIO 9WjBWqry SSCST_BIO9WjBWqrykBXbKQY2bb
添加
iqn.2018-11:dss.target2 10.92.30.151 3260
Targets > LUNs
LUNs > Targets
高级参数
确定
取消

图 4-4-5

图 4-4-6

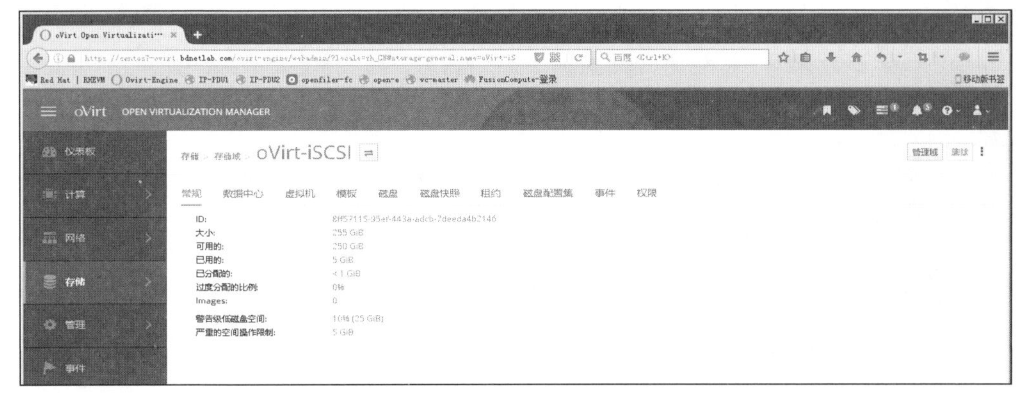

图 4-4-7

第 8 步，查看任务状态，添加进行中，如图 4-4-8 所示。

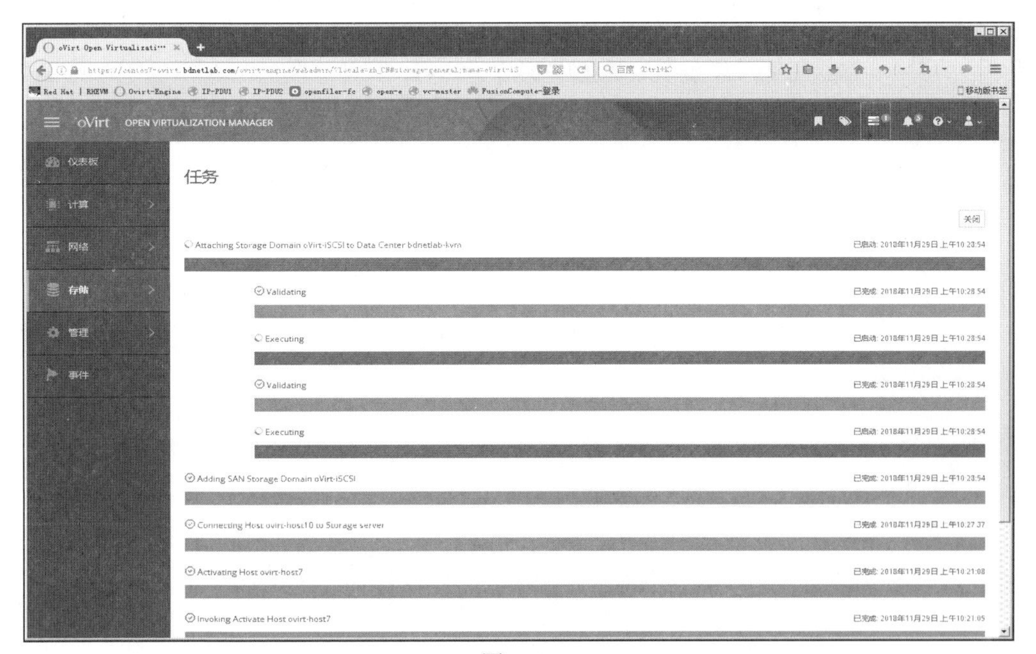

图 4-4-8

第 9 步，将 iSCSI 存储添加到 oVirt Engine 管理端完成，如图 4-4-9 所示，注意如果出现错误提示请检查配置。

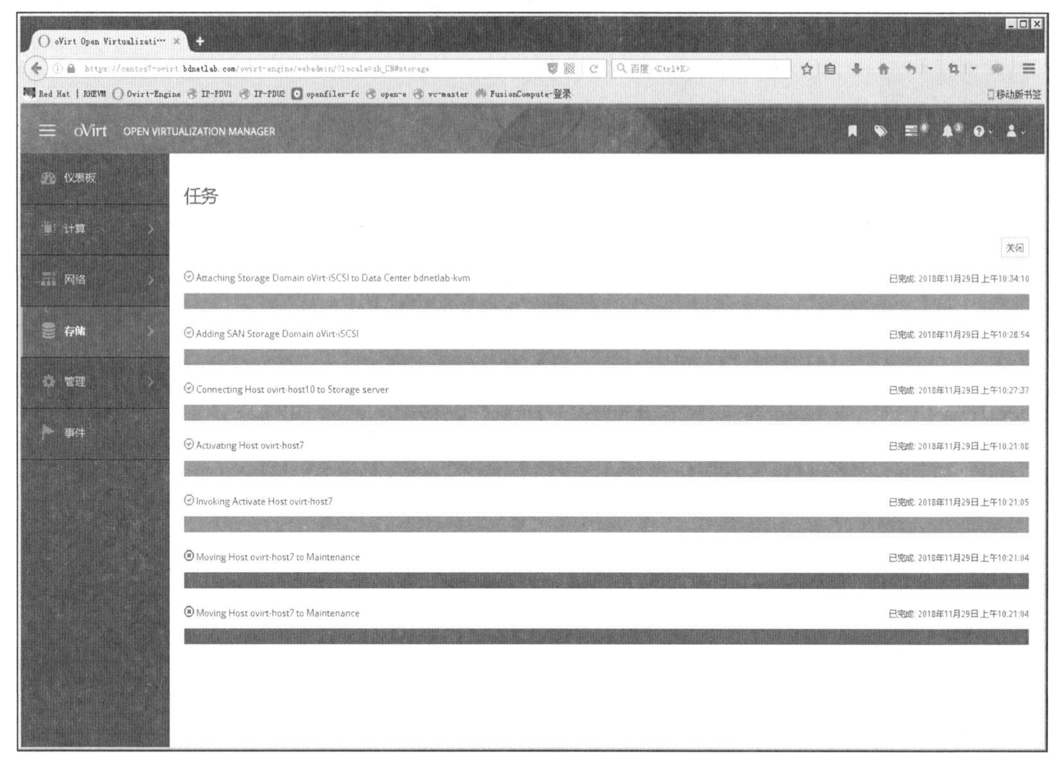

图 4-4-9

第 10 步，再次查看存储域 iSCSI 存储状态变为“活跃的”，如图 4-4-10 所示。

图 4-4-10

第 11 步，查看新建的数据中心状态已变为“Up”，如图 4-4-11 所示。

至此，oVirt Engine 管理端添加 iSCSI 存储完成。整体来说，oVirt Engine 管理端作为一个客户端配置比较简单，更多的是 iSCSI 存储服务器的配置。

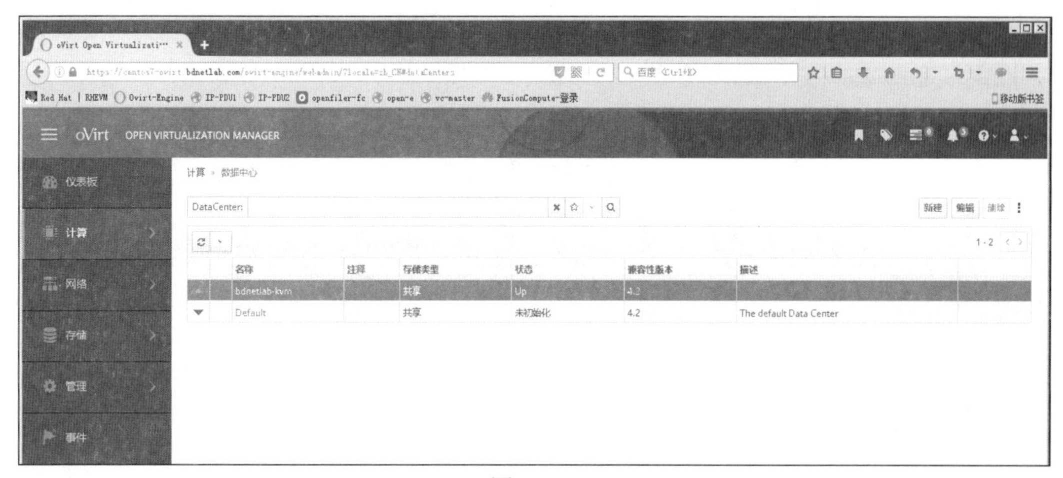

图 4-4-11

4.4.2 配置基于 NFS 存储的 ISO 域

oVirt Engine 管理端使用 ISO 域，其主要目的是上传后续用于安装操作系统的 ISO 镜像文件，在 oVirt Engine 管理端中使用特殊的域进行识别。

第 1 步，进入新建域界面，选择存储类型为 NFS，在“使用的主机”处输入名称，在“导出路径”处输入生产环境中的 NFS 存储服务器地址，如图 4-4-12 所示，单击“确定”按钮。

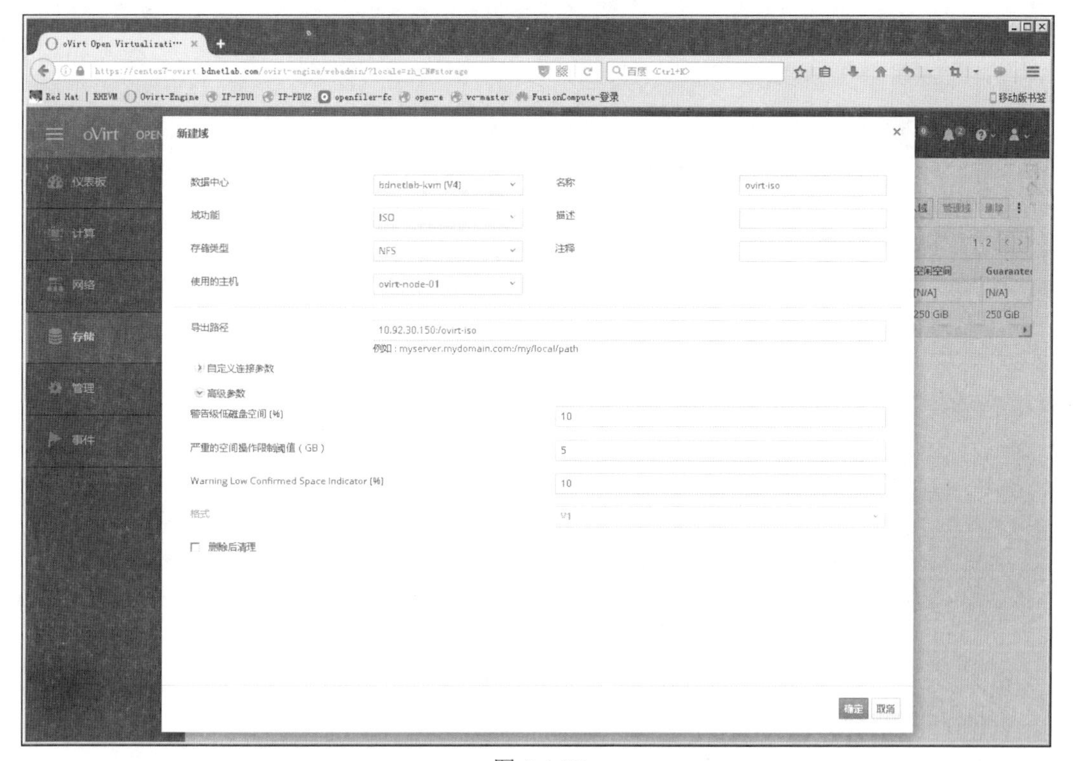

图 4-4-12

第 2 步，系统将 NFS 存储添加到 oVirt Engine 管理端，添加过程中存储域处于“不活跃的”状态，如图 4-4-13 所示。

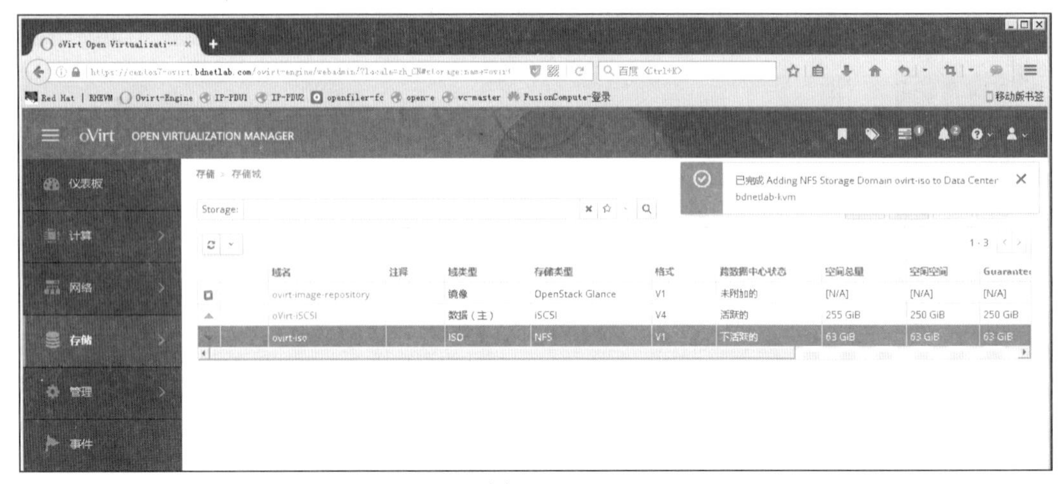

图 4-4-13

第 3 步，查看任务状态，添加进行中，如图 4-4-14 所示。

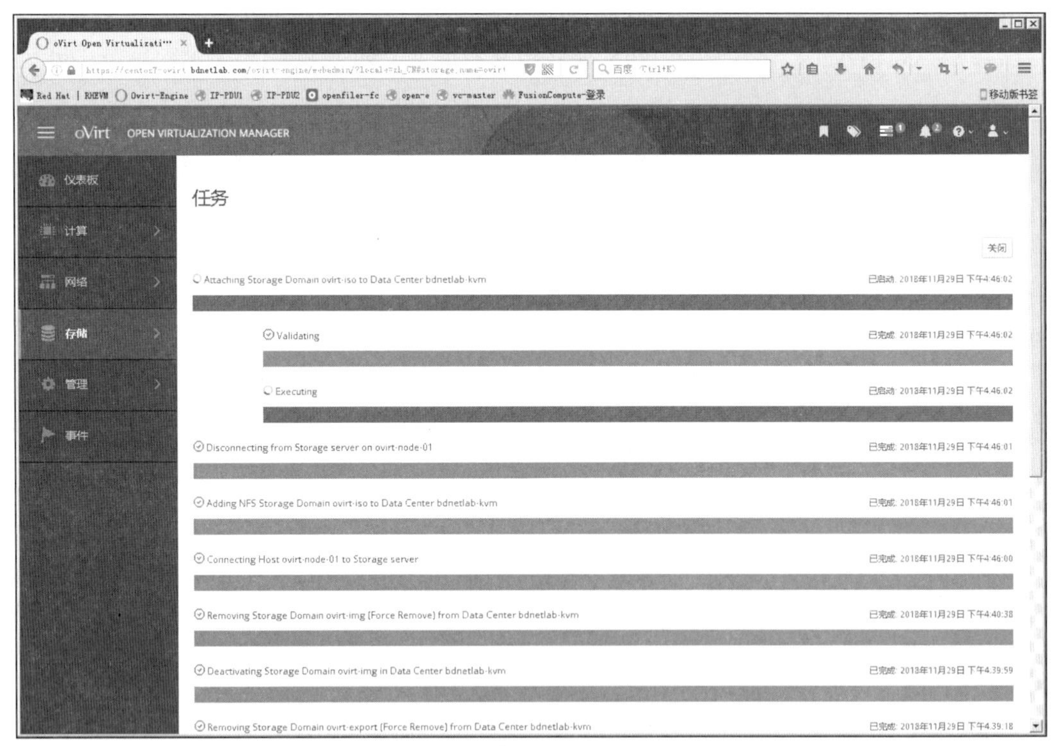

图 4-4-14

第 4 步，将 NFS 存储添加到 oVirt Engine 管理端完成，如图 4-4-15 所示，注意如果出现错误提示请检查配置。

第 5 步，再次查看存储域 NFS 存储状态，变为“活跃的”，如图 4-4-16 所示。

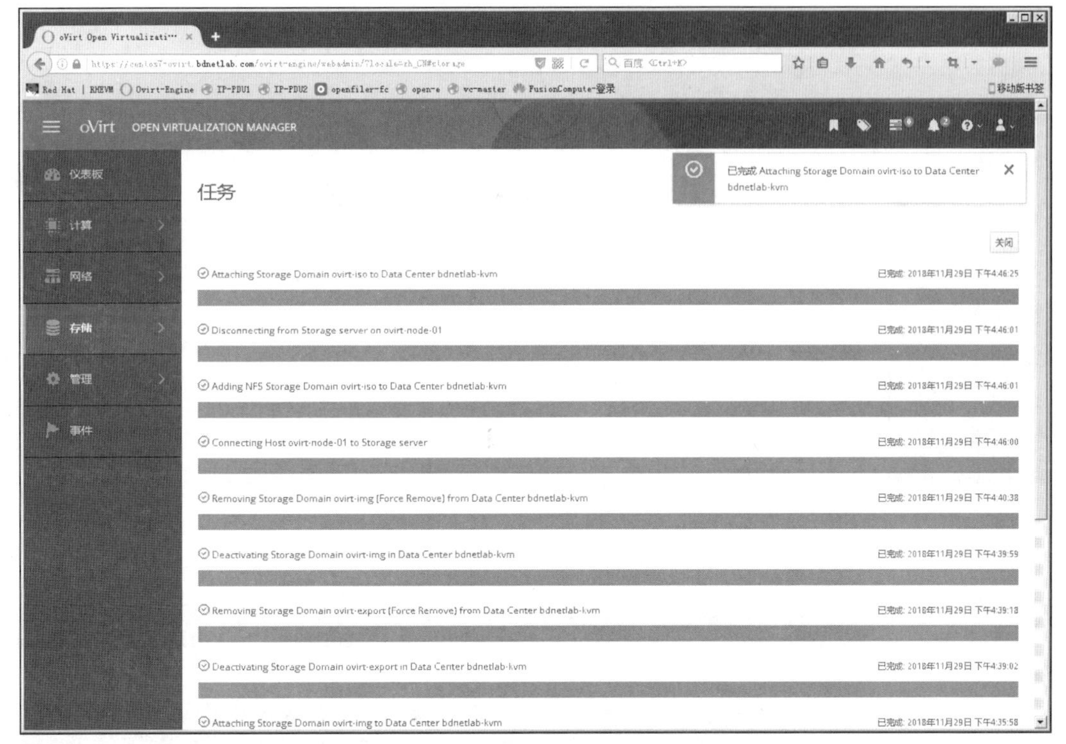

图 4-4-15

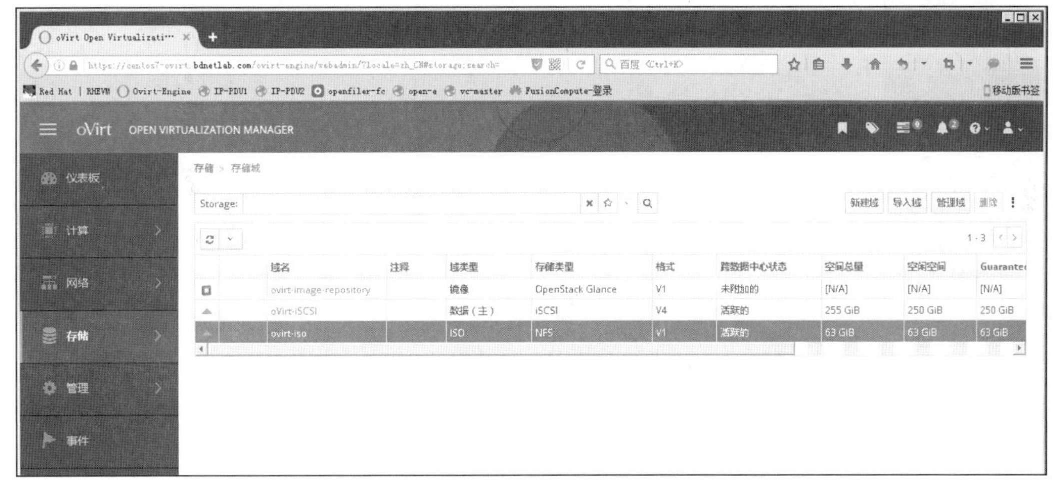

图 4-4-16

第 6 步，查看新建 NFS 存储的信息，如图 4-4-17 所示，新建的 ISO 域需要上传 ISO 文件后才能正常使用。

第 7 步，使用命令“ovirt-iso-uploader”将 ISO 文件上传至 ISO 域。

```
[root@centos7-ovirt tmp]# ovirt-iso-uploader -i ovirt-iso upload CentOS-7-x86_64-Minimal-1708.iso
Please provide the REST API password for the admin@internal oVirt Engine user (CTRL+D to abort):
Uploading, please wait...
INFO: Start uploading CentOS-7-x86_64-Minimal-1708.iso
Uploading: [########################################] 100%
INFO: CentOS-7-x86_64-Minimal-1708.iso uploaded successfully
```

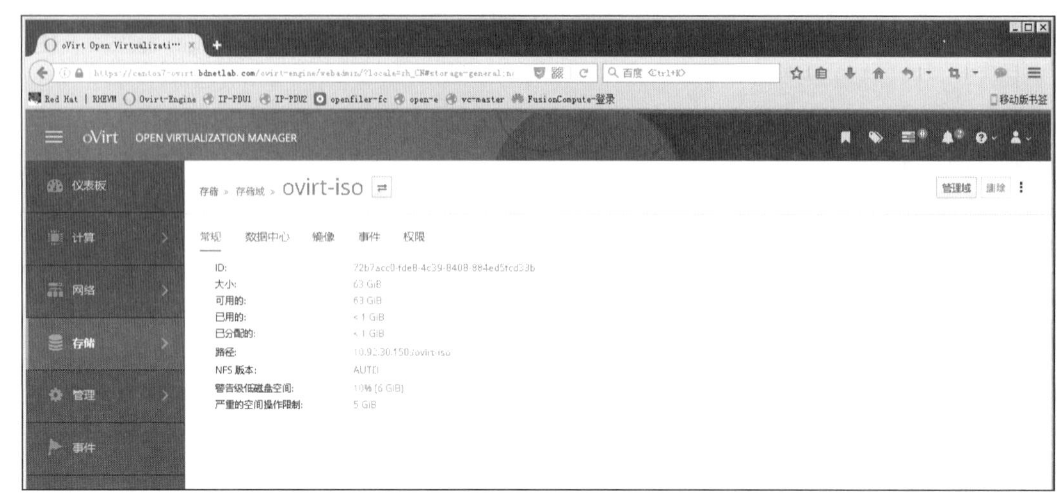

图 4-4-17

第 8 步，查看上传成功的 ISO 镜像文件，如图 4-4-18 所示。

图 4-4-18

至此，基于 NFS 存储的 ISO 域创建完成。需要注意 ISO 域的主要功能，新建完成后不上传 ISO 文件后续将无法使用。

4.4.3 配置基于 NFS 存储的导出域

oVirt Engine 管理端使用导出域，其主要目的是将虚拟机文件导出备份，在 oVirt Engine 管理端中使用特殊的域进行识别。

第 1 步，进入新建域界面，选择存储类型为 NFS，在“使用的主机”处输入名称，在“导出路径”处输入生产环境中的 NFS 存储服务器地址，如图 4-4-19 所示，单击“确定”。

第 2 步，系统将 NFS 存储添加到 oVirt Engine 管理端，添加过程中存储域处于“已锁定”状态，如图 4-4-20 所示。

第 3 步，查看任务状态，添加进行中，如图 4-4-21 所示。

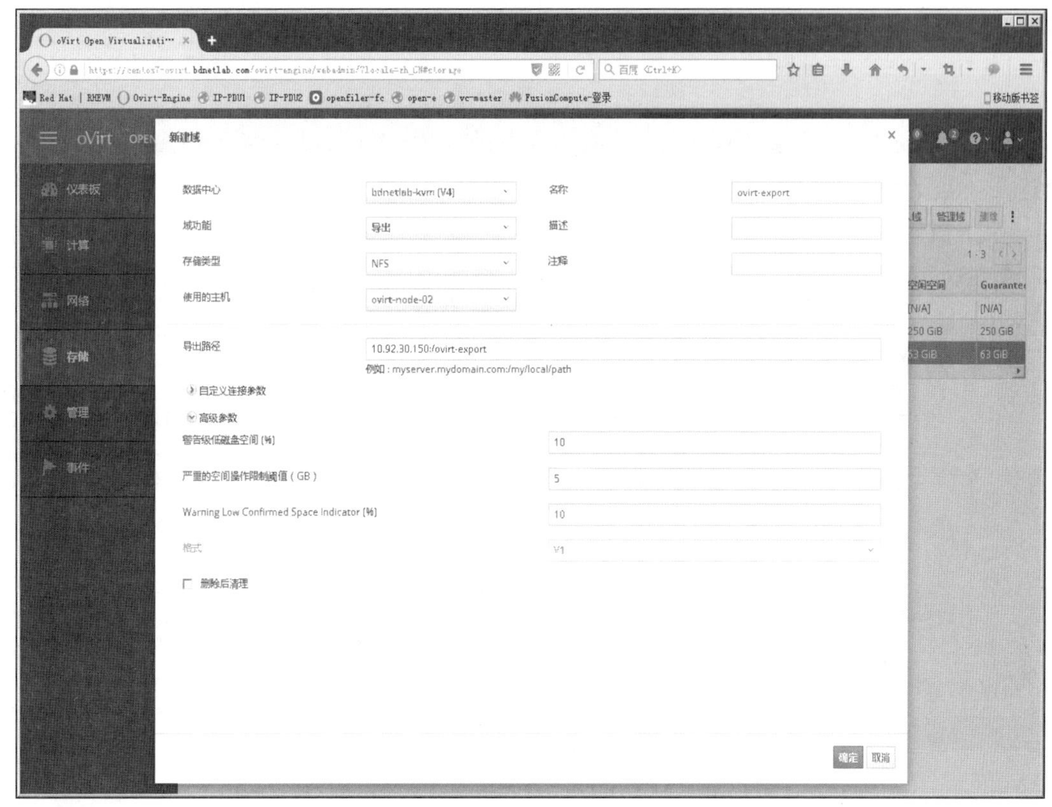

图 4-4-19

图 4-4-20

第 4 步，将 NFS 存储添加到 oVirt Engine 管理端完成，如图 4-4-22 所示，注意如果出现错误提示请检查配置。

第 5 步，再次查看存储域 NFS 存储状态变为“活跃的”，如图 4-4-23 所示。

第 6 步，查看新建 NFS 存储的信息，如图 4-4-24 所示，新建的导出域可在后续导出虚拟机文件的时候使用。

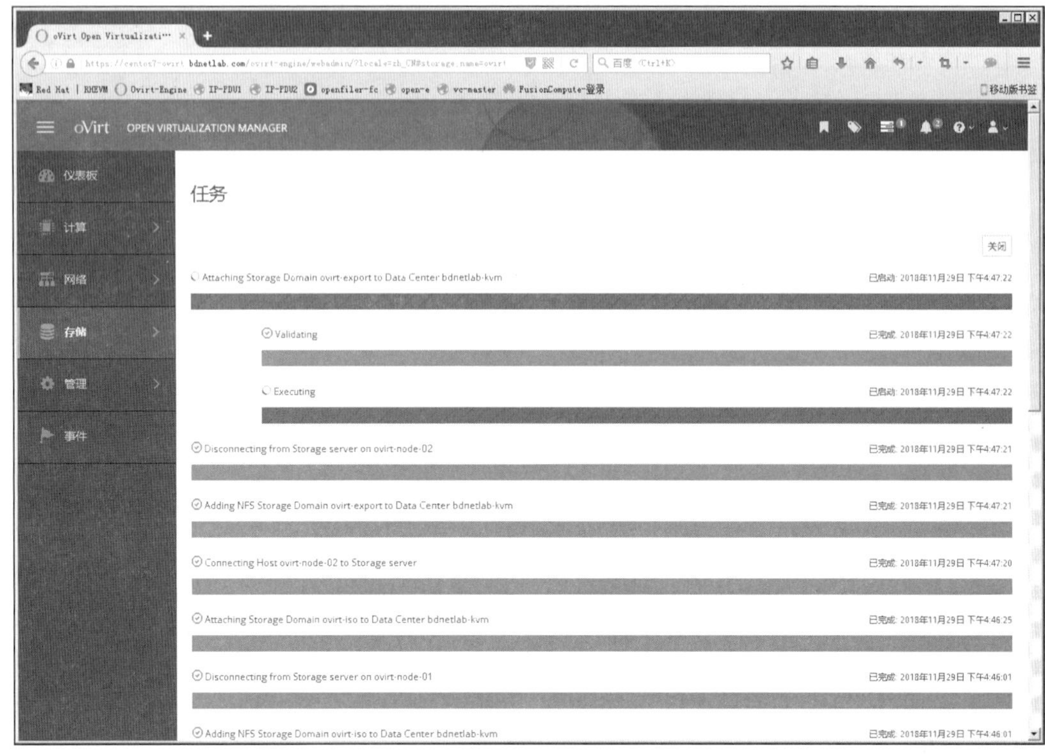
oVirt Open Virtualizati…
oVirt OPEN VIRTUALIZATION MANAGER
仪表板
计算
网络
存储
管理
事件
任务
关闭
Attaching Storage Domain ovirt-export to Data Center bdnetlab-kvm
Validating
Executing
Disconnecting from Storage server on ovirt-node-02
Adding NFS Storage Domain ovirt-export to Data Center bdnetlab-kvm
Connecting Host ovirt-node-02 to Storage server
Attaching Storage Domain ovirt-iso to Data Center bdnetlab-kvm
Disconnecting from Storage server on ovirt-node-01
Adding NFS Storage Domain ovirt-iso to Data Center bdnetlab-kvm

图 4-4-21

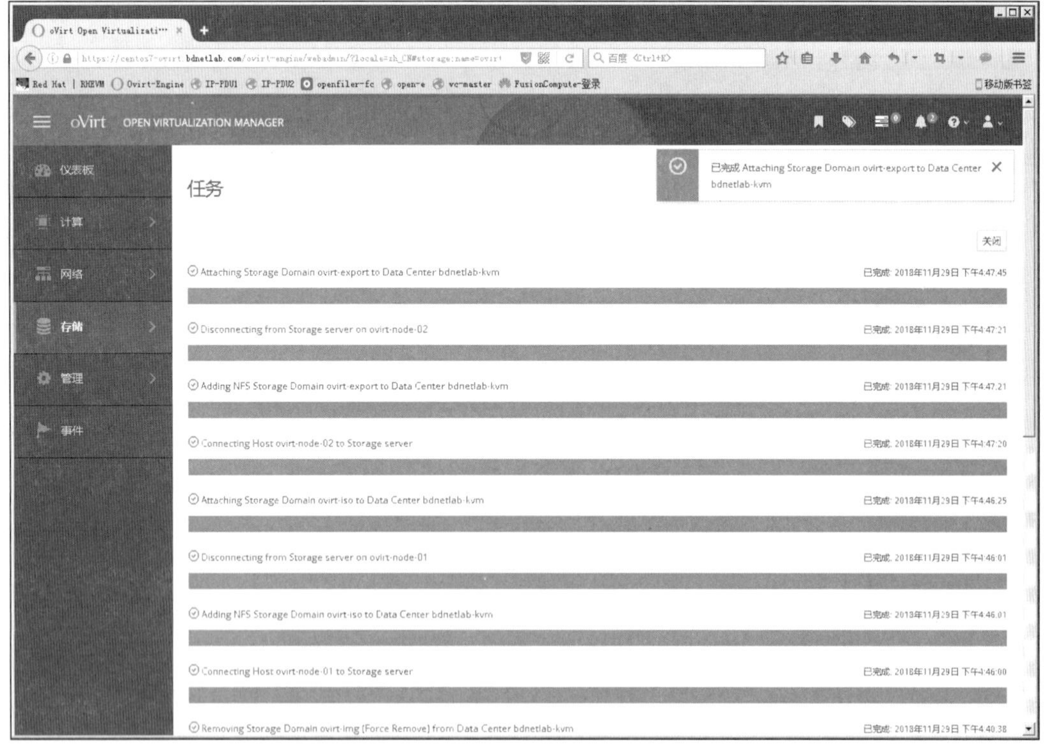
oVirt Open Virtualizati…
oVirt OPEN VIRTUALIZATION MANAGER
仪表板
计算
网络
存储
管理
事件
任务
已完成 Attaching Storage Domain ovirt-export to Data Center bdnetlab-kvm
关闭
Attaching Storage Domain ovirt-export to Data Center bdnetlab-kvm
Disconnecting from Storage server on ovirt-node-02
Adding NFS Storage Domain ovirt-export to Data Center bdnetlab-kvm
Connecting Host ovirt-node-02 to Storage server
Attaching Storage Domain ovirt-iso to Data Center bdnetlab-kvm
Disconnecting from Storage server on ovirt-node-01
Adding NFS Storage Domain ovirt-iso to Data Center bdnetlab-kvm
Connecting Host ovirt-node-01 to Storage server
Removing Storage Domain ovirt-img (Force Remove) from Data Center bdnetlab-kvm

图 4-4-22

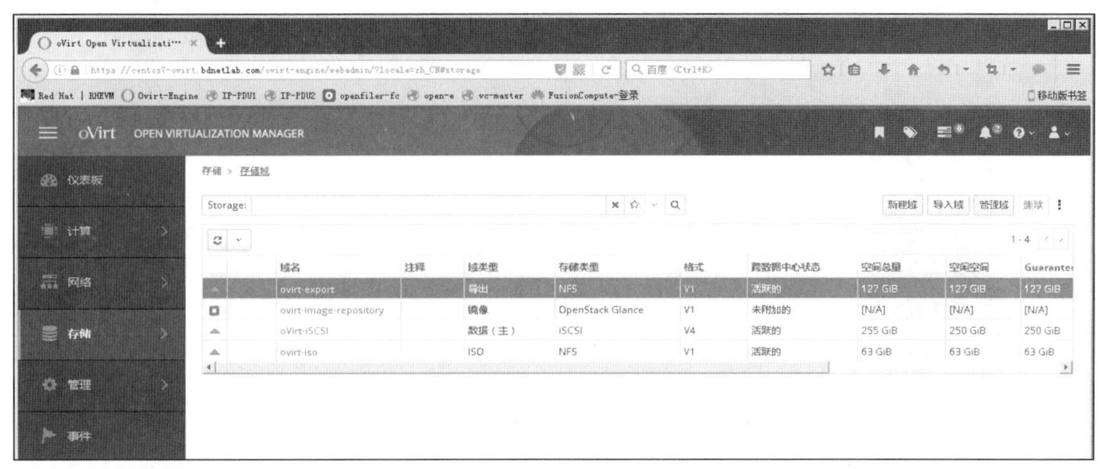

图 4-4-23

图 4-4-24

至此，基于 NFS 存储的 ISO 域和导出域均已创建完成。这也是在 oVirt 环境中比较特殊的部分，在生产环境中创建 ISO 域和导出域是必须的，可以根据实际情况确定容量。

4.5 创建使用虚拟机

通过 oVirt Engine 管理端也可以创建管理虚拟机，在创建管理过程中使用 GOI。本节介绍在生产环境中使用 oVirt Engine 管理端创建虚拟机。

4.5.1 创建 Linux 虚拟机

在 oVirt 环境中创建 Linux 虚拟机的兼容性非常好，底层操作系统就是基于 Linux 操作系统的。

第 1 步，使用浏览器登录 oVirt Engine 管理端，在“计算”选项中选择“虚拟机”，进入“新建虚拟机”界面，如图 4-5-1 所示，输入新建虚拟机名称，同时虚拟机需要“实例镜像”，也就是虚拟机使用的磁盘，单击“创建”按钮。

图 4-5-1

第 2 步，进入“新建虚拟磁盘”界面，推荐使用镜像，确定大小、别名、接口、存储域、分配策略以及磁盘配置集等信息，如图 4-5-2 所示，单击“确定”按钮。

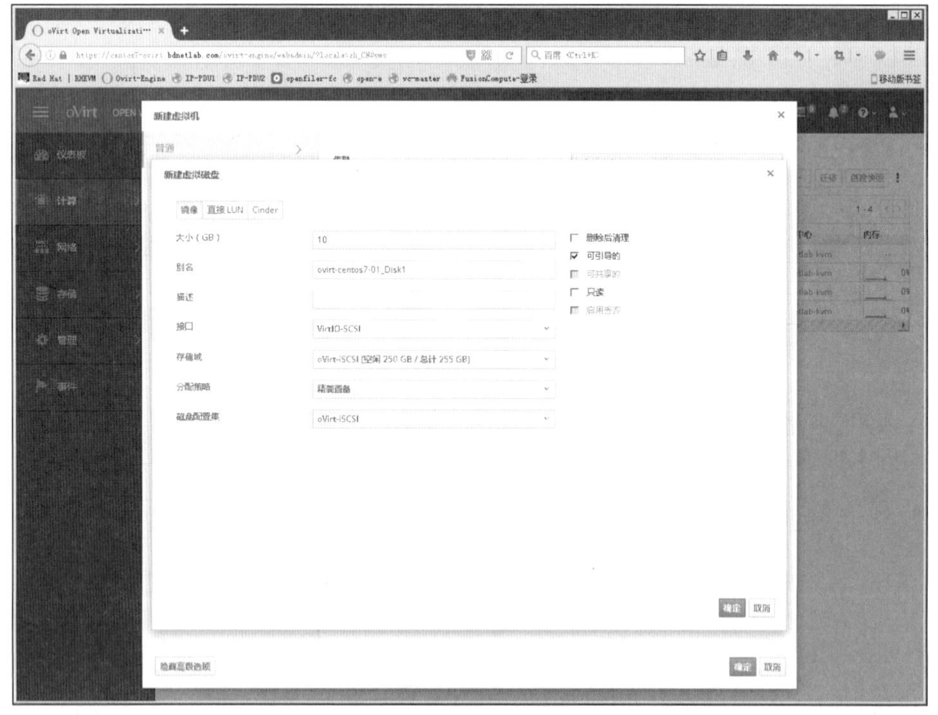

图 4-5-2

第 3 步，系统正在创建实例镜像，如图 4-5-3 所示。

图 4-5-3

第 4 步，查看新建虚拟机“系统”参数设置，如图 4-5-4 所示，可以根据生产环境的实际情况调整操作系统、实例类型、内存大小以及虚拟 CPU 的总数等参数。

图 4-5-4

第 5 步，查看新建虚拟机“初始运行”参数设置，如图 4-5-5 所示。

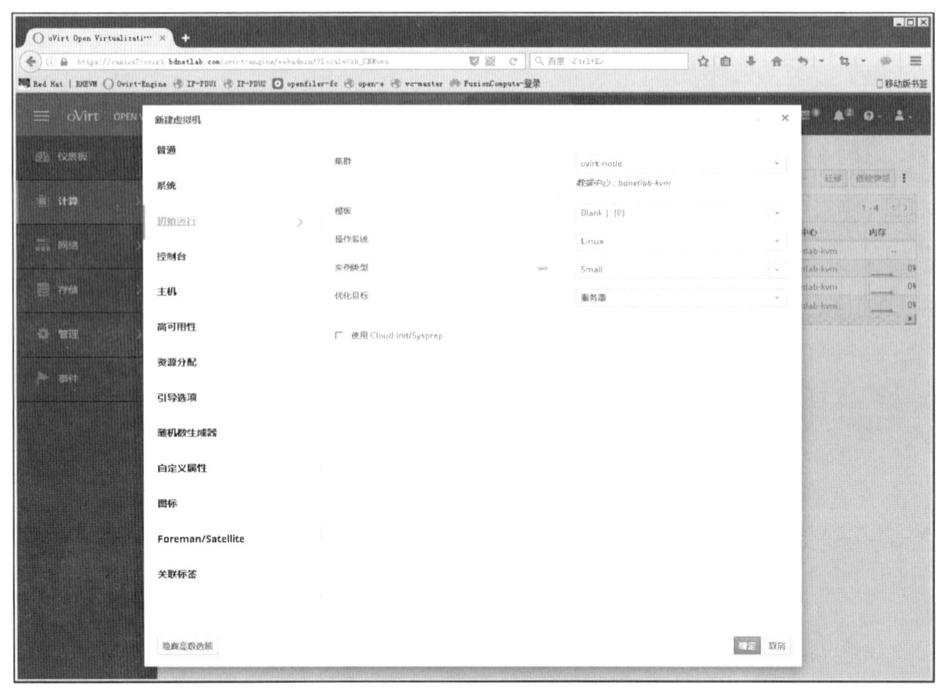

图 4-5-5

第 6 步，查看新建虚拟机“控制台”参数设置，如图 4-5-6 所示，可以根据生产环境的实际情况调整“图形控制台”的“图形界面协议”等参数。

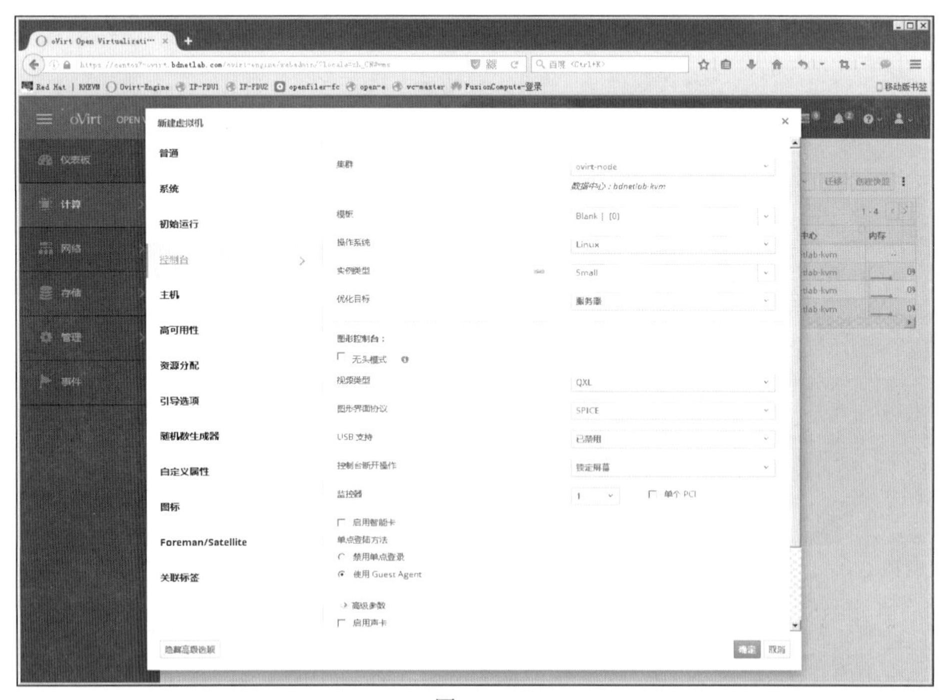

图 4-5-6

第 7 步，查看新建虚拟机“主机”参数设置，如图 4-5-7 所示，可以根据生产环境的实际情况调整虚拟机运行的主机等参数。

图 4-5-7

第 8 步，查看新建虚拟机“高可用性”参数设置，如图 4-5-8 所示，可以根据生产环境的实际情况调整虚拟机的高可用性参数。

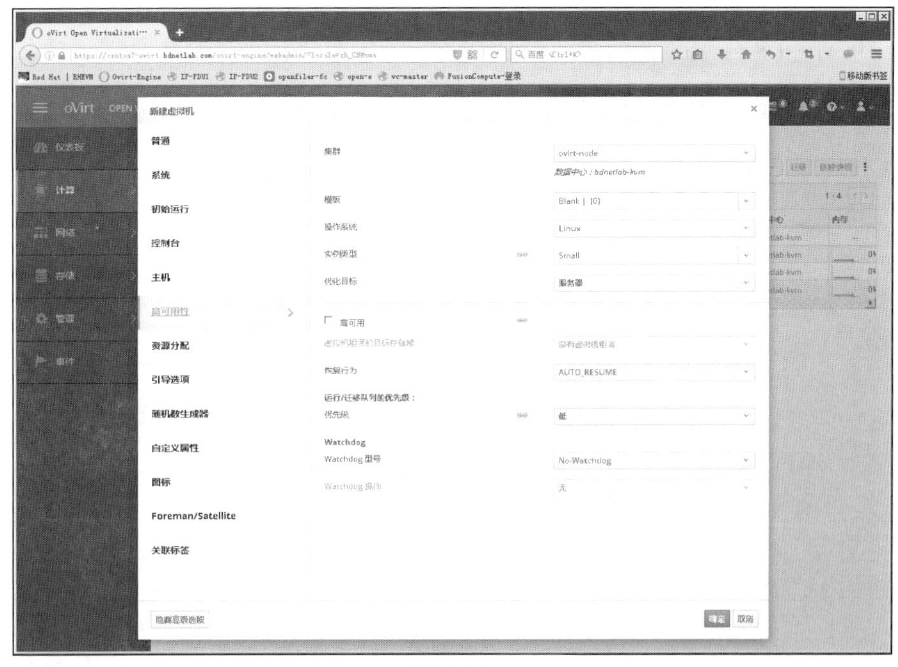

图 4-5-8

第 9 步，查看新建虚拟机“资源分配”参数设置，如图 4-5-9 所示，可以根据生产环境的实际情况分配 CPU 和内存等。

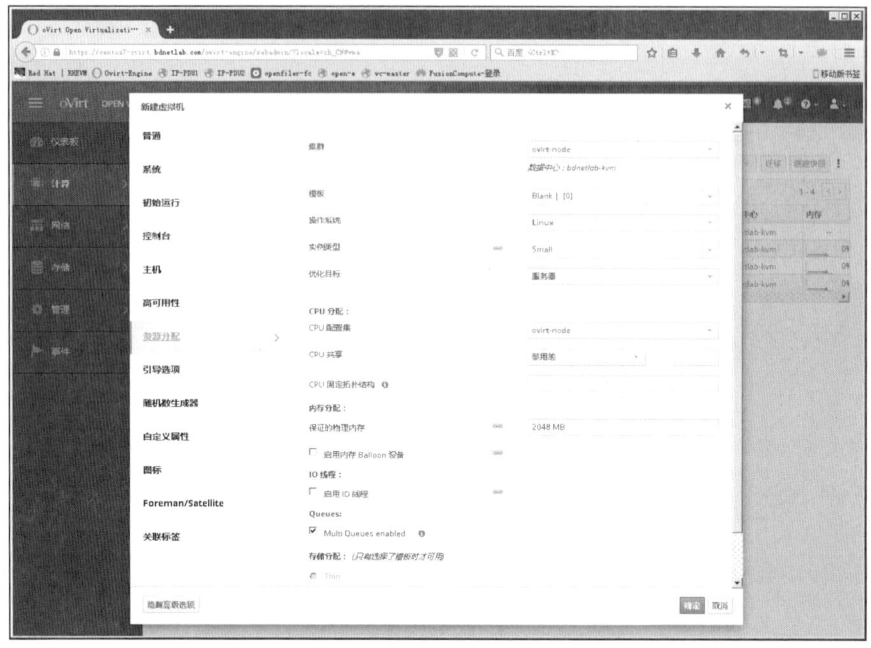

图 4-5-9

第 10 步，查看新建虚拟机“引导选项”参数设置，如图 4-5-10 所示，可以根据生产环境的实际情况配置第一个和第二个引导设备，如果安装虚拟机操作系统，需要设置“附加 CD”参数。

图 4-5-10

第 11 步，查看新建虚拟机“随机数生成器”参数设置，如图 4-5-11 所示。

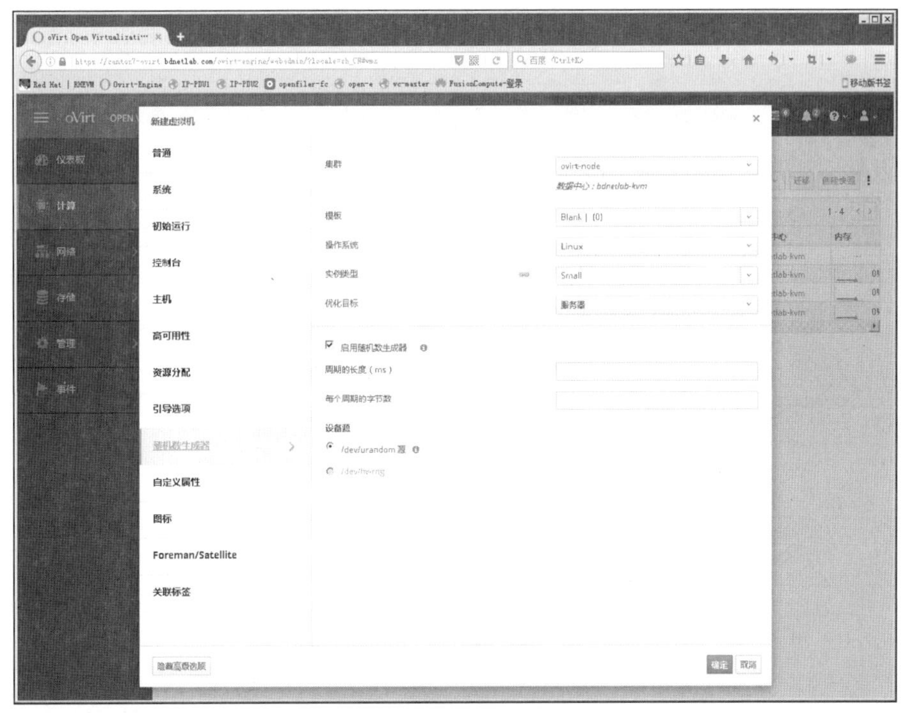

图 4-5-11

第 12 步，查看新建虚拟机“自定义属性”参数设置，如图 4-5-12 所示。

图 4-5-12

第 13 步，查看新建虚拟机“图标”参数设置，如图 4-5-13 所示。

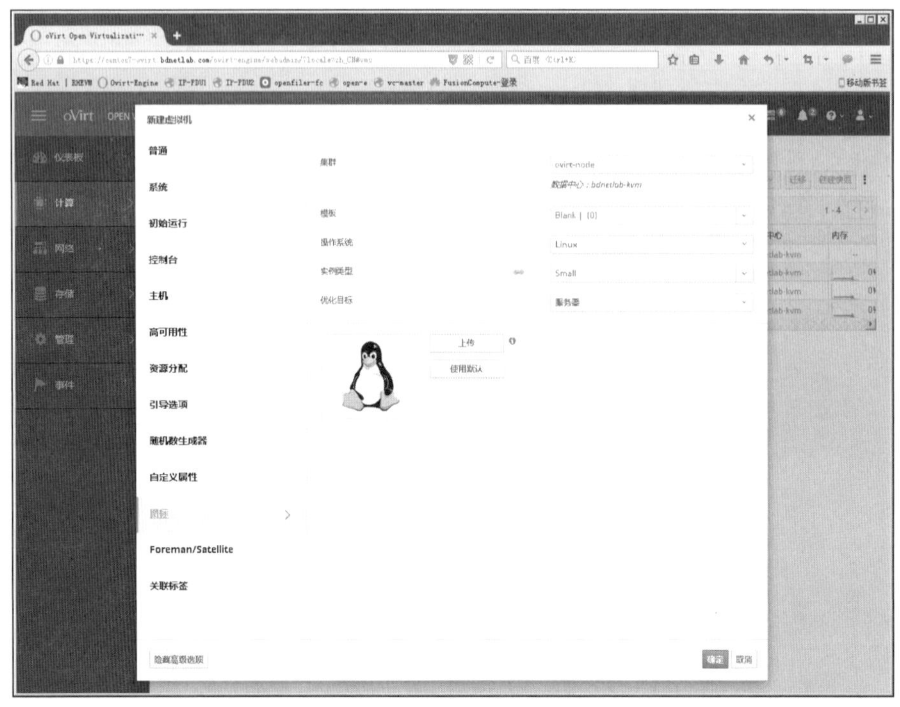

图 4-5-13

第 14 步，查看新建虚拟机“Foreman/Satellite”参数设置，如图 4-5-14 所示。

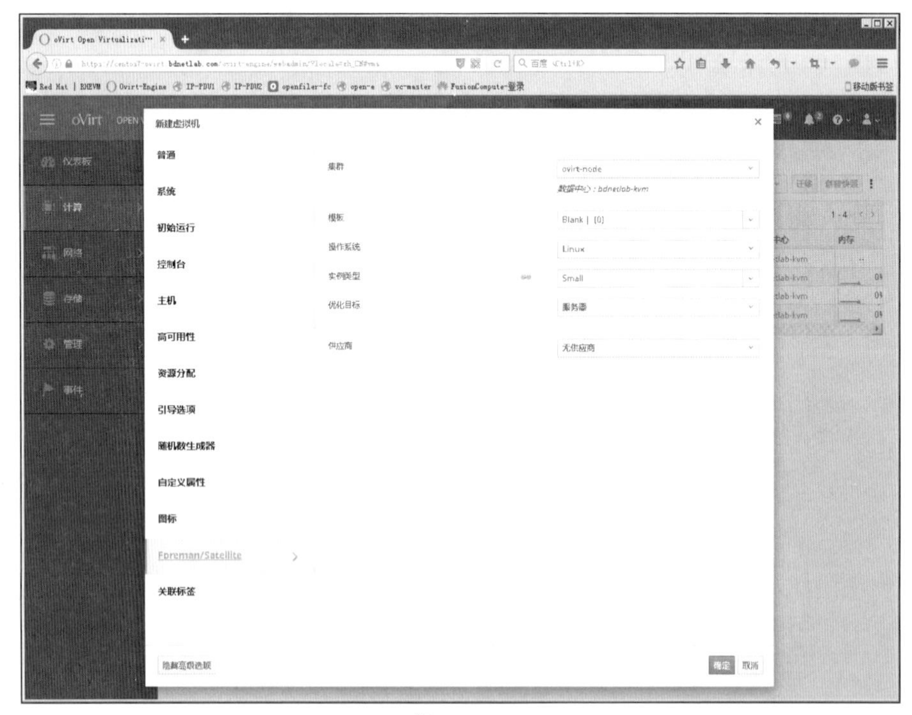

图 4-5-14

第 15 步，查看新建虚拟机“关联标签”参数设置，如图 4-5-15 所示。

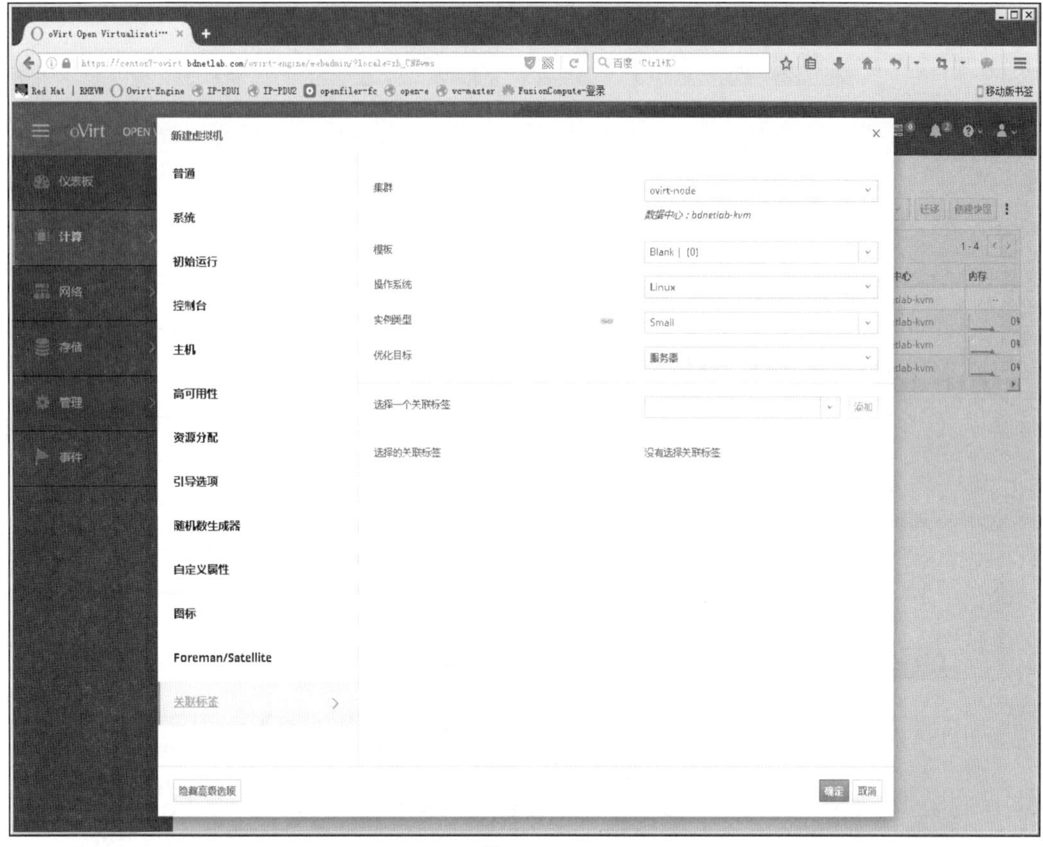

图 4-5-15

第 16 步，完成虚拟机创建，如图 4-5-16 所示，新建的虚拟机出现在列表中。

图 4-5-16

第 17 步，查看新建虚拟机的详细信息，如图 4-5-17 所示，单击“运行”按钮打开虚拟机电源。

图 4-5-17

第 18 步，虚拟机启动，进入 CentOS 7 操作系统安装界面，如图 4-5-18 所示。

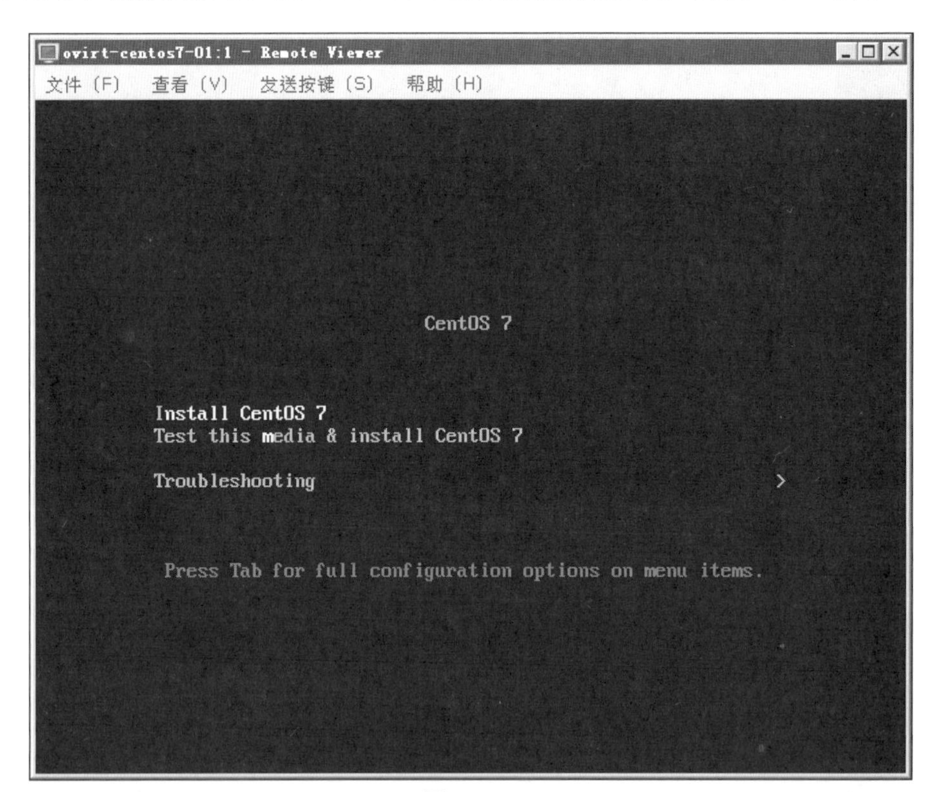

图 4-5-18

第 19 步，进入 CentOS 7 操作系统安装界面，如图 4-5-19 所示，单击“Continue”按钮。

第 20 步，虚拟机硬盘类型为 QEMU，如图 4-5-20 所示，说明虚拟机处于虚拟化模式。

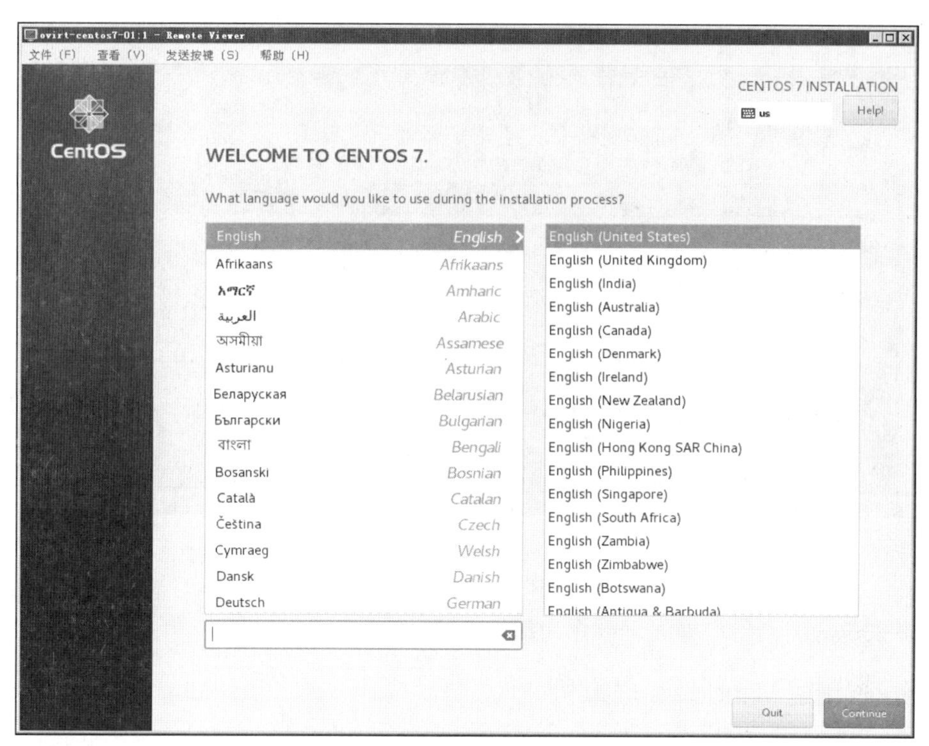

图 4-5-19

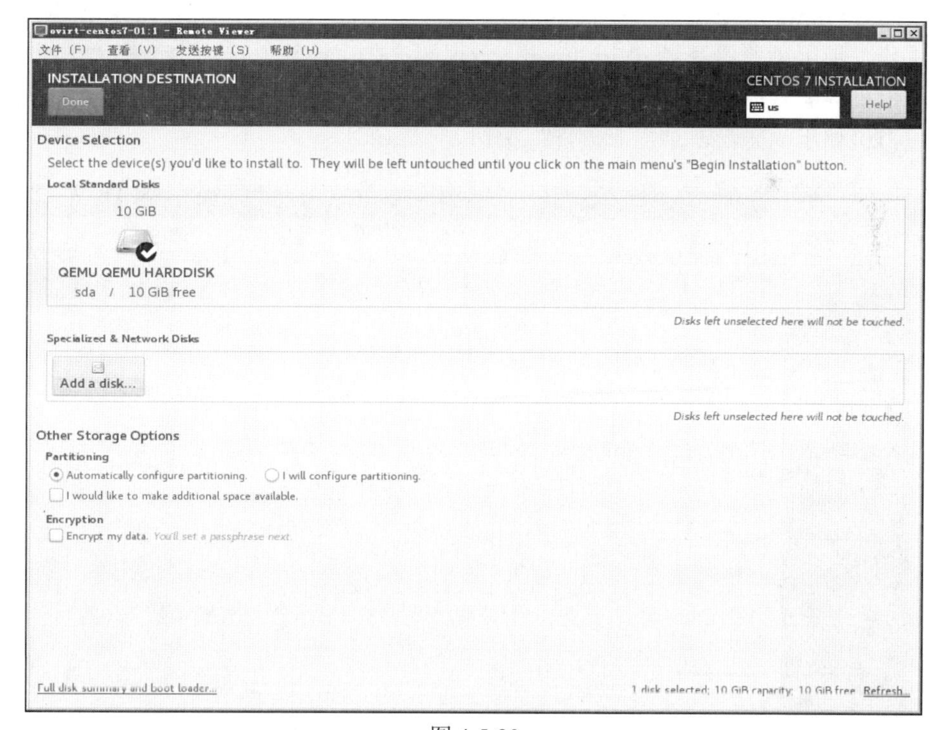

图 4-5-20

第 21 步，安装过程与前文介绍的相同，具体过程省略，安装结束后登录虚拟机并查看其网络的连通性，如图 4-5-21 所示。

```
ovirt-centos7-01:1 - Remote Viewer
文件 (F)   查看 (V)   发送按键 (S)   帮助 (H)

CentOS Linux 7 (Core)
Kernel 3.10.0-693.el7.x86_64 on an x86_64

localhost login: root
Password:
Last login: Thu Nov 29 17:34:51 on tty1
[root@localhost ~]# ip addr
1: lo: <LOOPBACK,UP,LOWER_UP> mtu 65536 qdisc noqueue state UNKNOWN qlen 1
    link/loopback 00:00:00:00:00:00 brd 00:00:00:00:00:00
    inet 127.0.0.1/8 scope host lo
       valid_lft forever preferred_lft forever
    inet6 ::1/128 scope host
       valid_lft forever preferred_lft forever
2: eth0: <BROADCAST,MULTICAST,UP,LOWER_UP> mtu 1500 qdisc pfifo_fast state UP qlen 1000
    link/ether 56:6f:fd:db:00:00 brd ff:ff:ff:ff:ff:ff
    inet 10.92.10.157/24 brd 10.92.10.255 scope global dynamic eth0
       valid_lft 86360sec preferred_lft 86360sec
    inet6 fe80::aa2d:6001:a257:9081/64 scope link
       valid_lft forever preferred_lft forever
3: bridge0: <NO-CARRIER,BROADCAST,MULTICAST,UP> mtu 1500 qdisc noqueue state DOWN qlen 1000
    link/ether ca:d8:80:19:79:0b brd ff:ff:ff:ff:ff:ff
[root@localhost ~]# ping qq.com
PING qq.com (111.161.64.40) 56(84) bytes of data.
64 bytes from dns40.online.tj.cn (111.161.64.40): icmp_seq=2 ttl=55 time=40.6 ms
64 bytes from dns40.online.tj.cn (111.161.64.40): icmp_seq=3 ttl=55 time=38.4 ms
64 bytes from dns40.online.tj.cn (111.161.64.40): icmp_seq=4 ttl=55 time=66.3 ms
64 bytes from dns40.online.tj.cn (111.161.64.40): icmp_seq=5 ttl=55 time=39.3 ms
^C
--- qq.com ping statistics ---
5 packets transmitted, 4 received, 20% packet loss, time 7346ms
rtt min/avg/max/mdev = 38.470/46.227/66.395/11.670 ms
[root@localhost ~]# _
```

图 4-5-21

至此，使用 oVirt Engine 管理端创建 Linux 虚拟机完成。对于不习惯或不喜欢命令行的运维人员来说，图形用户界面可以作为新的选择。

4.5.2 创建 Windows 虚拟机

在 oVirt 环境中创建 Windows 虚拟机兼容性相对于 Linux 来说要差一些，主要就是驱动程序问题，Windows 虚拟机需要加载驱动程序。

第 1 步，使用浏览器登录 oVirt Engine 管理端，在“计算”选项中选择“虚拟机”，进入新建虚拟机界面，如图 4-5-22 所示，输入新建虚拟机名称，同时虚拟机需要实例镜像，也就是虚拟机使用的磁盘，单击“创建”按钮。

图 4-5-22

第 2 步，进入“新建虚拟磁盘”界面，推荐使用镜像，确定大小、别名、接口、存储域、分配策略等信息，如图 4-5-23 所示，单击“确定”按钮，注意 Windows 虚拟机使用 VirtIO-SCSI 接口在安装过程中需要加载驱动程序，否则无法识别硬盘。

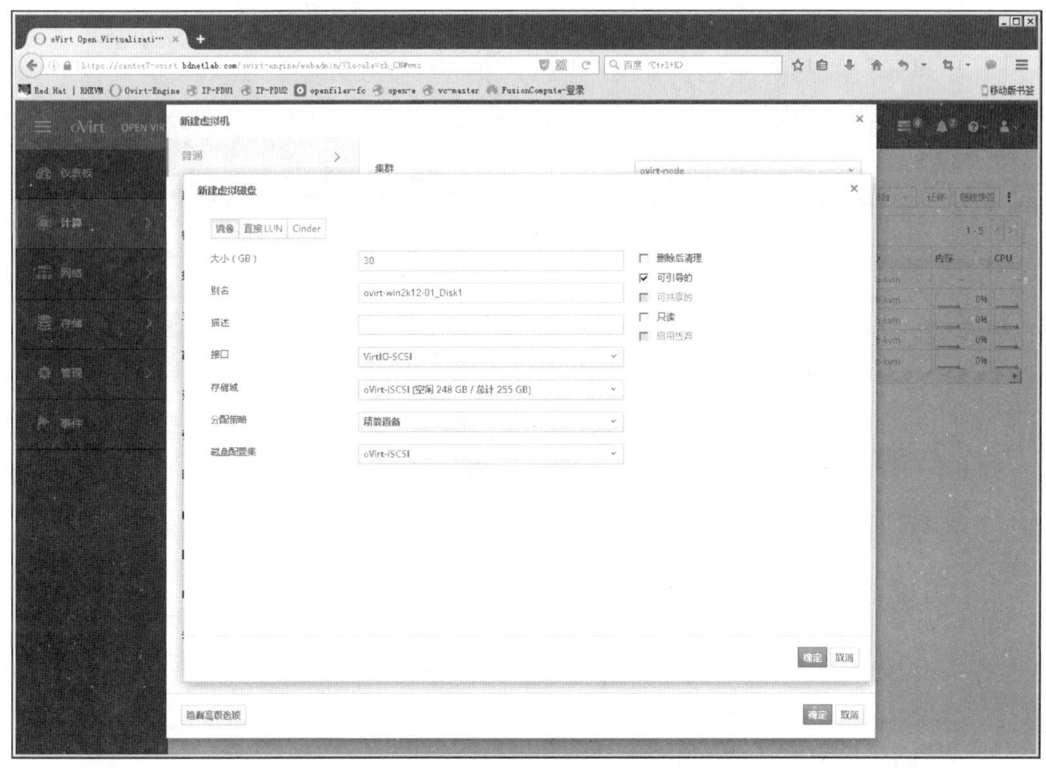

图 4-5-23

第 3 步，完成新建虚拟机，如图 4-5-24 所示，新建的虚拟机出现在列表中。

图 4-5-24

第 4 步，进入 Windows Server 2012 R2 操作系统安装界面，如图 4-5-25 所示，单击“下一步”按钮。

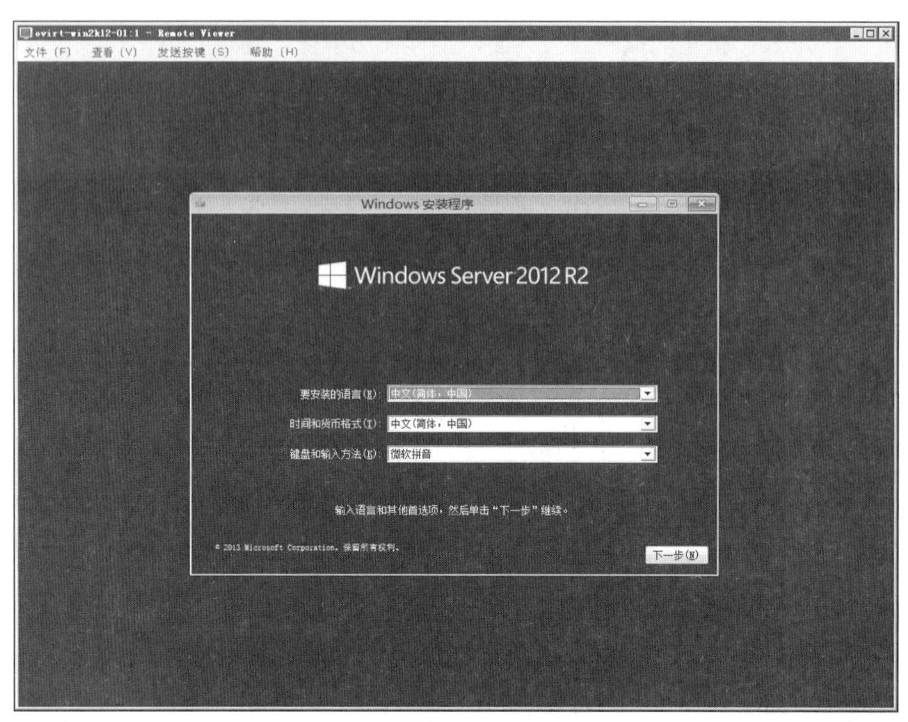

图 4-5-25

第 5 步，如果不加载驱动程序，Windows Server 2012 R2 操作系统不能识别 VirtIO-SCSI 驱动程序，如图 4-5-26 所示，单击“加载驱动程序”。

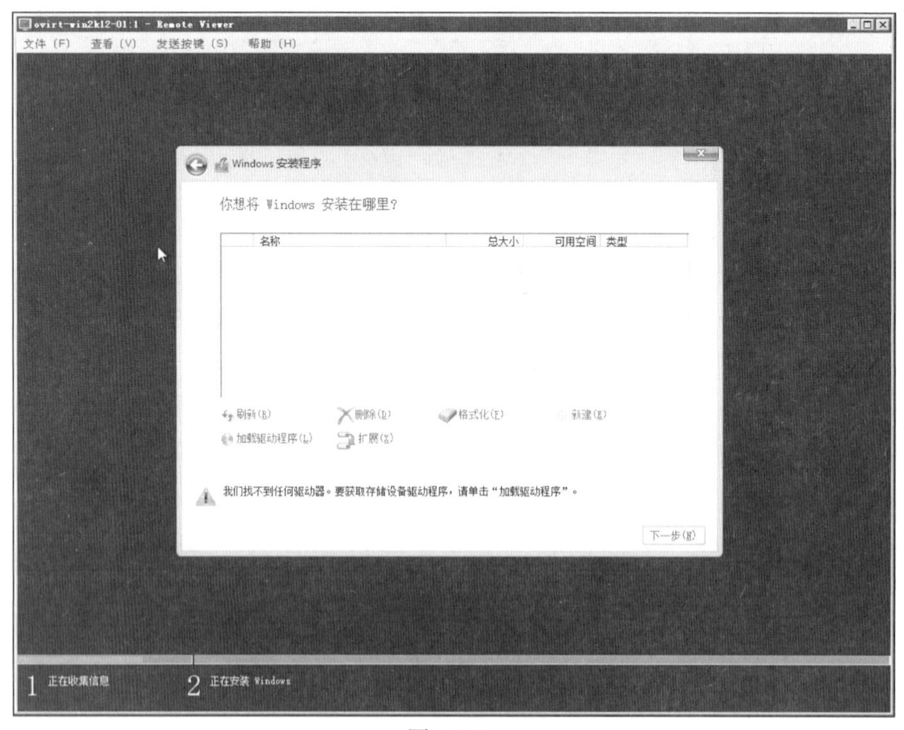

图 4-5-26

第 6 步，找到 VirtIO-SCSI 驱动程序，如图 4-5-27 所示，单击“下一步”按钮。

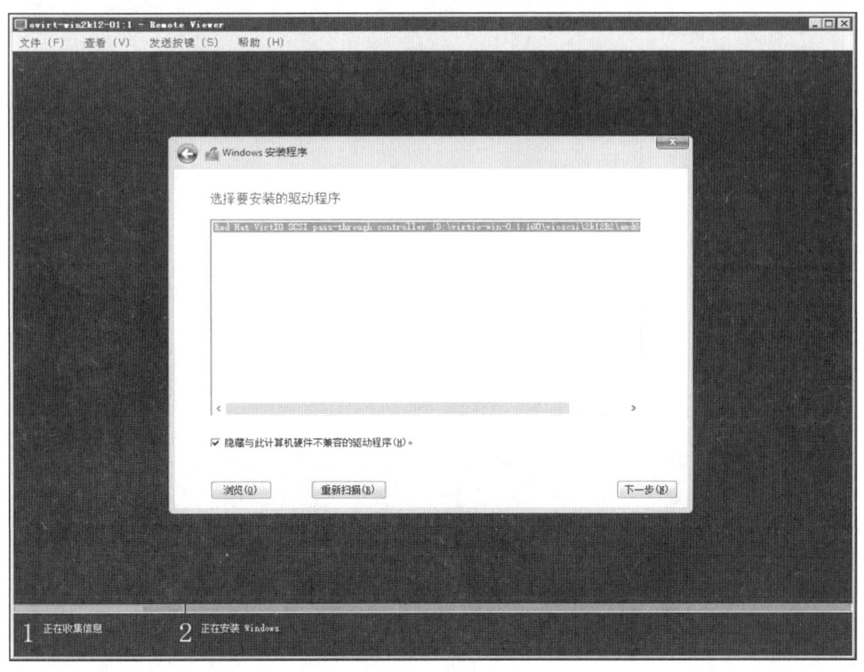

图 4-5-27

第 7 步，Windows Server 2012 R2 操作系统加载 VirtIO-SCSI 驱动程序后识别到硬盘，如图 4-5-28 所示，单击“下一步”按钮。

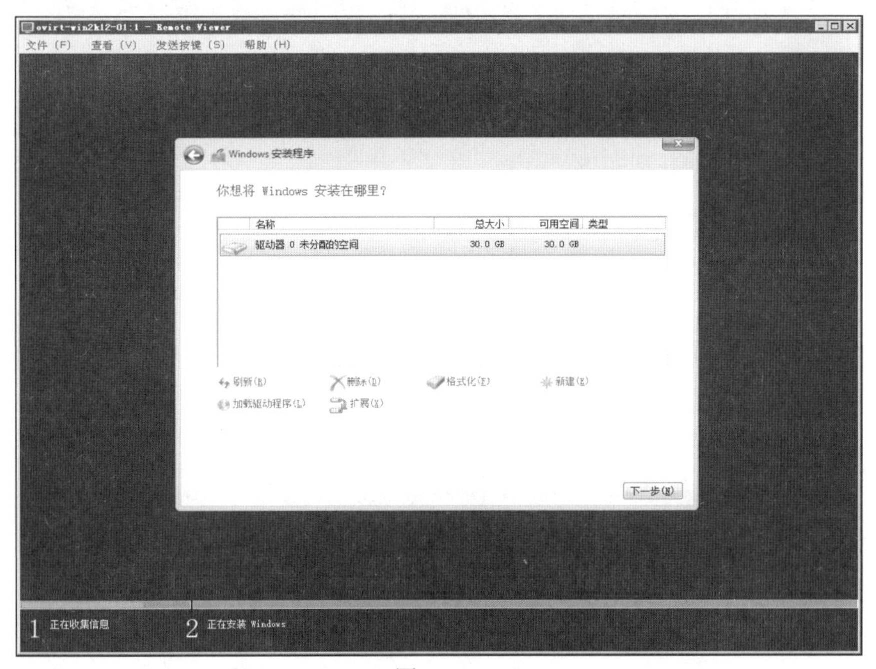

图 4-5-28

第 8 步，开始安装 Windows Server 2012 R2 操作系统，如图 4-5-29 所示。

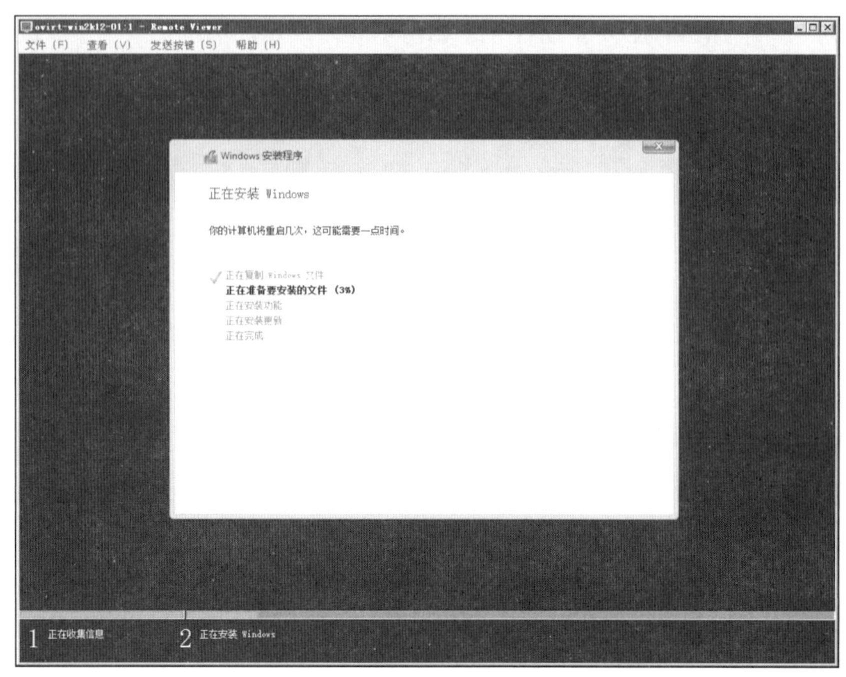

图 4-5-29

第 9 步，完成 Windows Server 2012 R2 操作系统的安装，查看虚拟机硬件信息，如图 4-5-30 所示，虚拟机使用 QEMU 硬盘。

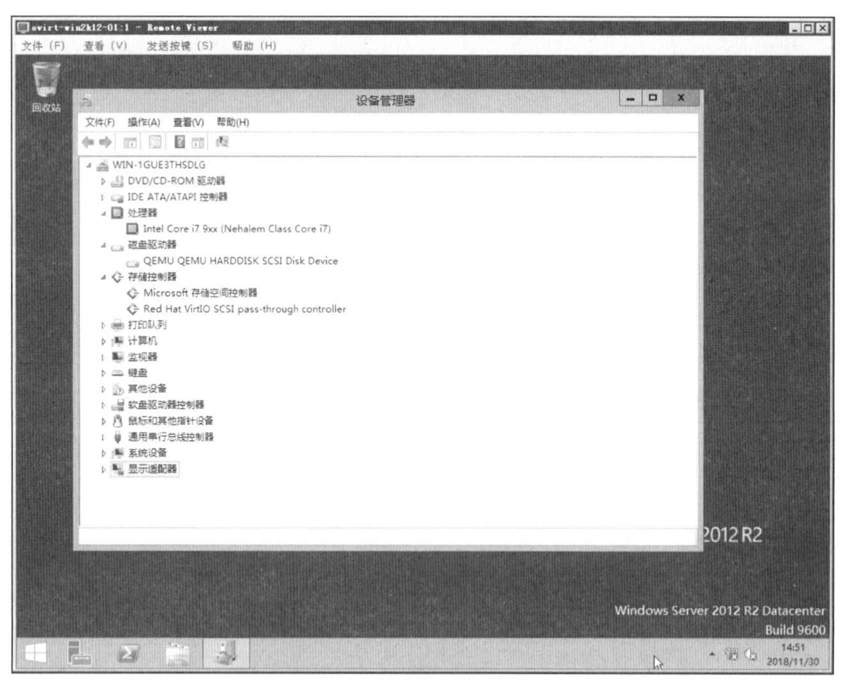

图 4-5-30

至此，使用 oVirt Engine 管理端创建 Windows 虚拟机完成，与创建 Linux 虚拟机相比，主要是驱动程序问题，解决完驱动程序问题，创建 Windows 虚拟机也比较容易。

4.6 配置 oVirt 平台高可用

生产环境中的高可用性意味着在出现问题的时候，虚拟机会重启。出现问题是指在除了通过虚拟机上的客户端或 oVirt Engine 管理端所发出的关机指令以外的情况下，造成虚拟机的停机。当出现问题时，具有高可用性的虚拟机会在它所在的主机或所在群集中的另外一台主机上自动进行重启。

高可用性的功能是 oVirt Engine 管理端通过实时监测主机和存储设备，并自动检测硬件故障来实现的。如果主机被发现出现故障，高可用性的虚拟机将会在群集中的另外一个主机上被重启。作为高可用性的虚拟机，它所提供的服务中断时间会被控制在很短的范围内。这是因为在出现问题时，虚拟机会在数秒之内重启，而不需要人工参与。高可用性的虚拟机会在当前资源使用率最低的主机上被重启，或根据用户事先配置的负载平衡和电源管理策略，在适当的主机上重启，从而保证资源利用率的平衡。这还可以使整个环境有足够的资源来保证可以随时重启虚拟机。oVirt 虚拟化架构提供了多种高可用性特性，可采用颗粒的方式，从一个单一的虚拟机级别应对多个主机故障情况。此外，oVirt 虚拟化架构可以保护虚拟机应对各种故障，配置电源管理，oVirt Engine 管理端的故障检测和故障恢复解决方案相结合。本节介绍配置使用虚拟机的高可用性。

4.6.1 使用高可用注意事项

在生产环境中使用高可用和迁移需要考虑多个问题。高可用性的主机需要一个电源管理设备，并且需要配置它的隔离参数，但是很多环境的主机电源未被配置，包括本书中使用的主机。另外，当高可用性虚拟机所在的主机无法工作时，虚拟机需要在群集中的其他主机上被重启，要保证可以迁移高可用性虚拟机，整体来说需要考虑以下问题。

1）运行高可用性虚拟机的主机的电源管理被配置。

2）运行高可用性虚拟机的主机所在的群集必须有其他可用的主机。

3）迁移的目标主机必须正在运行。

4）源和目标主机必须都可以访问虚拟机所在的数据域。

5）源和目标主机必须都可以访问相同的虚拟网络和 VLAN。

6）目标主机必须有足够的可用 CPU 资源来支持虚拟机的需求。

7）目标主机必须有足够的可用内存来支持虚拟机的需求。

在生产环境中虚拟机迁移也需要考虑多个问题，整体来说需要考虑以下问题。

1. 设置虚拟机迁移优先级

oVirt Engine 管理端会把一个主机上的虚拟机迁移请求放入一个队列中。当这个队列中有一个迁移请求，而且群集中有可用的主机时，一个迁移事件就会根据群集中的负载均衡策略被触发。每一分钟负载均衡处理都会被运行，那些正在处理迁移事件的主机在它们的迁移事件完成前不会被包括在负载均衡处理中。oVirt Engine 管理端允许用户通过为每个虚拟机设置优先级来改变它们在迁移队列中的顺序，具有高优先级的虚拟机会被先迁移。

2. 取消正在进行的虚拟机迁移

一个正在进行的虚拟机，迁移所用的时间比预期的要长，需要对系统进行一些改变。因此，需要取消正在进行的虚拟机迁移。

3. 查看虚拟机自动迁移事件和日志

当一台虚拟机因为高可用性设置发生自动迁移操作时，自动迁移的详细信息会被记录在事件标签页和引擎的日志中。这些信息可以被用来进行故障排除。

4. 防止虚拟机自动迁移

oVirt Engine 管理端允许禁用虚拟机自动迁移。另外，也可以通过设定虚拟机只能在一个特定主机上运行的方式，来禁用虚拟机自动迁移。禁用虚拟机自动迁移并指定虚拟机只能在一个特定的主机上运行，对于使用高可用性产品非常有用。

5. 虚拟机热迁移需要考虑的问题

热迁移也称为实时迁移，就是虚拟机在运行的时候可以在不同的物理主机间进行迁移，不需要停止虚拟机所提供的服务。实时迁移对于最终用户是透明的：在虚拟机被迁移到一台新的主机的过程中，这个虚拟机仍然保持运行状态，它上面的用户应用程序仍然可以被使用。

6. 选择其他问题

在使用实时迁移前，请确定 oVirt 虚拟化架构环境已经被正确配置为可以进行实时迁移。要使实时迁移可以成功进行，至少需要满足以下条件。

1）源主机和目标主机必须在同一个群集中，并且它们的 CPU 必须兼容。

2）源主机和目标主机的状态必须都为“Up”。

3）源主机和目标主机必须都可以访问相同的虚拟网络。

4）源主机和目标主机必须都可以访问虚拟机所在的数据存储域。

5）目标主机必须有足够的 CPU 资源来支持虚拟机的需求。

6）目标主机必须有足够的可用内存来支持虚拟机的需求。

7）要迁移的虚拟机必须没有“cache!=none”这个自定义属性设置。

最后，为了获得较好的性能，存储网络和管理网络应该被分开，从而避免出现网络饱和的情况，因为虚拟机的迁移会在主机间传输大量的数据。

4.6.2 配置虚拟机高可用

在生产环境中配置高可用主要分为启用高可用和迁移两个选项。

1. 启用虚拟机高可用特性

在 oVirt 虚拟化架构中，虚拟机的高可用性需要在虚拟机上单独进行配置。

第 1 步，使用浏览器登录 oVirt Engine 管理端，选择需要配置高可用性的虚拟机，单击“编辑”按钮，弹出图 4-6-1 所示窗口。

第 2 步，在“编辑虚拟机”窗口选择“高可用性”，勾选“高可用”启用虚拟机的高可用性，根据实际情况选择“运行/迁移队列的优先级”，如图 4-6-2 所示，当需要进行虚拟机迁移的时候，一个队列会被创建。有高优先级的虚拟机会排在队列的前面被优先处理。如果群集的可用资源较少，只有高可用性虚拟机会被迁移。

图 4-6-1

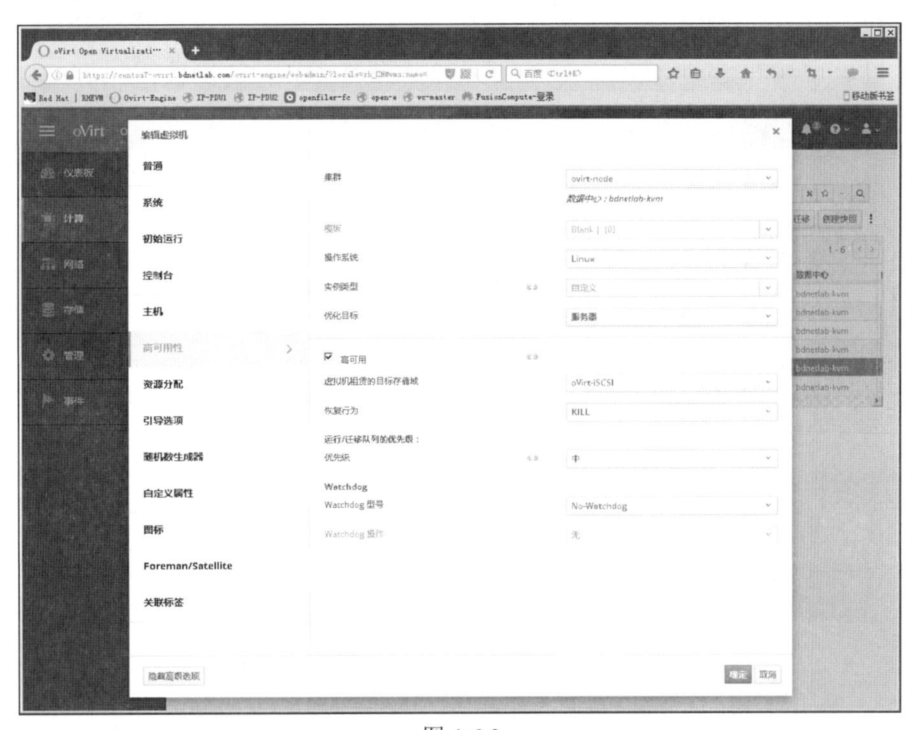

图 4-6-2

第 3 步，调整虚拟机“高可用性”参数会出现提示，一些改变需要重启虚拟机才能使用，如图 4-6-3 所示。

第 4 步，如果虚拟机未重启，虚拟机名称前的图标上会出现提示，如图 4-6-4 所示。

图 4-6-3

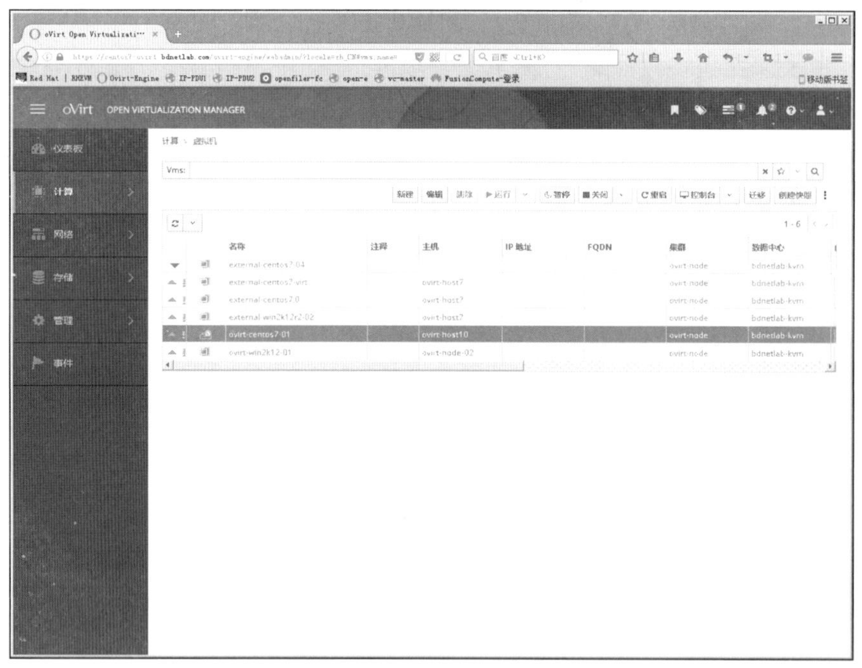

图 4-6-4

2. 调整虚拟机迁移模式

当一个主机被设为维护模式后，oVirt Engine 管理端将会自动启动虚拟机的实时迁移，它会把这个主机上运行的所有虚拟机迁移到这个群集中的其他主机上。oVirt Engine 管理端会根据群集策略中所规定的负载平衡或电源管理级别来自动进行虚拟机的实时迁移。在默认的情况下，群集策略没有被定义，但是如果需要使用实时迁移的功能，推荐设置一个适用于具体情况的群集策略。也可以在需要的时候，在特定虚拟机上禁用自动（甚至手动）实时迁移的功能。

第 1 步，默认情况下，虚拟机“迁移模式”为“允许手动和自动迁移”，如图 4-6-5 所示。

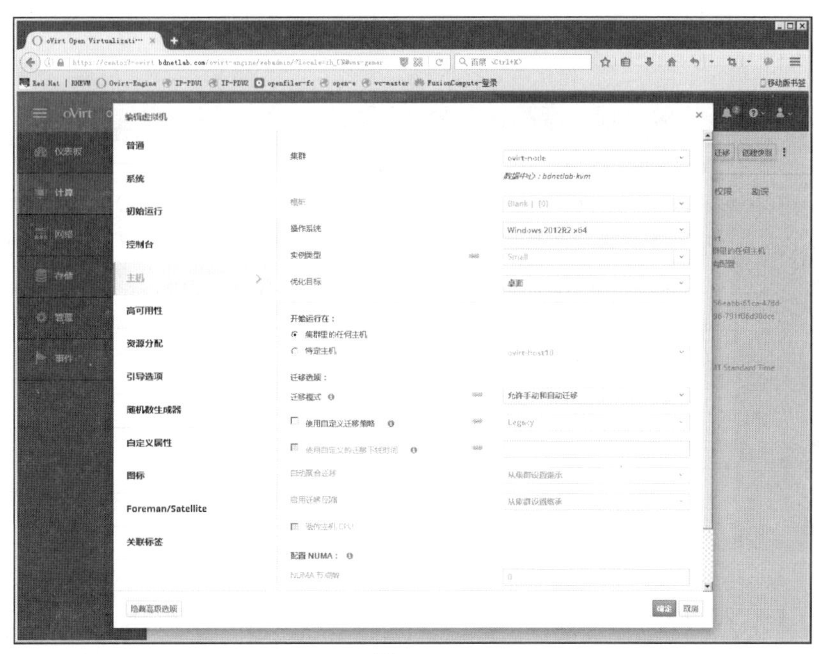

图 4-6-5

第 2 步，如果在生产环境中不允许该虚拟机发生自动迁移，通过下拉列表，将“迁移模式”修改为“不允许迁移”，如图 4-6-6 所示，单击“确定”按钮。

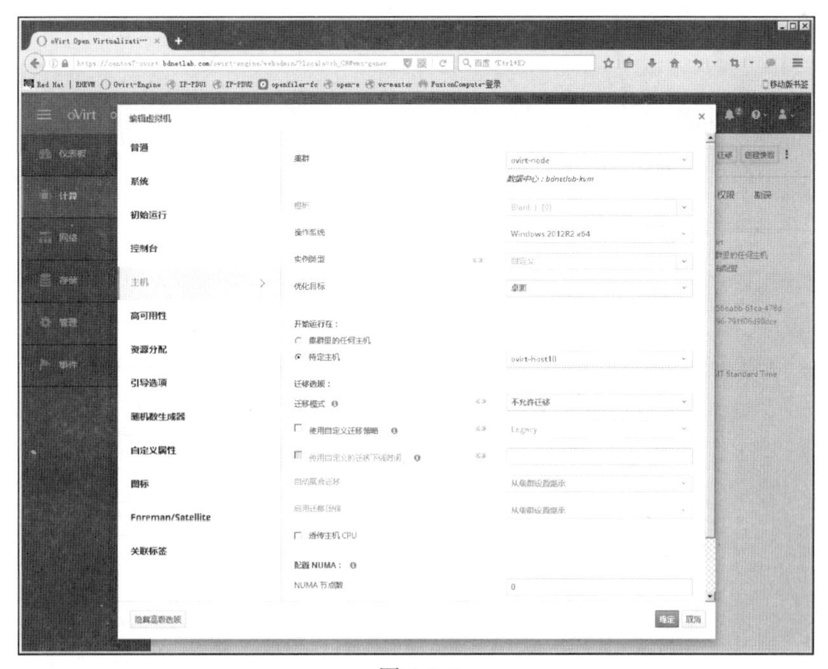

图 4-6-6

3. 调整手动迁移虚拟机

在生产环境中，可能由于对服务器进行一些调整或其他原因，需要手动对虚拟机进行迁移，具体操作如下。

第 1 步，打开虚拟机电源，虚拟机运行的主机为 ovirt-node-02，单击“迁移”按钮，如图 4-6-7 所示。

图 4-6-7

第 2 步，默认选项为“自动选择主机”，手动将选项改为“选择目的地主机”，如图 4-6-8 所示，单击“确定”按钮。

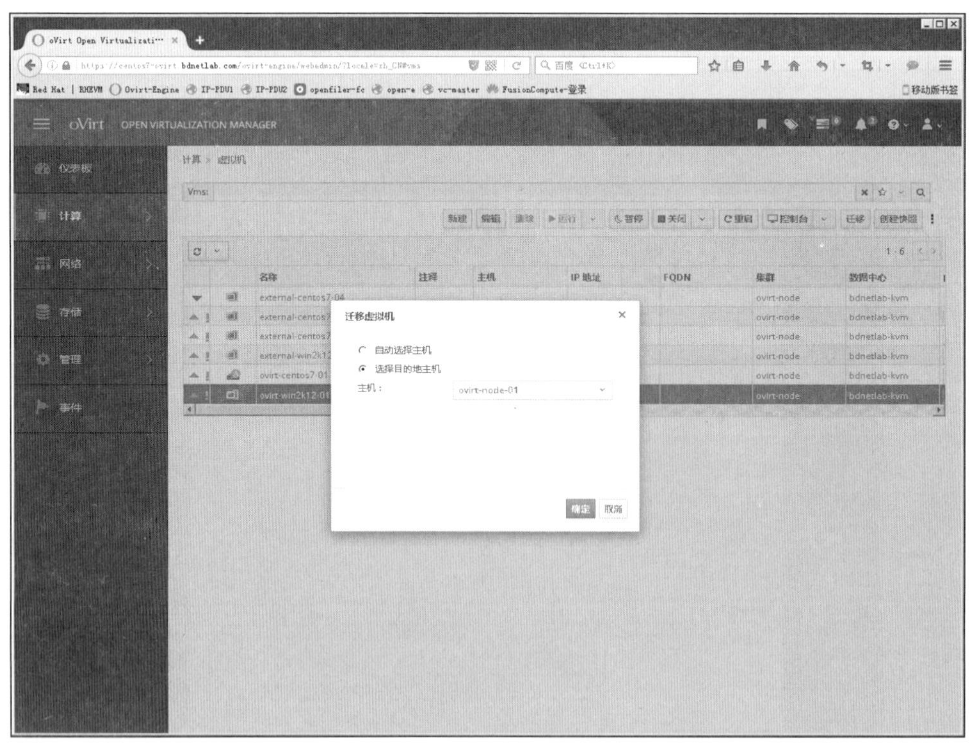

图 4-6-8

第 3 步，查看“任务”中手动迁移的情况如图 4-6-9 所示。

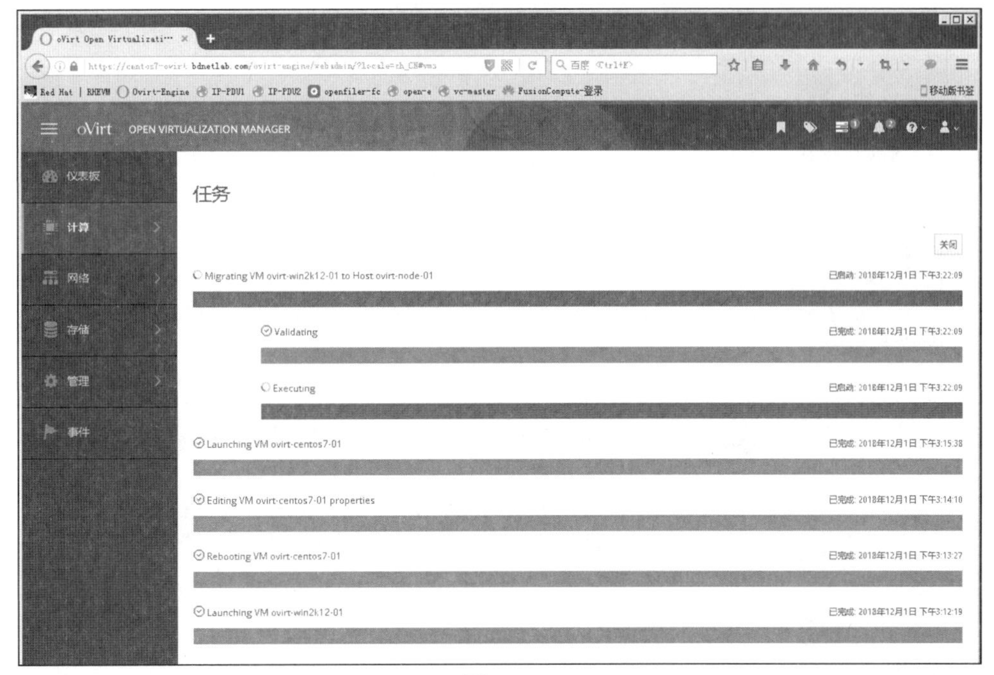

图 4-6-9

第 4 步，手动迁移完成，如图 4-6-10 所示，注意“任务”中的提示。

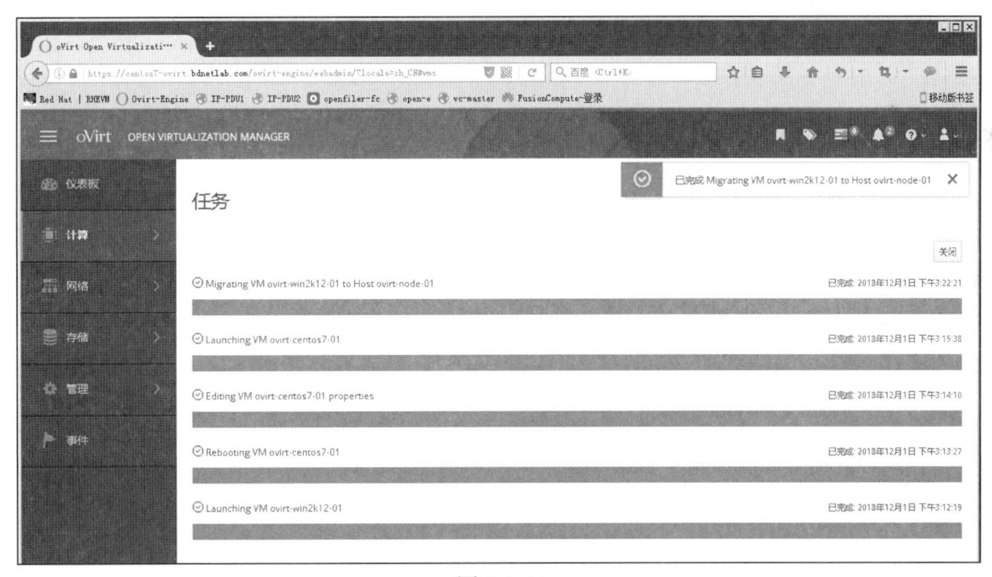

图 4-6-10

第 5 步，查看虚拟机运行的情况，主机变为 ovirt-node-01，如图 4-6-11 所示，说明手动迁移成功。

第 6 步，查看主机 ovirt-node-01 虚拟机运行情况，如图 4-6-12 所示。

图 4-6-11

图 4-6-12

4. 将主机调整为维护模式自动迁移虚拟机

在生产环境中，如果需要对主机进行维护，需要将主机设置为维护模式，处于维护模式的主机会将运行的虚拟机迁移到集群中正常运行的主机。

第 1 步，虚拟机 ovirt-centos7-01 运行在 ovirt-host10 主机，如图 4-6-13 所示。

图 4-6-13

第 2 步，将主机 ovirt-host10 调整为维护模式，如图 4-6-14 所示，单击“确定”按钮。

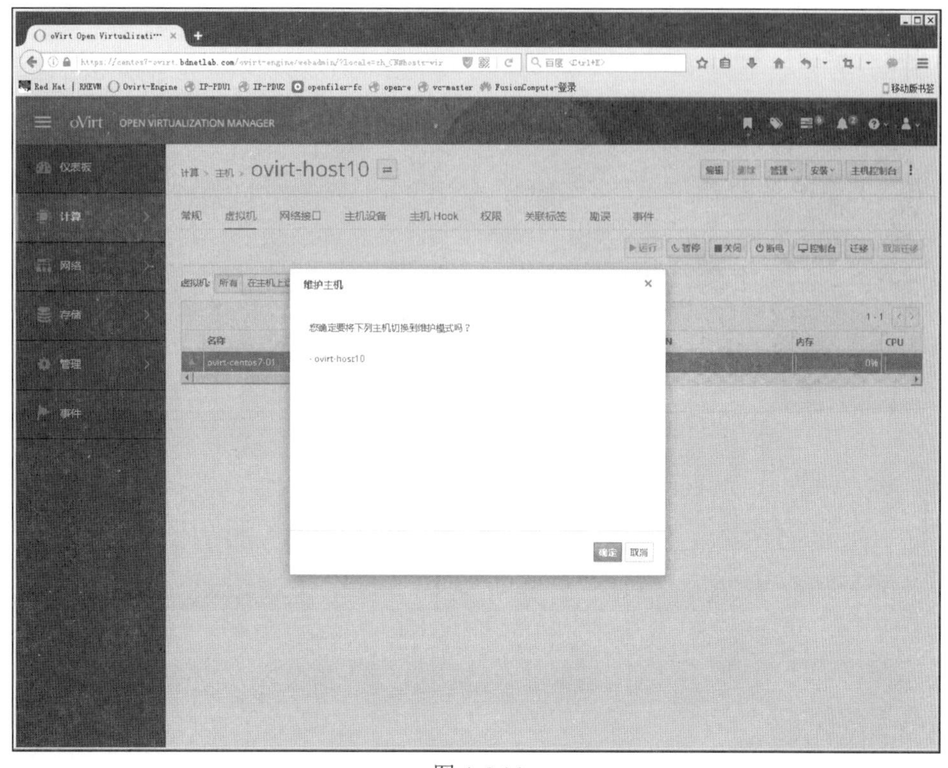

图 4-6-14

第 3 步，查看“任务”中主机处于维护模式的情况，如图 4-6-15 所示，此时虚拟机会迁移到其他正常运行的主机。

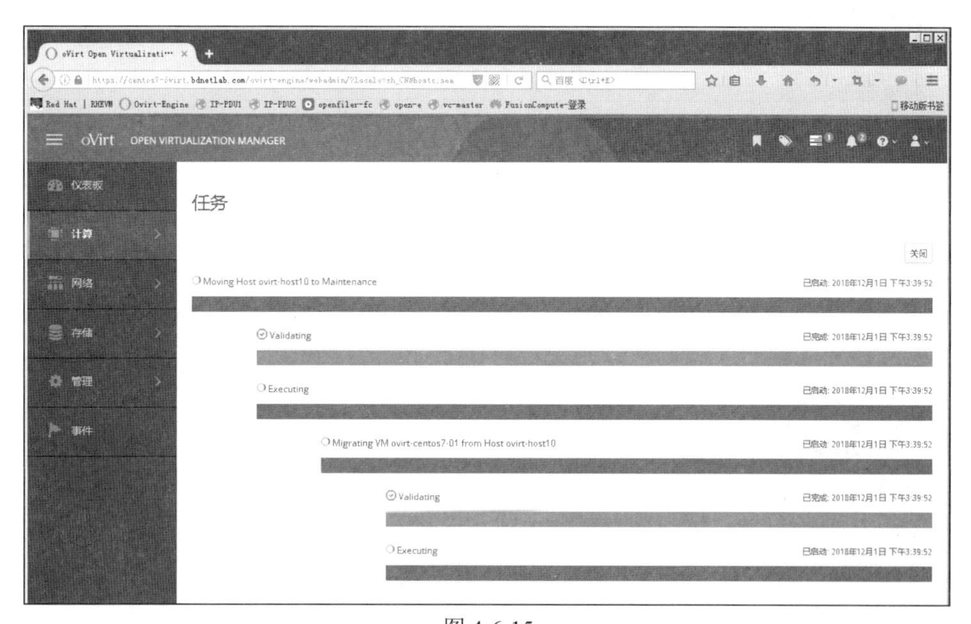

图 4-6-15

第 4 步，将主机 ovirt-host10 调整为维护模式完成，如图 4-6-16 所示。

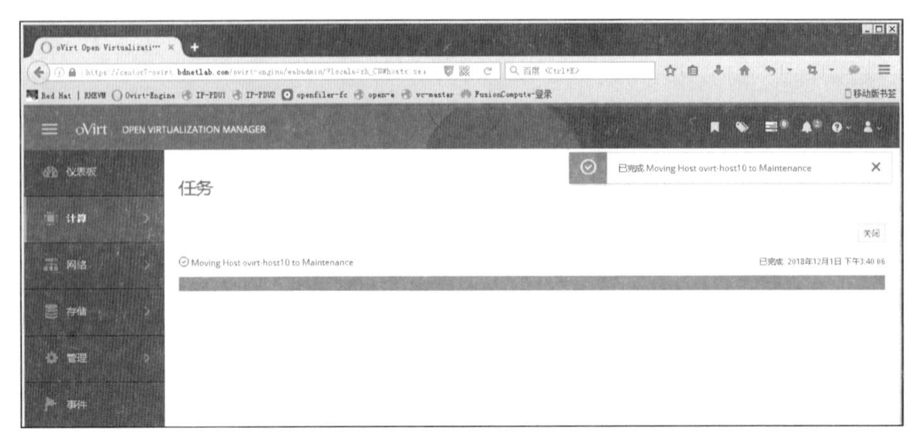

图 4-6-16

第 5 步，虚拟机 ovirt-centos7-01 运行在 ovirt0-node-02 主机，如图 4-6-17 所示，说明迁移成功。

图 4-6-17

第 6 步，查看主机 ovirt-node-02 虚拟机运行情况，如图 4-6-18 所示。

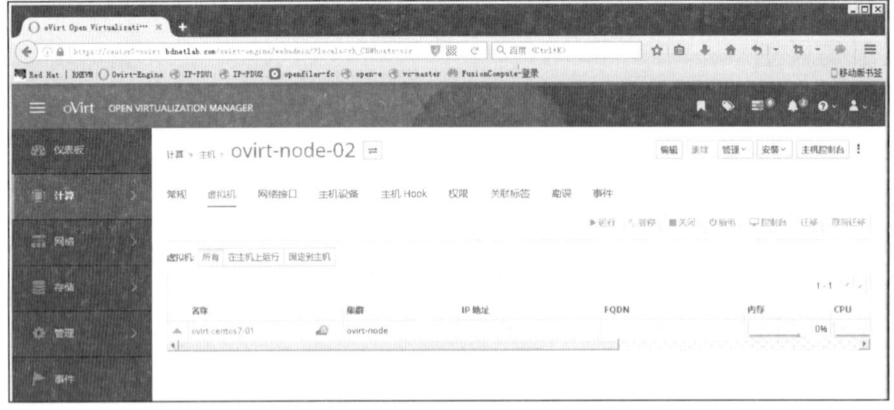

图 4-6-18

至此，使用 oVirt Engine 管理端配置虚拟机迁移完成。相对于使用命令行操作来说，使用 GUI 具有很多优势。在生产环境中保证虚拟机的高可用是必须的操作，但是一定要结合实际情况来配置虚拟机的高可用和迁移等选项，不适当的配置或错误的配置可能会影响生产环境中虚拟机的运行。

4.7 备份和恢复虚拟机

无论生产环境使用什么虚拟化平台，对虚拟机的备份都是非常重要的，不同的虚拟化平台对于虚拟机备份和恢复支持是不一样的。如 VMware vSphere 平台官方提供 VDP 备份工具，常用第三方 veeam backup 对其支持也非常好，对于 KVM 和 oVirt 平台来说，备份工具相对薄弱，但是依旧需要对虚拟机进行备份。本节介绍在生产环境备份恢复虚拟机。

4.7.1 使用导出域备份虚拟机

前文介绍过通过 oVirt Engine 管理端创建导出域，本节介绍如何使用导出域备份恢复虚拟机。导出操作相当于备份虚拟机，导入操作相当于恢复虚拟机。

第 1 步，使用浏览器登录 oVirt Engine 管理端，查看导出域中虚拟机导入信息，列表为空，如图 4-7-1 所示，因为目前没有对虚拟机进行过导出操作。

图 4-7-1

第 2 步，选择需要备份的虚拟机进行导出操作，“导出到导出域”为灰色状态无法操作，如图 4-7-2 所示，其原因是虚拟机处于“开机”状态。

第 3 步，关闭虚拟机电源，“导出到导出域”正常，如图 4-7-3 所示。

第 4 步，导出虚拟机会有两个参数供选择：如果导出域已经有原导出的信息，必须勾选“强制覆盖”，否则无法导出；“Collapse 快照”是整合虚拟机原有快照，建议勾选，如图 4-7-4 所示，单击“确定”按钮。

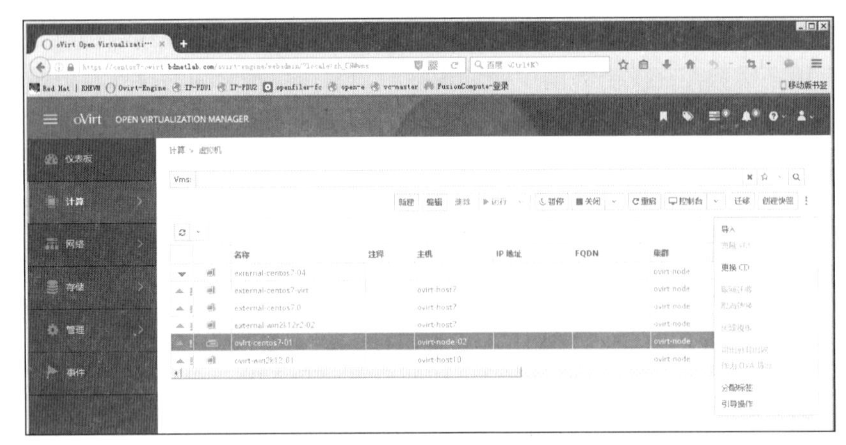

图 4-7-2

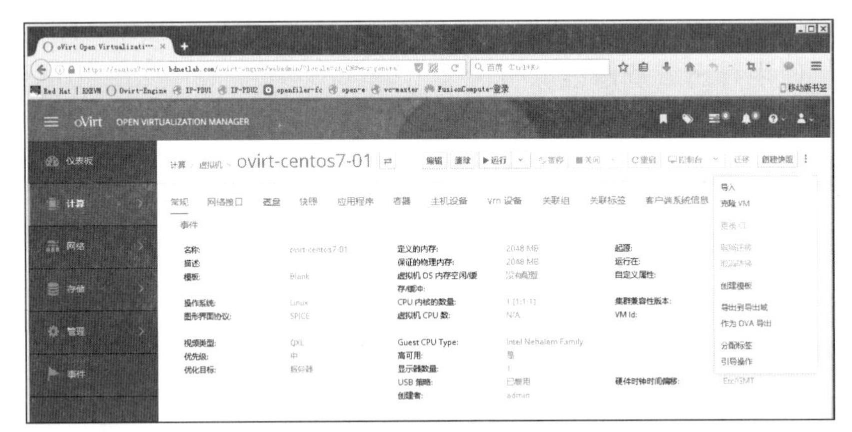

图 4-7-3

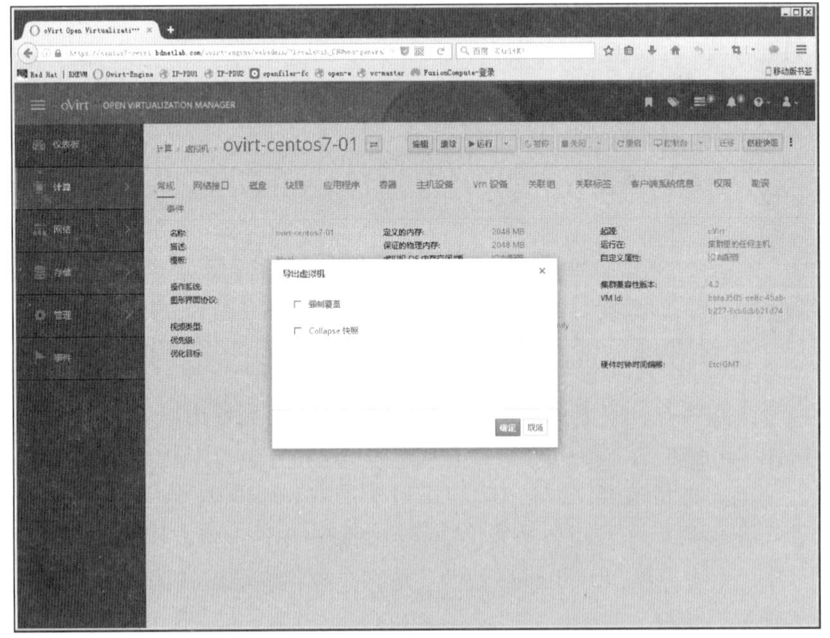

图 4-7-4

第 5 步，开始执行虚拟机导出操作，相当于备份虚拟机，如图 4-7-5 所示。

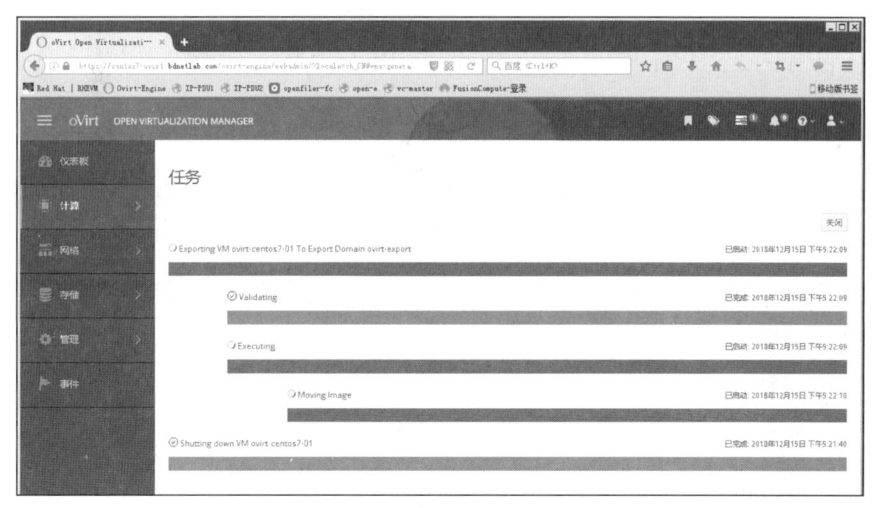

图 4-7-5

第 6 步，完成虚拟机导出操作，如图 4-7-6 所示。

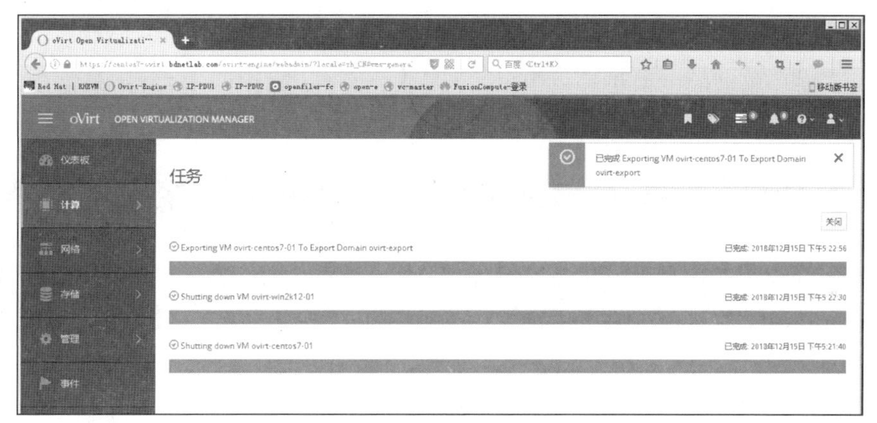

图 4-7-6

第 7 步，导出域的虚拟机导出完成后出现虚拟机的相关信息，如图 4-7-7 所示。

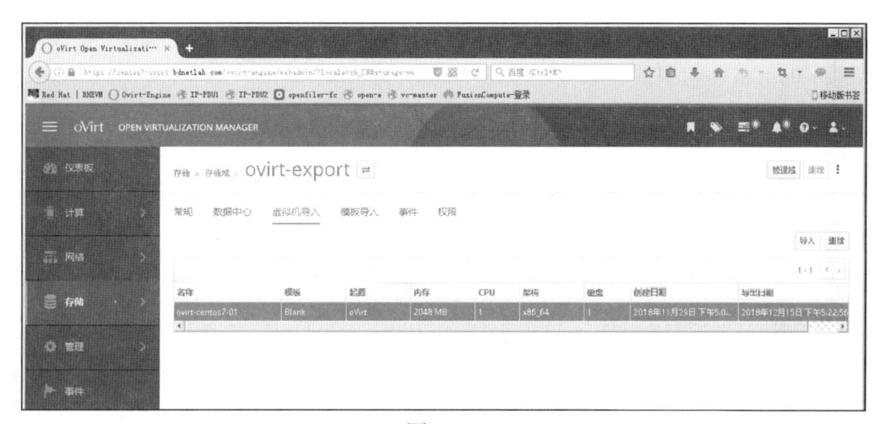

图 4-7-7

4.7.2　使用导出域恢复虚拟机

前文介绍过通过 oVirt Engine 管理端创建导出域，本节介绍如何使用导出域备份恢复虚拟机。

第 1 步，为还原真实环境，模拟虚拟机故障，将虚拟机从列表中删除，如图 4-7-8 所示。

图 4-7-8

第 2 步，在导出域的虚拟机导出中单击“导入”按钮，弹出图 4-7-9 所示窗口，单击“确定”按钮。

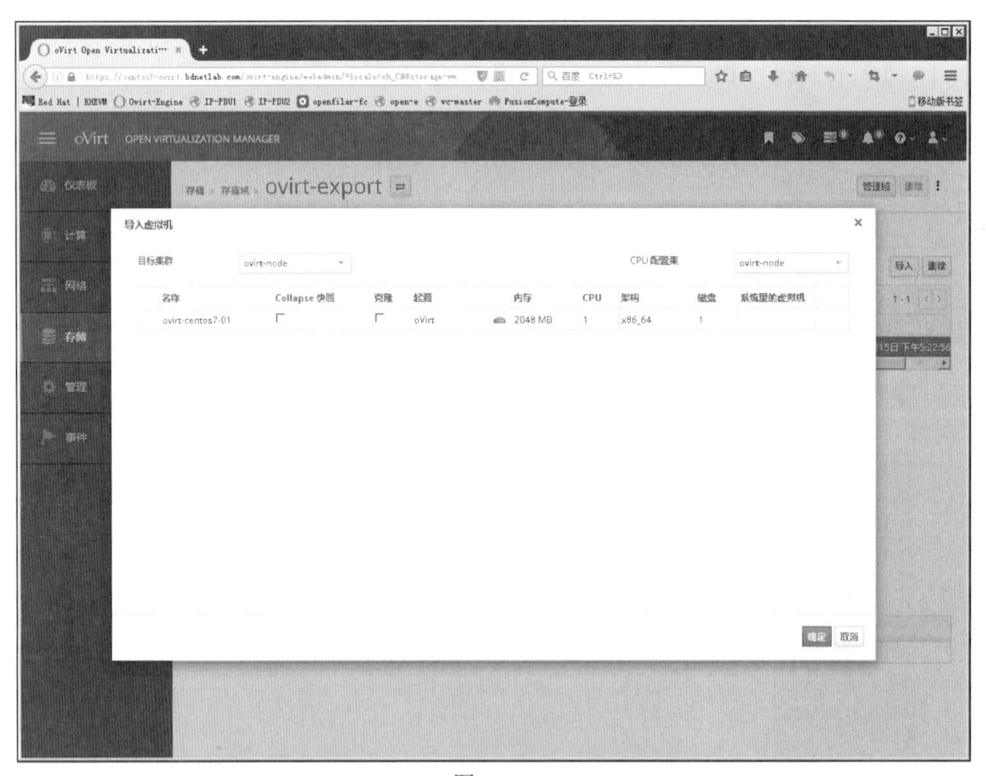

图 4-7-9

第 3 步，系统提示“导入虚拟机”，如图 4-7-10 所示，单击“关闭”按钮。

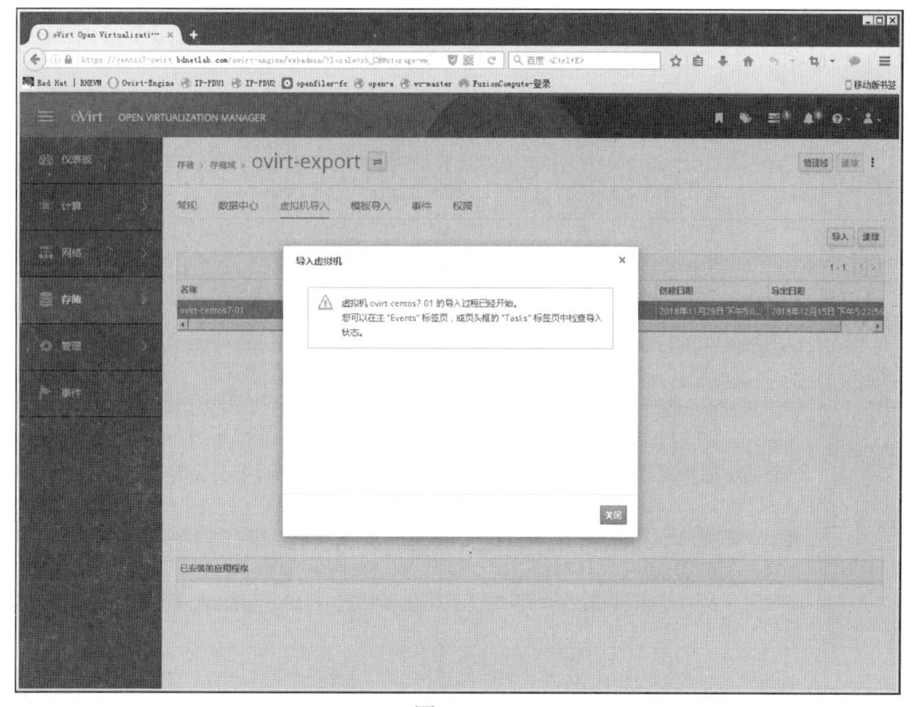

图 4-7-10

第 4 步，开始导入虚拟机，相当于恢复虚拟机，如图 4-7-11 所示。

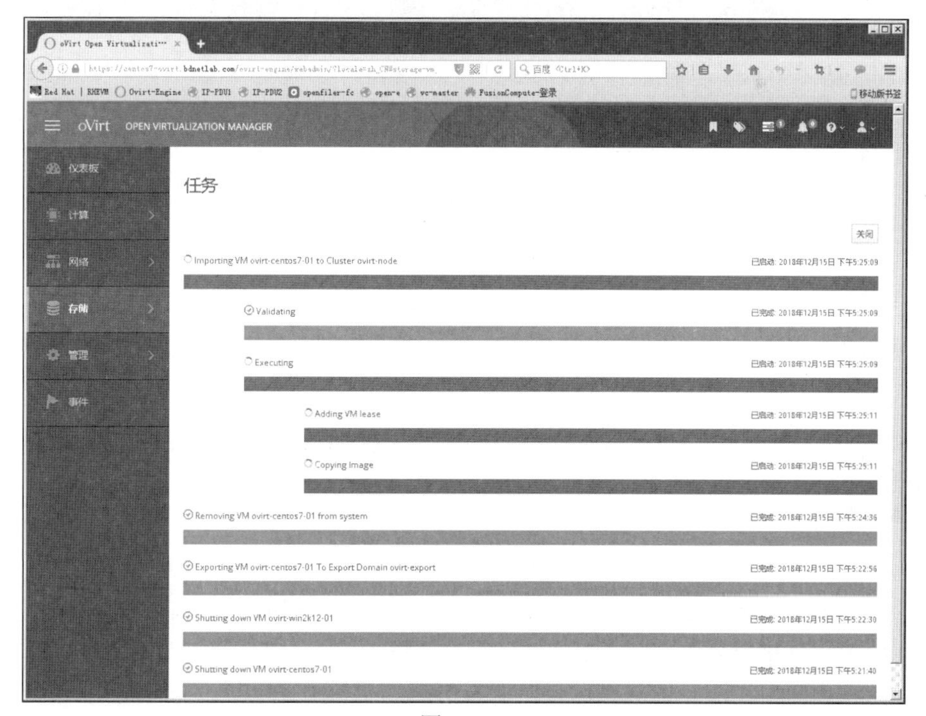

图 4-7-11

第 5 步，完成虚拟机导入操作，如图 4-7-12 所示。

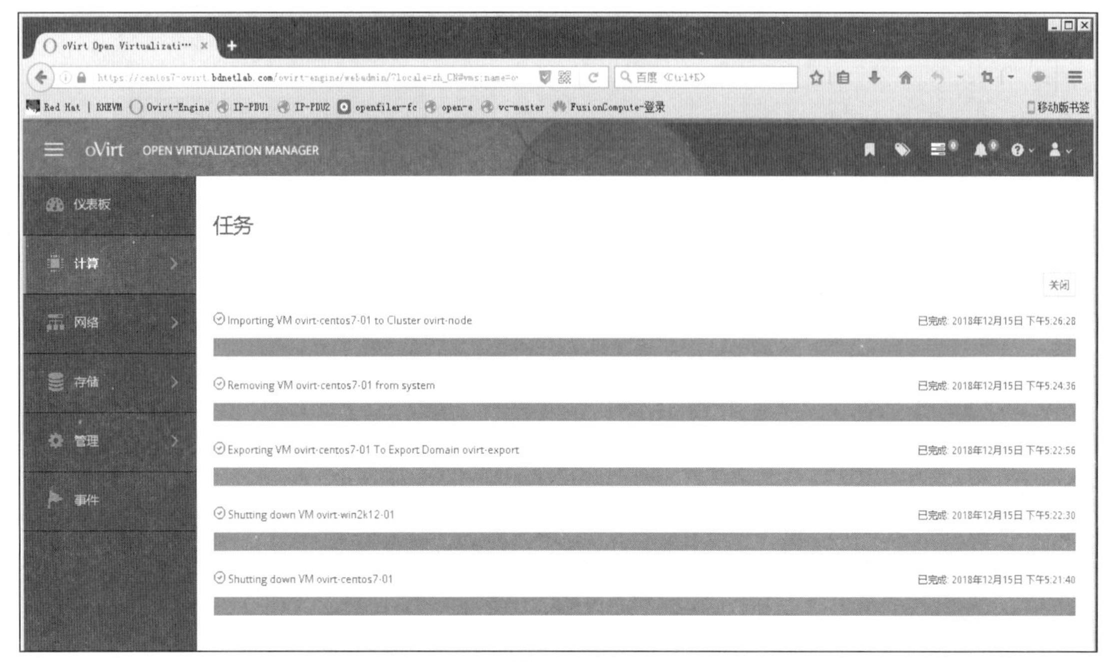

图 4-7-12

第 6 步，虚拟机出现在列表中，如图 4-7-13 所示。

图 4-7-13

第 7 步，虚拟机运行正常，说明导入成功，如图 4-7-14 所示。

至此，使用导出域备份恢复虚拟机完成。整体操作也是比较简单的，但需要提前创建导出域，需注意的是无法在虚拟机运行的时候导出，只能在虚拟机关机状态下导出。

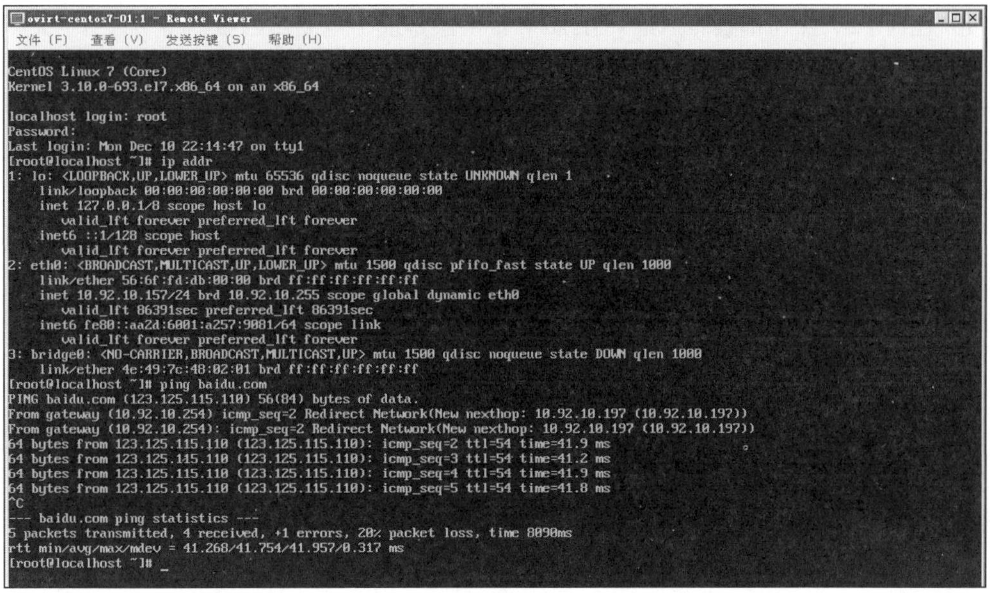

```
ovirt-centos7-01:1 - Remote Viewer
文件 (F)  查看 (V)  发送按键 (S)  帮助 (H)

CentOS Linux 7 (Core)
Kernel 3.10.0-693.el7.x86_64 on an x86_64

localhost login: root
Password:
Last login: Mon Dec 10 22:14:47 on tty1
[root@localhost ~]# ip addr
1: lo: <LOOPBACK,UP,LOWER_UP> mtu 65536 qdisc noqueue state UNKNOWN qlen 1
    link/loopback 00:00:00:00:00:00 brd 00:00:00:00:00:00
    inet 127.0.0.1/8 scope host lo
       valid_lft forever preferred_lft forever
    inet6 ::1/128 scope host
       valid_lft forever preferred_lft forever
2: eth0: <BROADCAST,MULTICAST,UP,LOWER_UP> mtu 1500 qdisc pfifo_fast state UP qlen 1000
    link/ether 56:6f:fd:db:00:00 brd ff:ff:ff:ff:ff:ff
    inet 10.92.10.157/24 brd 10.92.10.255 scope global dynamic eth0
       valid_lft 86391sec preferred_lft 86391sec
    inet6 fe80::aa2d:6001:a257:9081/64 scope link
       valid_lft forever preferred_lft forever
3: bridge0: <NO-CARRIER,BROADCAST,MULTICAST,UP> mtu 1500 qdisc noqueue state DOWN qlen 1000
    link/ether 4e:49:7c:48:02:01 brd ff:ff:ff:ff:ff:ff
[root@localhost ~]# ping baidu.com
PING baidu.com (123.125.115.110) 56(84) bytes of data.
From gateway (10.92.10.254) icmp_seq=2 Redirect Network(New nexthop: 10.92.10.197 (10.92.10.197))
From gateway (10.92.10.254): icmp_seq=2 Redirect Network(New nexthop: 10.92.10.197 (10.92.10.197))
64 bytes from 123.125.115.110 (123.125.115.110): icmp_seq=2 ttl=54 time=41.9 ms
64 bytes from 123.125.115.110 (123.125.115.110): icmp_seq=3 ttl=54 time=41.2 ms
64 bytes from 123.125.115.110 (123.125.115.110): icmp_seq=4 ttl=54 time=41.9 ms
64 bytes from 123.125.115.110 (123.125.115.110): icmp_seq=5 ttl=54 time=41.8 ms
^C
--- baidu.com ping statistics ---
5 packets transmitted, 4 received, +1 errors, 20% packet loss, time 8090ms
rtt min/avg/max/mdev = 41.268/41.754/41.957/0.317 ms
[root@localhost ~]# _
```

图 4-7-14

4.8 将物理服务器迁移到 oVirt 平台

生产环境中将物理服务器转换为虚拟机是常见的操作，特别是一些数据中心的改造项目，可能需要将大量的物理服务器转换为虚拟机。比较常见的工具有 virt-p2v 和 clonezilla。对于 Windows 物理服务器来说因其存在一些特殊性，还有使用 VMware 转换工具将其转换后再导入 KVM 平台的方式，这种方式过于复杂烦琐，不推荐使用。本节介绍如何将 Windows 物理服务器转换为虚拟机。

4.8.1 迁移方式

从物理机到虚拟机（Physical to Virtual，P2V）的实质是迁移物理服务器上的操作系统、应用软件及其数据到虚拟机中。P2V 迁移方式主要有以下几种。

1. 手动迁移

一般来说，虚拟化平台厂商或第三方会提供 P2V 工具，KVM 平台常见的是 virt-p2v，VMware vSphere 平台常见的是 VMware vCenter Converter Standalone Client。使用这些工具进行 P2V 操作相对容易。

2. 热迁移

某些生产环境或项目要求在 P2V 操作中不能停止服务。这对于 VMware vSphere 平台来说实现难度不大；但对于 KVM 平台来说有较大的难度，由于其特殊性，P2V 操作基本都是在物理服务器关机状态下进行。

4.8.2 迁移物理服务器的注意事项

在生产环境或项目中将 Windows 操作系统的物理机转换为虚拟机，需要注意几个问题。

1）任何 P2V 操作都具有风险，在 P2V 转换操作前一定要做好数据备份。

2）基于 Windows 域控制器不推荐使用 P2V 转换，建议在 KVM 平台新建虚拟机后通过域角色切换的方式进行操作。

3）基于 Windows Exchange 的服务器不推荐使用 P2V 转换，建议在 KVM 平台新建虚拟机后通过数据迁移的方式进行操作。

4）基于 Windows SQL 的服务器不推荐使用 P2V 转换，建议在 KVM 平台新建虚拟机后通过数据库备份恢复的方式进行操作。

在生产环境中将 Windows 操作系统的物理机转换为虚拟机存在过多的不确定因素，不局限于上述问题，在操作前一定要评估备份后再进行操作。特别注意：P2V 转换操作会中断物理服务器提供的各种服务。

与 Windows 物理服务器相比较，在生产环境中将 Linux 物理服务器转换为虚拟机相对容易得多，KVM 本身就是使用的 Linux 操作系统。但是，在生产环境中将 Linux 物理服务器转换为虚拟机依旧需要注意几个问题。

1）虽然都是 Linux 操作系统，但是任何 P2V 操作都具有风险，在 P2V 转换操作前一定要做好数据备份。

2）运行 Oracle RAC 的物理服务器不推荐直接使用 P2V 转换，建议在 KVM 平台新建虚拟机后通过配置的方式进行操作。

3）运行其他 Linux 集群的物理服务器不推荐直接使用 P2V 转换，建议在 KVM 平台新建虚拟机后通过配置的方式进行操作。

4.8.3 迁移 Windows 物理服务器

KVM 平台常用的 P2V 工具是 virt-p2v，是 Red Hat 公司开发用于将物理服务器转换为虚拟机的工具。将 Windows 物理服务器转换为虚拟机分两个阶段，本节介绍如何迁移 Windows 物理服务器。

1. 配置 KVM 主机

KVM 主机用于运行转换后的虚拟机，需要安装 virt-v2v 和 virtio-win 等工具，使用命令“yum install virt-v2v libguestfs-winsupport virtio-win”进行安装。

```
[root@host11 ~]# yum install virt-v2v libguestfs-winsupport virtio-win
已加载插件：fastestmirror, langpacks
Loading mirror speeds from cached hostfile
 * base: mirrors.nwsuaf.edu.cn
 * extras: mirrors.163.com
 * updates: mirrors.cqu.edu.cn
正在解决依赖关系
--> 正在检查事务
---> 软件包 libguestfs-winsupport.x86_64.0.7.2-2.el7 将被安装
--> 正在处理依赖关系 libguestfs >= 1:1.28.1，它被软件包 libguestfs-winsupport-7.2-2.el7.x86_64 需要
……（省略）
--> 解决依赖关系完成
依赖关系解决
================================================================================
 Package                     架构         版本                                源          大小
================================================================================
正在安装：
 libguestfs-winsupport       x86_64       7.2-2.el7                           base        2.1 M
```

```
 virt-v2v                  x86_64    1:1.38.2-12.el7_4.4           updates    12 M
……（省略）
为依赖而安装：
 OVMF                      noarch    20180508-3.gitee3198e672e2.el7   base       1.6 M
……（省略）
 事务概要
=====================================================================================
安装  2 软件包 (+15 依赖软件包)
总下载量：25 M
安装大小：56 M
Is this ok [y/d/N]: y
Downloading packages:
(1/17): hexedit-1.2.13-5.el7.x86_64.rpm        | 39 kB       00:00:00
(2/17): hivex-1.3.10-6.9.el7.x86_64.rpm        | 101 kB      00:00:03
……（省略）
已安装：
  libguestfs-winsupport.x86_64 0:7.2-2.el7         virt-v2v.x86_64 1:1.38.2-12.el7_4.4
作为依赖被安装：
  OVMF.noarch 0:20180508-3.gitee3198e672e2.el7     hexedit.x86_64 0:1.2.13-5.el7
  hivex.x86_64 0:1.3.10-6.9.el7                    libguestfs.x86_64 1:1.38.2-12.el7_4.4
完毕！
```

2. 将 Windows 物理服务器转换为虚拟机操作

在正式进行 P2V 操作之前，将下载好的 virt-p2v-1.34.4-4.el7.iso 刻录为光盘或制作为 U 盘，用于物理服务器引导操作。

第 1 步，为保证操作的真实性，使用一台 HP 服务器安装 Windows Server 2008 R2 操作系统，如图 4-8-1 所示。

图 4-8-1

第 2 步，关闭物理服务器电源，使用 virt-p2v 进行引导操作，如图 4-8-2 所示，选择“Start Virt P2V”。

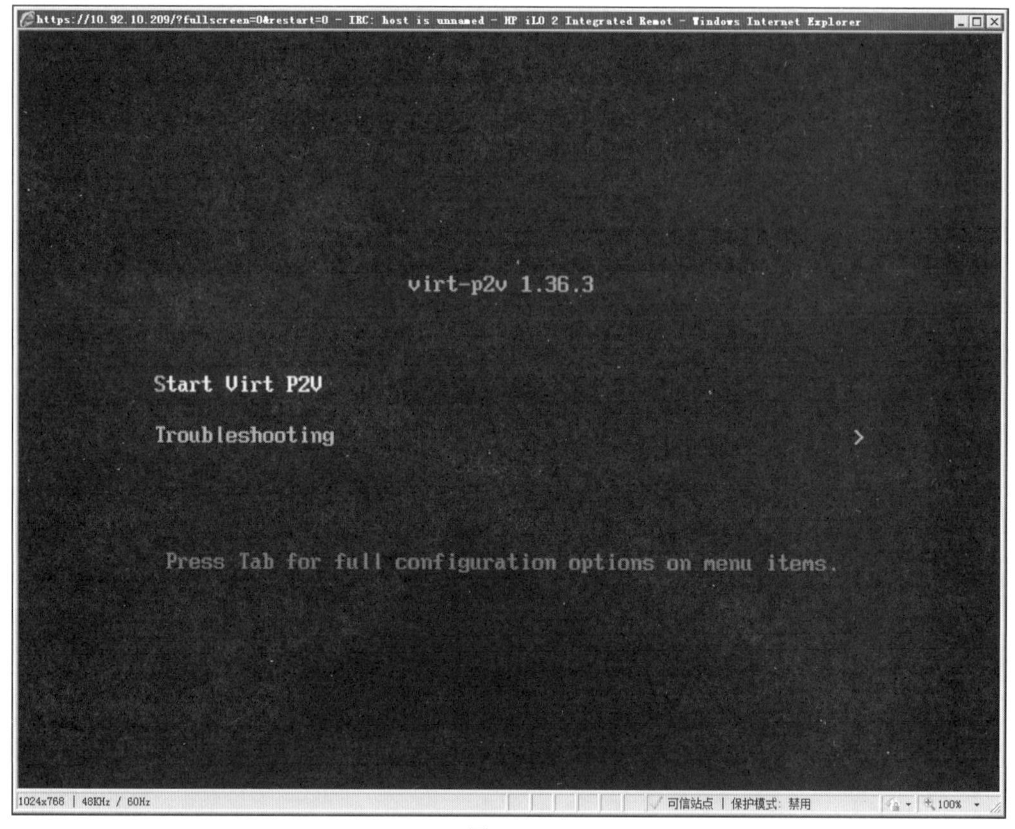

图 4-8-2

第 3 步，引导成功，进入 P2V 操作界面，如果没有自动进入 P2V 操作界面，可以使用命令“launch-virt-p2v”进入，如图 4-8-3 所示。

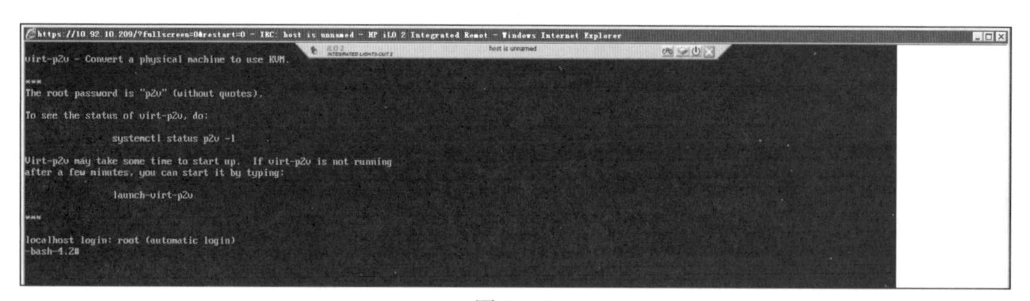

图 4-8-3

第 4 步，进入 P2V 操作界面，如图 4-8-4 所示。

第 5 步，输入 KVM 主机 IP 地址和用户名密码，如图 4-8-5 所示，单击“Test connection”按钮进行连接测试。

第 6 步，测试成功后根据提示单击“Next”进行下一步，如图 4-8-6 所示。注意：如果出现其他提示，请根据提示处理后才能进行下一步。

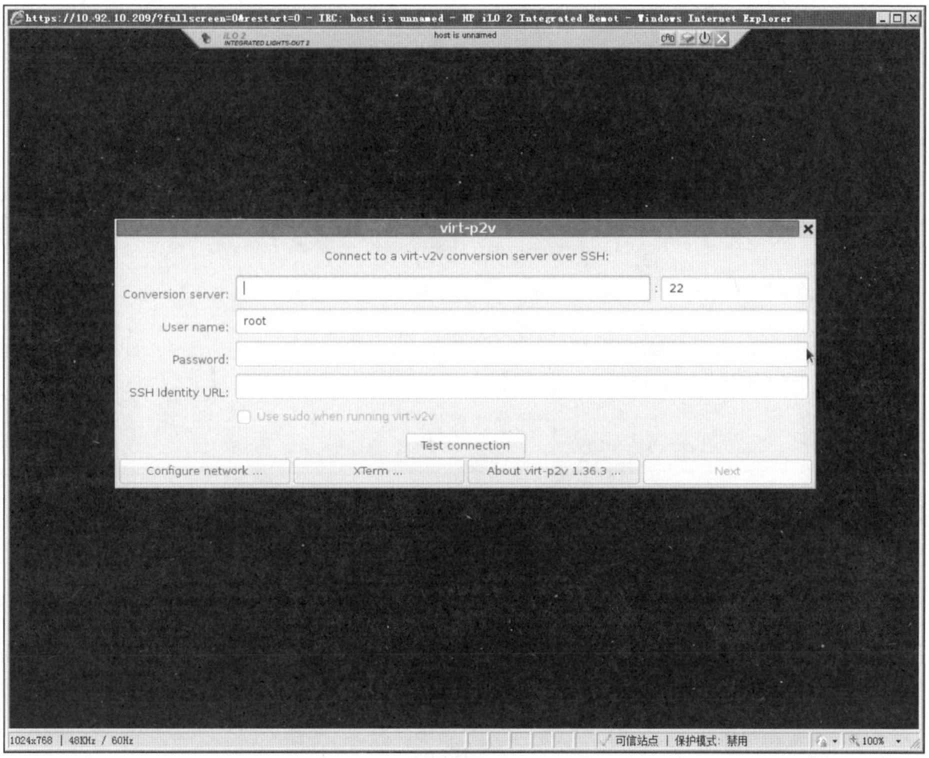

https://10.92.10.209/?fullscreen=0&restart=0 - IRC: host is unnamed - HP iLO 2 Integrated Remot - Windows Internet Explorer

iLO 2 INTEGRATED LIGHTS-OUT 2 host is unnamed

virt-p2v

Connect to a virt-v2v conversion server over SSH:

Conversion server: : 22

User name: root

Password:

SSH Identity URL:

Use sudo when running virt-v2v

Test connection

Configure network ... XTerm ... About virt-p2v 1.36.3 ... Next

1024x768 | 48KHz / 60Hz 可信站点 | 保护模式: 禁用 100%

图 4-8-4

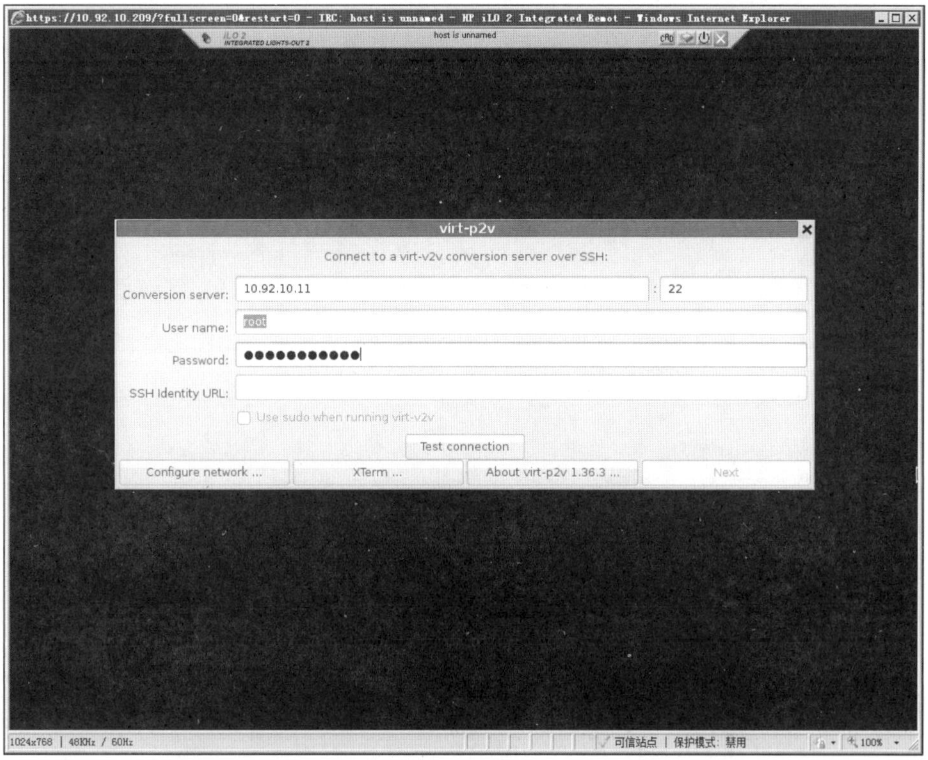

https://10.92.10.209/?fullscreen=0&restart=0 - IRC: host is unnamed - HP iLO 2 Integrated Remot - Windows Internet Explorer

iLO 2 INTEGRATED LIGHTS-OUT 2 host is unnamed

virt-p2v

Connect to a virt-v2v conversion server over SSH:

Conversion server: 10.92.10.11 : 22

User name: root

Password: ●●●●●●●●●●●●

SSH Identity URL:

Use sudo when running virt-v2v

Test connection

Configure network ... XTerm ... About virt-p2v 1.36.3 ... Next

1024x768 | 48KHz / 60Hz 可信站点 | 保护模式: 禁用 100%

图 4-8-5

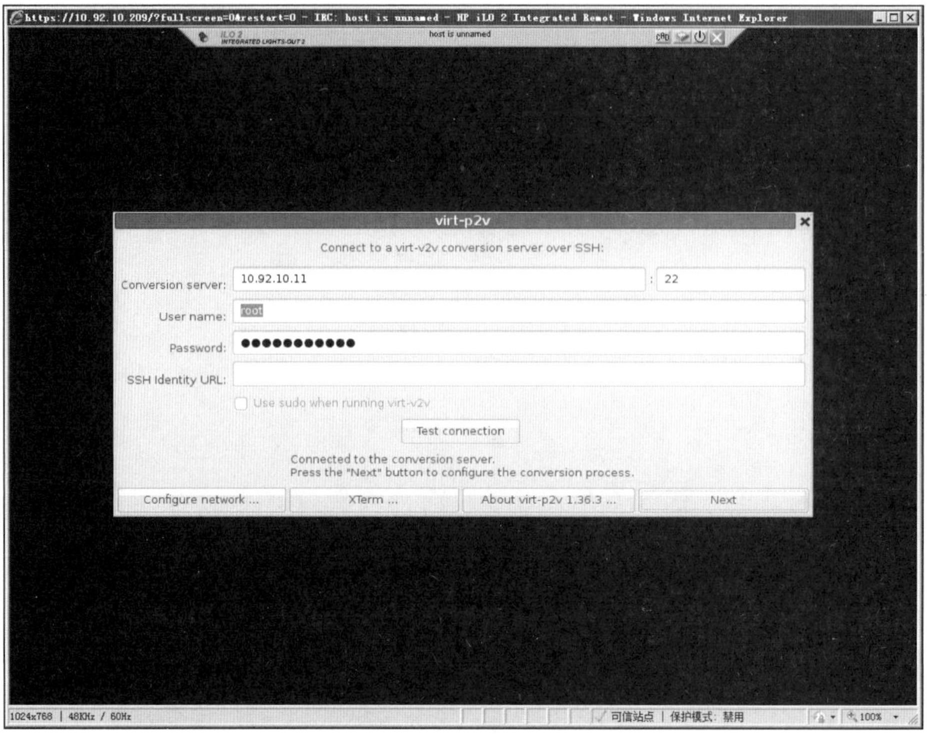

https://10.92.10.209/?fullscreen=0&restart=0 - IRC: host is unnamed - HP iLO 2 Integrated Remot - Windows Internet Explorer
iLO 2 INTEGRATED LIGHTS-OUT 2
host is unnamed

virt-p2v

Connect to a virt-v2v conversion server over SSH:

Conversion server: 10.92.10.11 : 22

User name: root

Password: ●●●●●●●●●●●●

SSH Identity URL:

Use sudo when running virt-v2v

Test connection

Connected to the conversion server.
Press the "Next" button to configure the conversion process.

Configure network ... | XTerm ... | About virt-p2v 1.36.3 ... | Next

1024x768 | 48KHz / 60Hz
可信站点 | 保护模式: 禁用
100%

图 4-8-6

第 7 步，进入 P2V 工具参数配置界面，如图 4-8-7 所示。

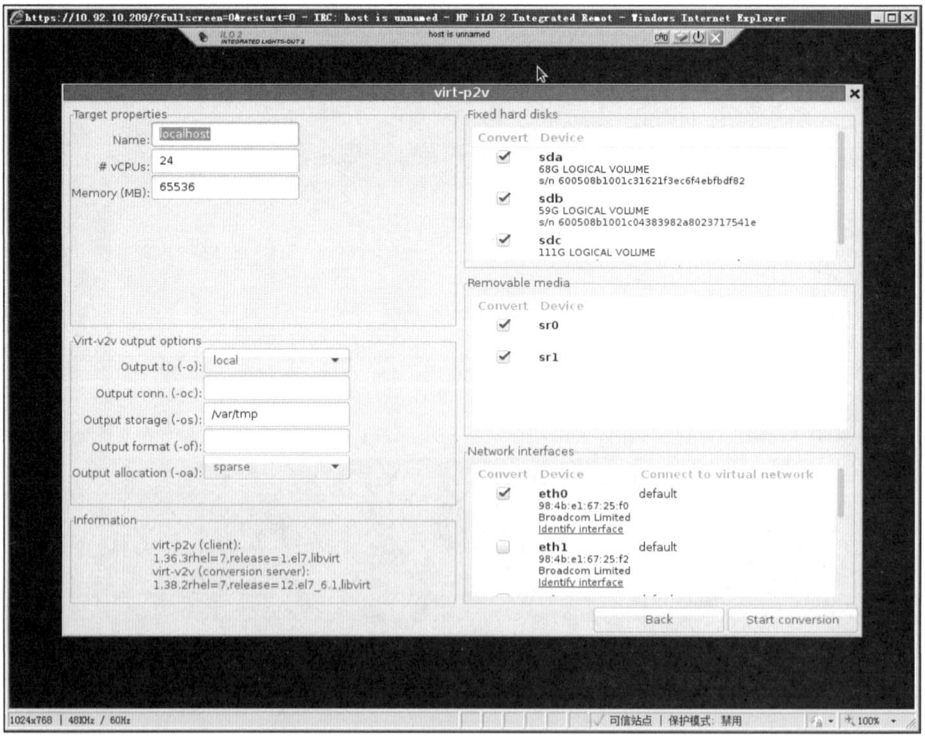

https://10.92.10.209/?fullscreen=0&restart=0 - IRC: host is unnamed - HP iLO 2 Integrated Remot - Windows Internet Explorer
iLO 2 INTEGRATED LIGHTS-OUT 2
host is unnamed

virt-p2v

Target properties
Name: localhost
vCPUs: 24
Memory (MB): 65536

Virt-v2v output options
Output to (-o): local
Output conn. (-oc):
Output storage (-os): /var/tmp
Output format (-of):
Output allocation (-oa): sparse

Information
virt-p2v (client):
1.36.3rhel=7,release=1.el7,libvirt
virt-v2v (conversion server):
1.38.2rhel=7,release=12.el7_6.1,libvirt

Fixed hard disks
Convert Device
✔ sda
68G LOGICAL VOLUME
s/n 600508b1001c31621f3ec6f4ebfbdf82
✔ sdb
59G LOGICAL VOLUME
s/n 600508b1001c04383982a8023717541e
✔ sdc
111G LOGICAL VOLUME

Removable media
Convert Device
✔ sr0
✔ sr1

Network interfaces
Convert Device Connect to virtual network
✔ eth0 default
98:4b:e1:67:25:f0
Broadcom Limited
Identify interface
eth1 default
98:4b:e1:67:25:f2
Broadcom Limited
Identify interface

Back | Start conversion

1024x768 | 48KHz / 60Hz
可信站点 | 保护模式: 禁用
100%

图 4-8-7

第 8 步，配置转换后的虚拟机名、vCPU 数以及内存大小，指定 KVM 存储路径和转换后的硬盘格式，对于 Windows 服务器来说，网卡可以等完成转换后再增加，参数如图 4-8-8 所示，单击“Start Conversion”。

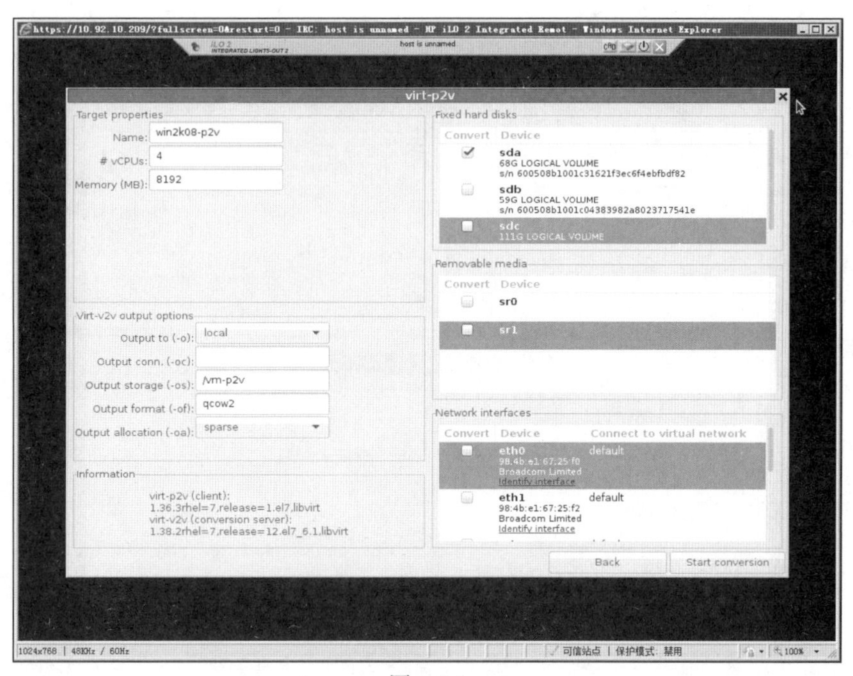

图 4-8-8

第 9 步，P2V 工具将 Windows 物理服务器转换为虚拟机，如图 4-8-9 所示。

图 4-8-9

第 10 步，完成 P2V 操作，如图 4-8-10 所示。注意：一定要出现成功的提示才说明 P2V 操作成功，如果出现错误提示说明 P2V 操作出现问题。

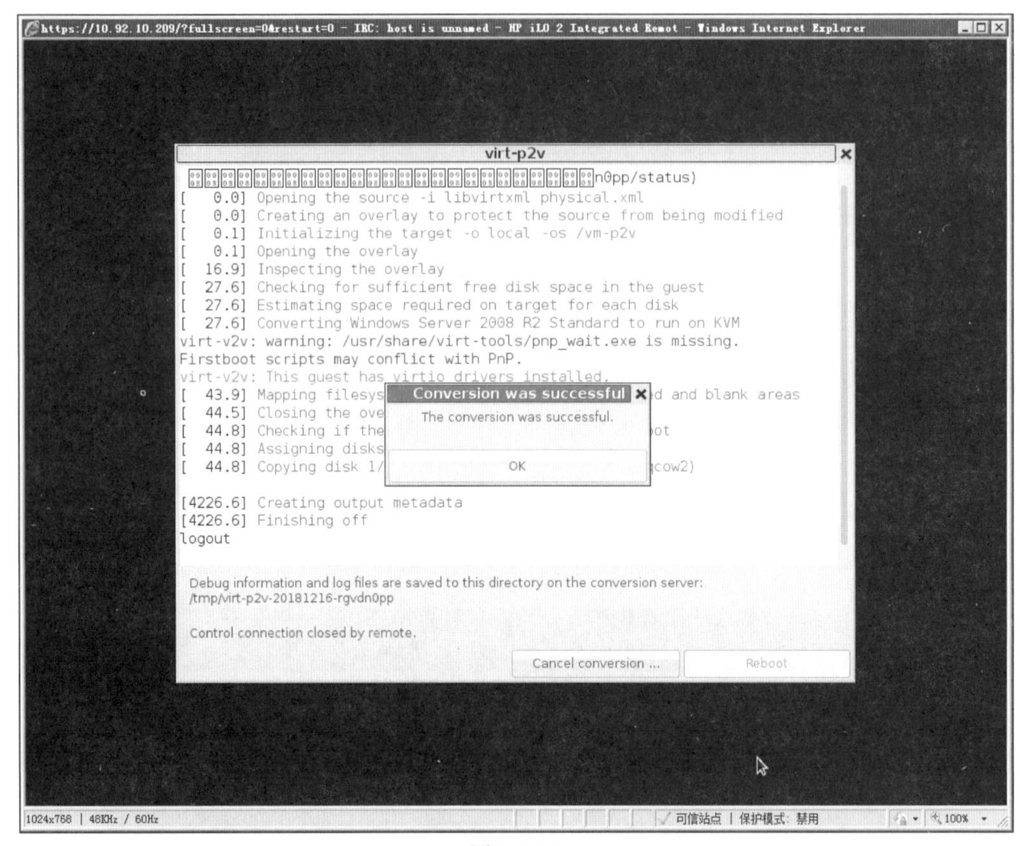

图 4-8-10

第 11 步，由于 Windows 虚拟机需要使用 GUI，因此登录 KVM 主机 GUI，使用命令查看转换后的虚拟机硬盘文件和配置文件，虚拟机处于未注册状态，如图 4-8-11 所示。

图 4-8-11

第 12 步，使用命令“virsh define win2k08-p2v.xml”将虚拟机注册到 KVM 主机，然后使用命令将其启动，如图 4-8-12 所示。

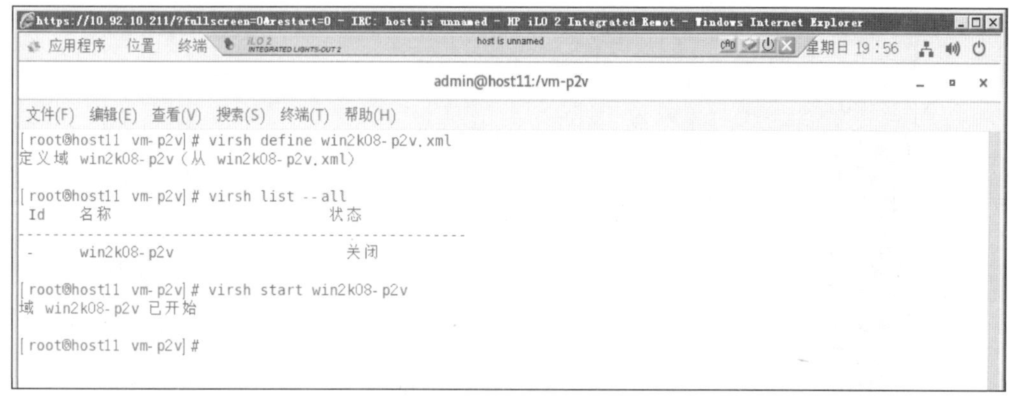

图 4-8-12

第 13 步，P2V 后虚拟机启动成功，如图 4-8-13 所示。

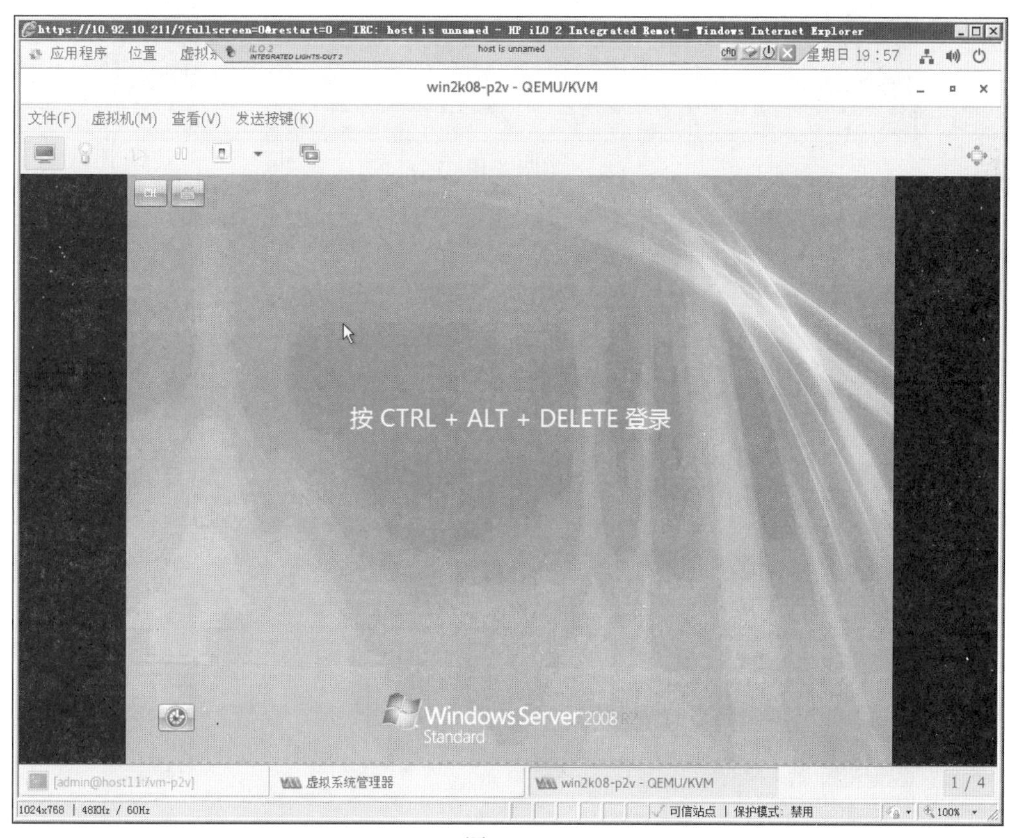

图 4-8-13

第 14 步，登录虚拟机查看配置，如图 4-8-14 所示，由于在 P2V 过程中未添加网卡，所以网络处于未连接状态。

第 15 步，为虚拟机增加网卡，查看网络连接情况，如图 4-8-15 所示。

至此，将 Windows 物理服务器转换为虚拟机操作完成。整体来说，使用工具难度系数并不大，如果原物理服务器数据较多，P2V 过程会比较慢。最后再说明一下，任何 P2V 操作都是有风险的，也不一定能够一次成功，P2V 出现问题请根据提示进行操作。

图 4-8-14

图 4-8-15

4.8.4 迁移 Linux 物理服务器

对于 Linux 物理服务器的迁移，依旧推荐使用常用的 P2V 工具 virt-p2v。将 Linux 物理服务器转换为虚拟机一般可以分两个阶段，本节介绍如何迁移 Linux 物理服务器。

第 1 步，为保证操作的真实性，使用一台 HP 服务器安装 CentOS 操作系统，如图 4-8-16 所示。

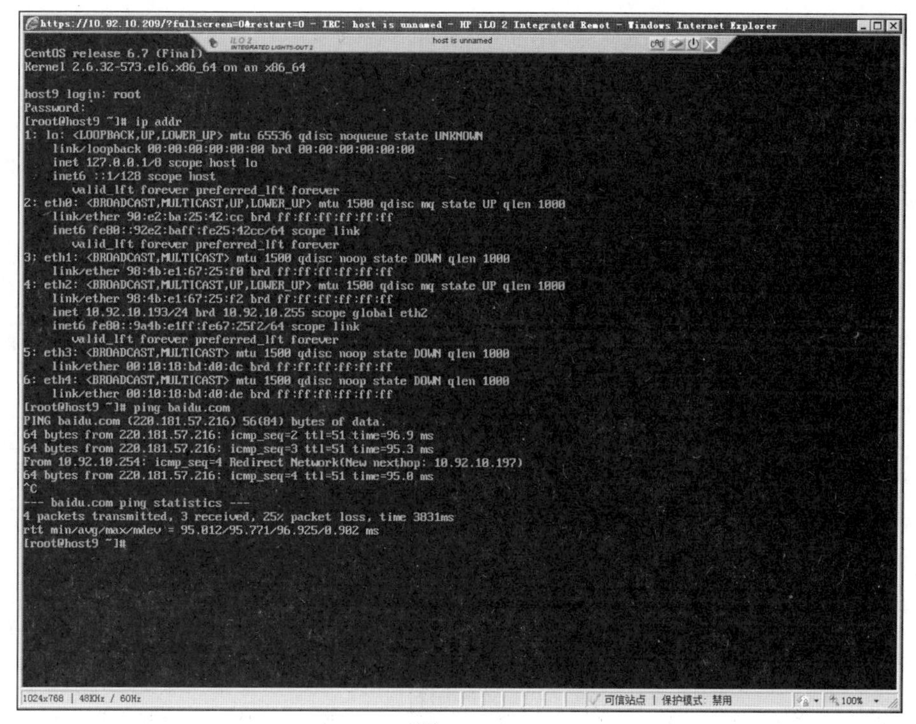

图 4-8-16

第 2 步，关闭物理服务器电源，使用 virt-p2v 进行引导操作，如图 4-8-17 所示，选择“Start Virt P2V”。

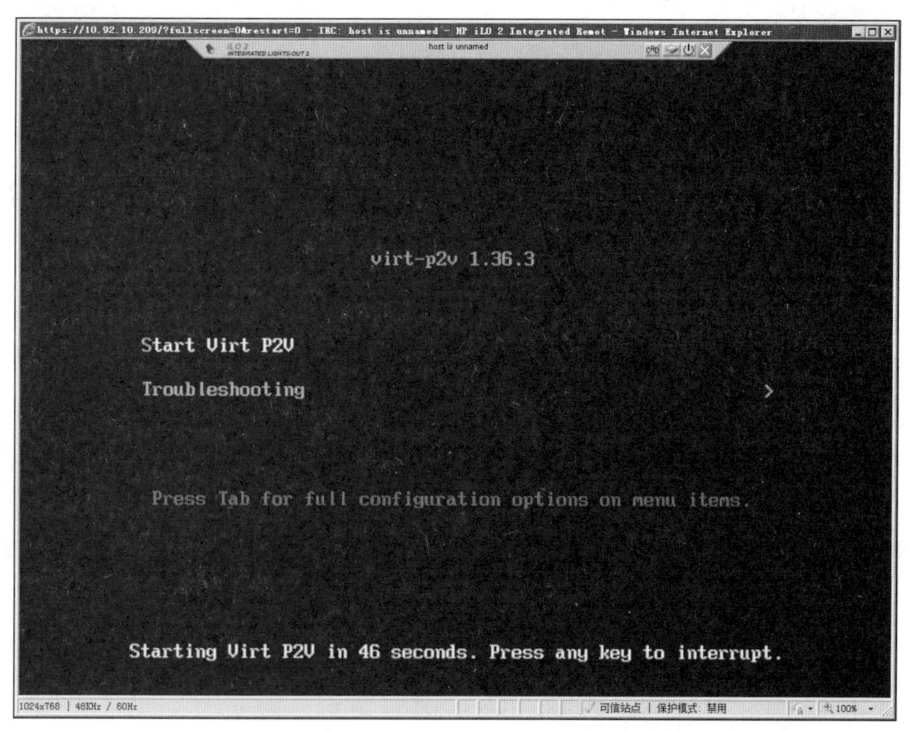

图 4-8-17

第 3 步，进入 P2V 操作界面，如图 4-8-18 所示。

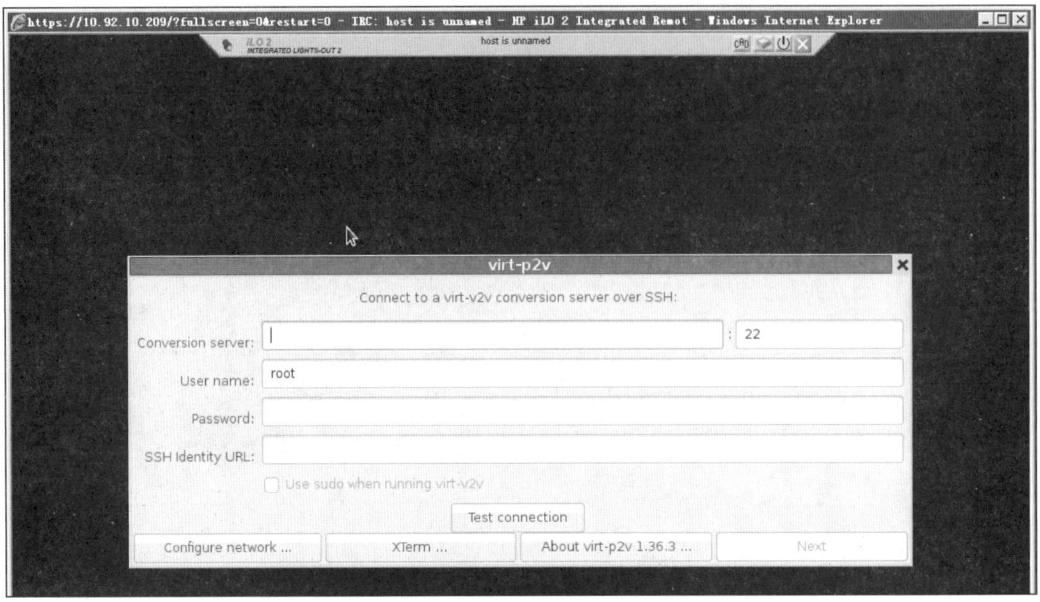

图 4-8-18

第 4 步，输入 KVM 主机 IP 地址和用户名、密码，如图 4-8-19 所示，单击“Test connection”按钮进行连接测试。

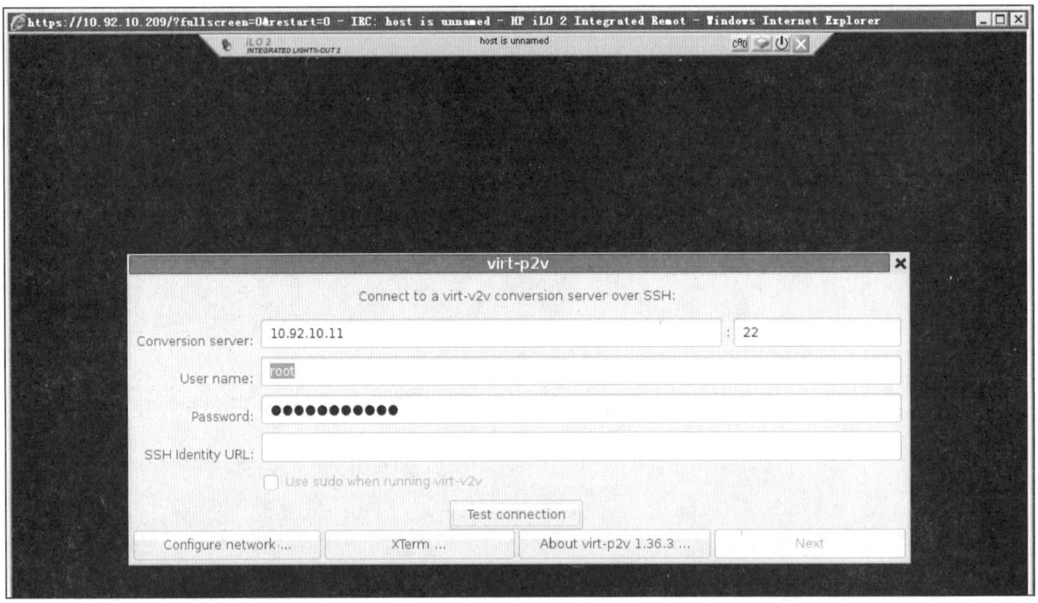

图 4-8-19

第 5 步，测试成功后根据提示单击“Next”进行下一步，如图 4-8-20 所示。注意：如果出现其他提示，请根据提示处理后才能进行下一步。

第 6 步，进入 P2V 工具参数配置界面，如图 4-8-21 所示。

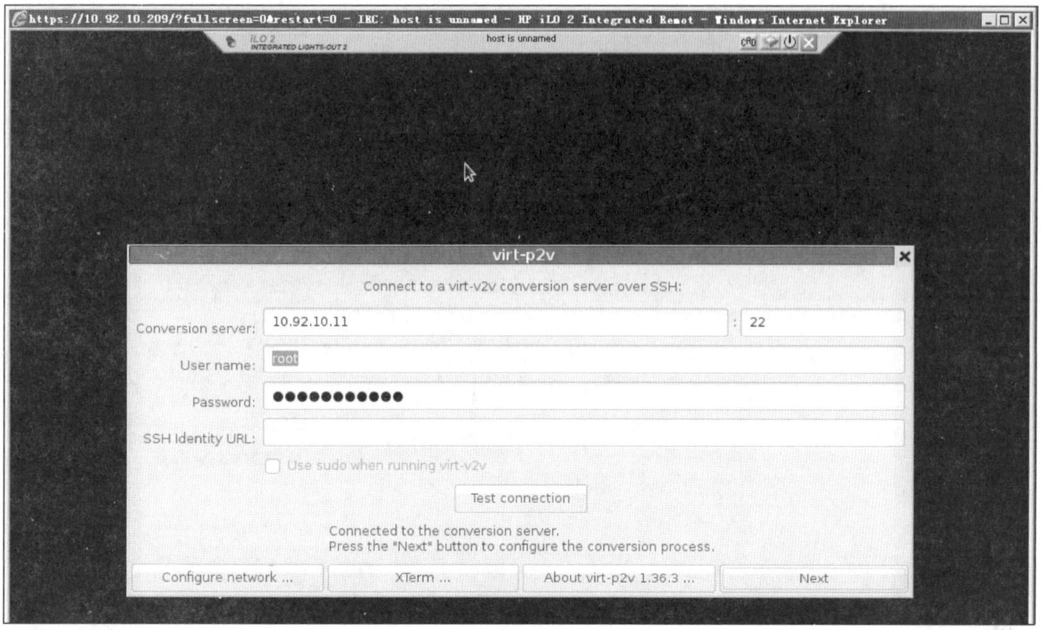

https://10.92.10.209/?fullscreen=0&restart=0 - IRC: host is unnamed - HP iLO 2 Integrated Remot - Windows Internet Explorer
iLO 2 INTEGRATED LIGHTS-OUT 2
host is unnamed
virt-p2v
Connect to a virt-v2v conversion server over SSH:
Conversion server: 10.92.10.11 : 22
User name: root
Password: ●●●●●●●●●●●●
SSH Identity URL:
Use sudo when running virt-v2v
Test connection
Connected to the conversion server.
Press the "Next" button to configure the conversion process.
Configure network ...
XTerm ...
About virt-p2v 1.36.3 ...
Next

图 4-8-20

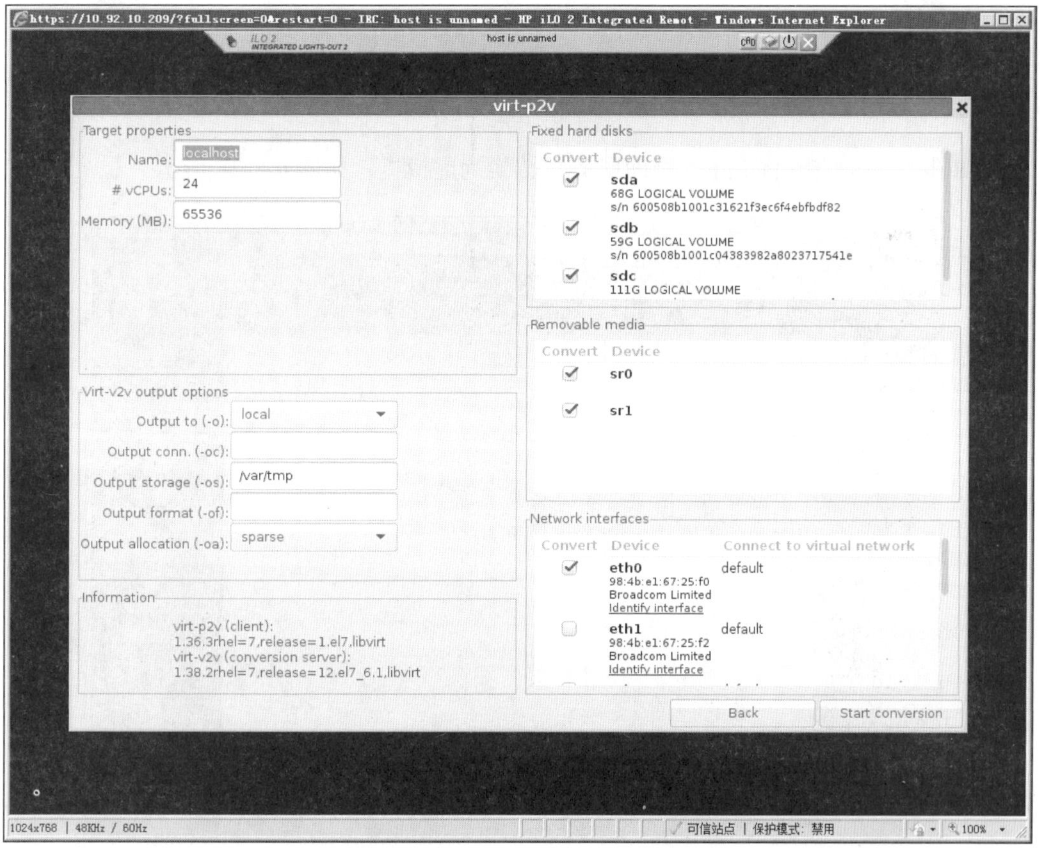

https://10.92.10.209/?fullscreen=0&restart=0 - IRC: host is unnamed - HP iLO 2 Integrated Remot - Windows Internet Explorer
iLO 2 INTEGRATED LIGHTS-OUT 2
host is unnamed
virt-p2v
Target properties
Name: localhost
vCPUs: 24
Memory (MB): 65536
Virt-v2v output options
Output to (-o): local
Output conn. (-oc):
Output storage (-os): /var/tmp
Output format (-of):
Output allocation (-oa): sparse
Information
virt-p2v (client):
1.36.3rhel=7,release=1.el7,libvirt
virt-v2v (conversion server):
1.38.2rhel=7,release=12.el7_6.1,libvirt
Fixed hard disks
Convert Device
sda
68G LOGICAL VOLUME
s/n 600508b1001c31621f3ec6f4ebfbdf82
sdb
59G LOGICAL VOLUME
s/n 600508b1001c04383982a8023717541e
sdc
111G LOGICAL VOLUME
Removable media
Convert Device
sr0
sr1
Network interfaces
Convert Device Connect to virtual network
eth0 default
98:4b:e1:67:25:f0
Broadcom Limited
Identify interface
eth1 default
98:4b:e1:67:25:f2
Broadcom Limited
Identify interface
Back
Start conversion
1024x768 | 48KHz / 60Hz
可信站点 | 保护模式: 禁用
100%

图 4-8-21

第 7 步，配置转换后的虚拟机名、vCPU 数以及内存大小，指定 KVM 存储路径和转换后的硬盘格式，对于 Linux 服务器来说，网卡可以等完成转换后再增加，参数如图 4-8-22 所示，单击“Start Conversion”。

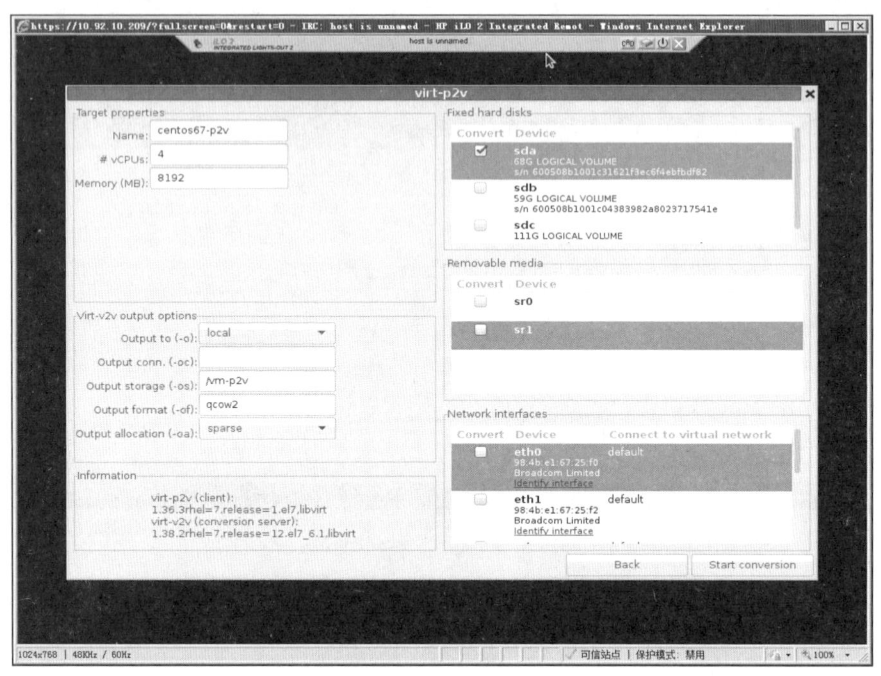

图 4-8-22

第 8 步，P2V 工具将 Linux 物理服务器转换为虚拟机，如图 4-8-23 所示。

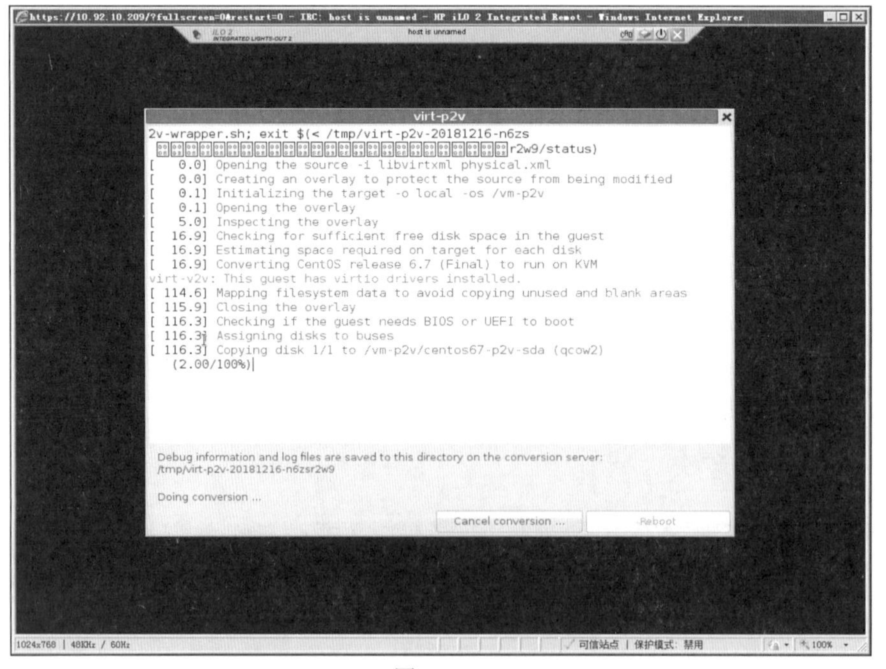

图 4-8-23

第 9 步，完成 P2V 操作，如图 4-8-24 所示。注意：一定要出现成功的提示才说明 P2V 操作成功，如果出现错误提示说明 P2V 操作出现问题。

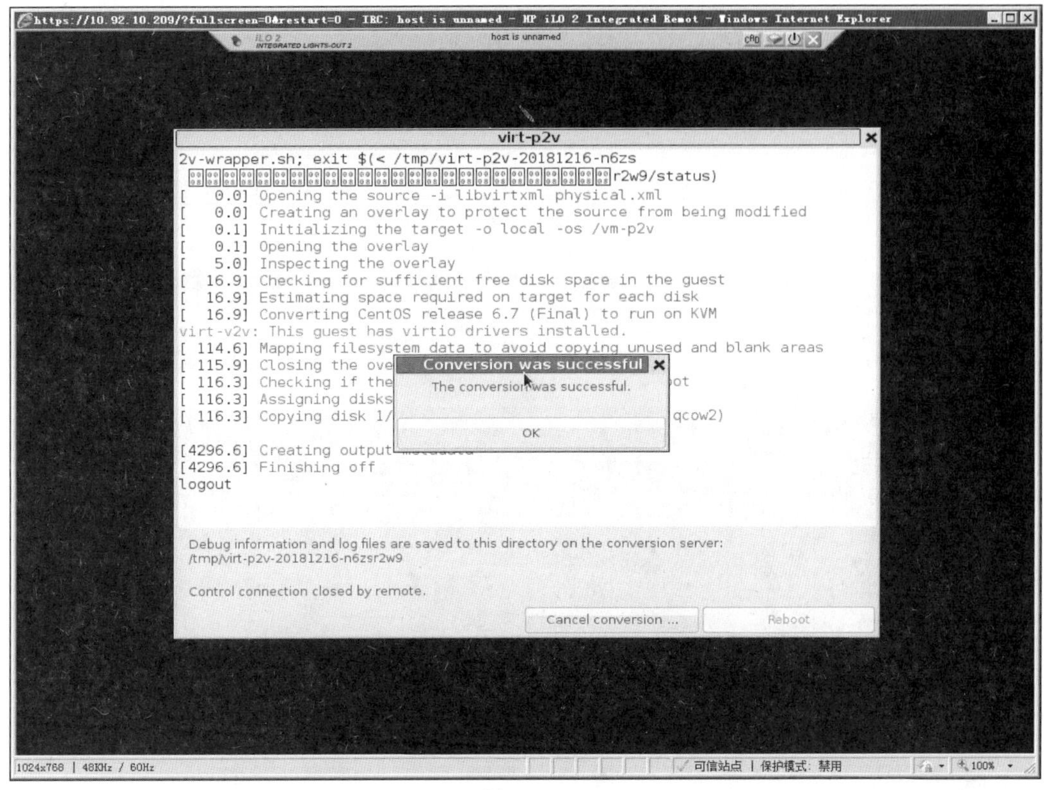

图 4-8-24

第 10 步，登录 KVM 主机 GUI，使用命令查看转换后的虚拟机硬盘文件和配置文件，虚拟机处于未注册状态，如图 4-8-25 所示。

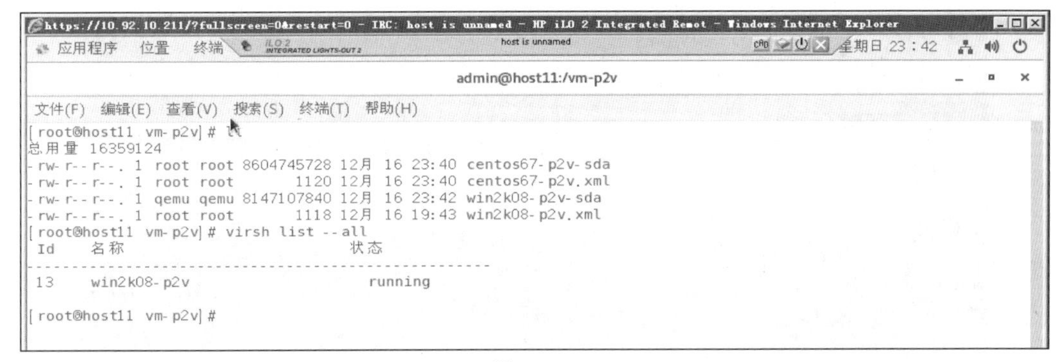

图 4-8-25

第 11 步，使用命令“virsh define win2k08-p2v.xml”将虚拟机注册到 KVM 主机，然后使用命令将其启动，如图 4-8-26 所示。

第 12 步，P2V 后虚拟机启动成功，如图 4-8-27 所示。

第 13 步，为虚拟机添加网卡后登录虚拟机，测试网络的连通性，如图 4-8-28 所示。

```
[root@host11 vm-p2v]# virsh define centos67-p2v.xml
定义域 centos67-p2v（从 centos67-p2v.xml）

[root@host11 vm-p2v]# virsh list --all
 Id    名称                           状态
----------------------------------------------------
 13    win2k08-p2v                    running
 -     centos67-p2v                   关闭

[root@host11 vm-p2v]# virsh start centos67-p2v
域 centos67-p2v 已开始

[root@host11 vm-p2v]#
```

图 4-8-26

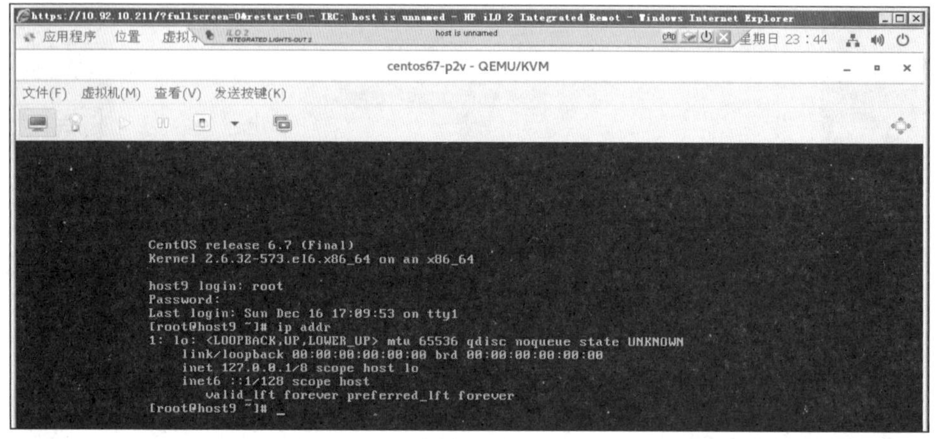

图 4-8-27

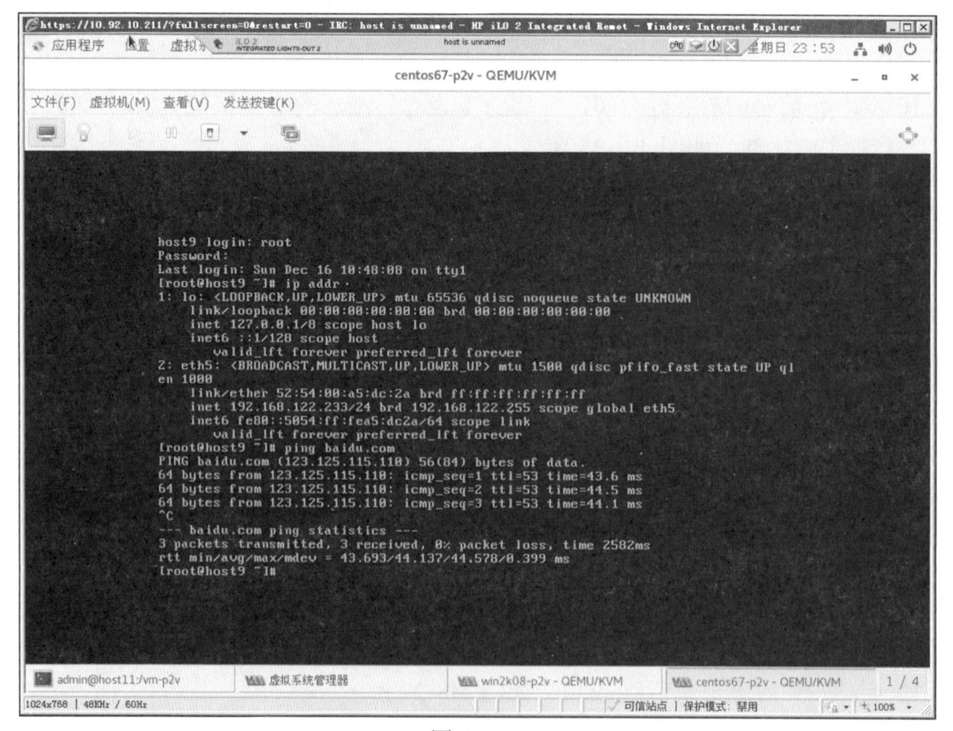

图 4-8-28

第 14 步，使用命令“virsh list--all”查看虚拟机运行状态，如图 4-8-29 所示，可以看到 2 台 P2V 的虚拟机均处于运行状态。

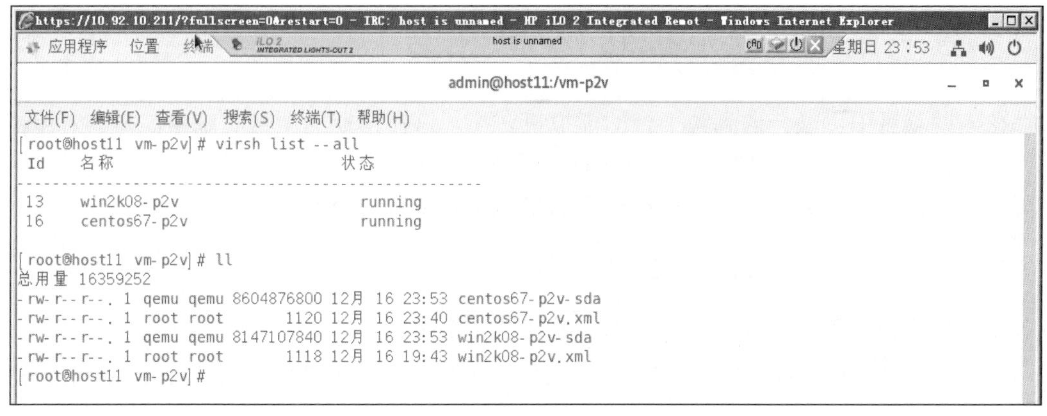

图 4-8-29

第 15 步，使用虚拟系统管理器查看虚拟机运行状态，如图 4-8-30 所示。

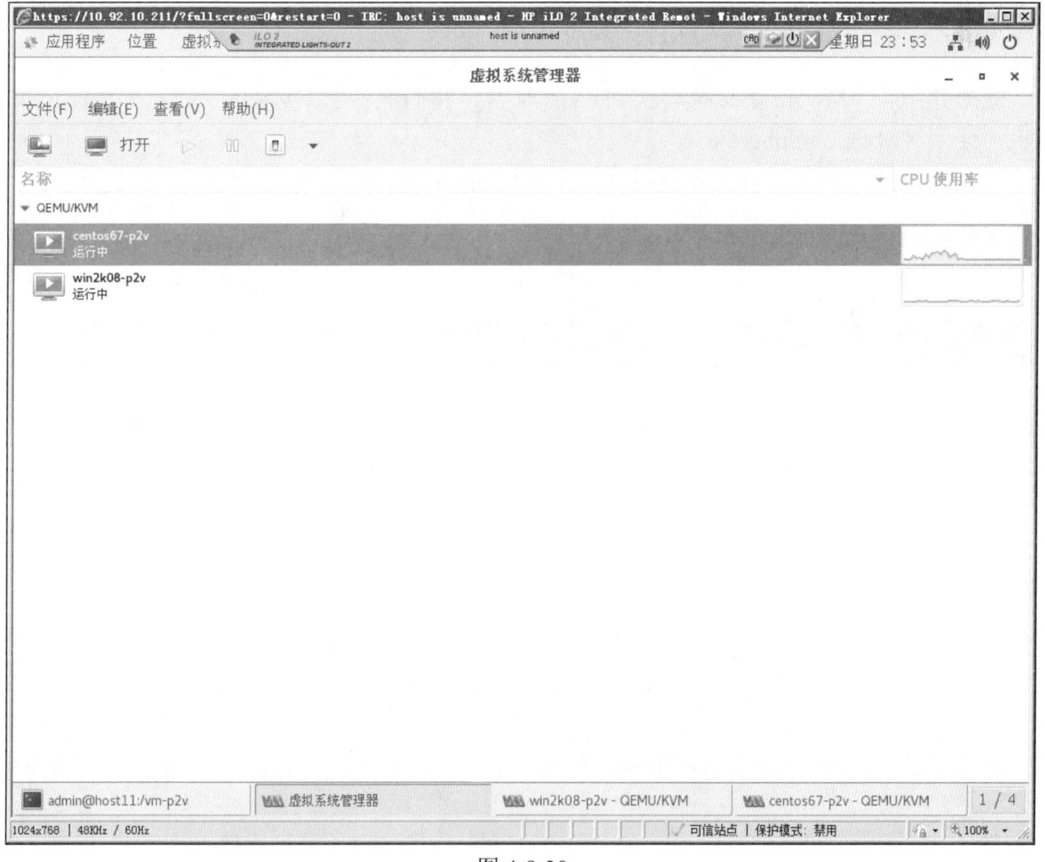

图 4-8-30

至此，将 Linux 物理服务器转换为虚拟机操作完成。整体来说，由于都是 Linux 操作系统，将 Linux 物理服务器转换成虚拟化问题相对较少。但是任何 P2V 操作都是有风险的，也不一定能够一次成功，P2V 出现问题请根据提示进行操作。部分 Linux 物理服务器添加网卡后无法生成配置文件，需要手动配置网络参数。

4.9 跨平台迁移虚拟机到 oVirt 平台

在一些项目中可能会遇到跨平台的虚拟机转换。这在早期的项目实施中，是有难度的，因为平台的兼容性问题。但现在各平台都会提供一些相应的工具来进行转换，难度系数相对降低很多。本节介绍如何将 VMware 平台虚拟机迁移到 oVirt 平台。

4.9.1 跨平台迁移虚拟机的注意事项

生产环境中跨平台的虚拟机迁移，需要注意几个问题。

1）跨平台转换与传统 P2V 操作相比具有更大的风险，转换操作前一定要做好数据备份和评估工作，以便出现问题及时回退。

2）推荐使用工具进行转换，前期通过导出 OVA 文件的方式可能导致虚拟机文件损坏，特别是容量大的 OVA 文件更容易出现问题。

4.9.2 将 VMware 虚拟机迁移到 oVirt 平台

此处介绍将 VMware 虚拟机转换到 oVirt 平台，我们准备了 VMware vCenter Server 和虚拟机，使用 VMware vSphere 6.7。

第 1 步，使用浏览器登录 VMware vCenter Server 查看整体情况，如图 4-9-1 所示。

图 4-9-1

第 2 步，确定将 WIN2K08-01 虚拟机转换到 oVirt 平台，查看虚拟机信息，如图 4-9-2 所示。

第 3 步，使用浏览器登录 oVirt Engine 管理端，在虚拟机选项中选择“导入”，如图 4-9-3 所示。

图 4-9-2

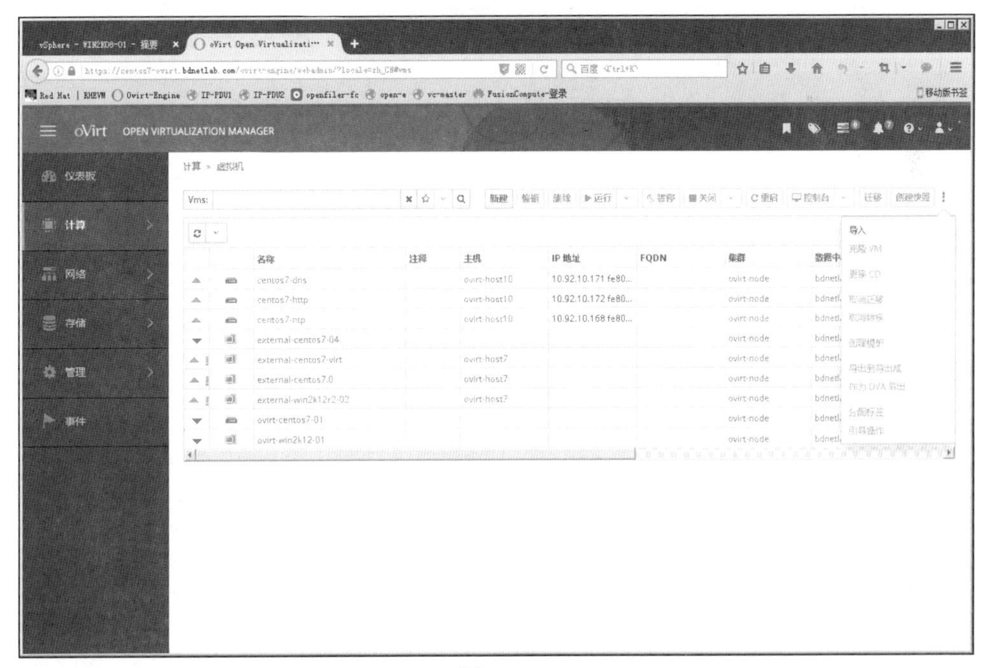

图 4-9-3

第 4 步，进入“导入虚拟机”界面，在“源”中选择“VMware”，如图 4-9-4 所示。

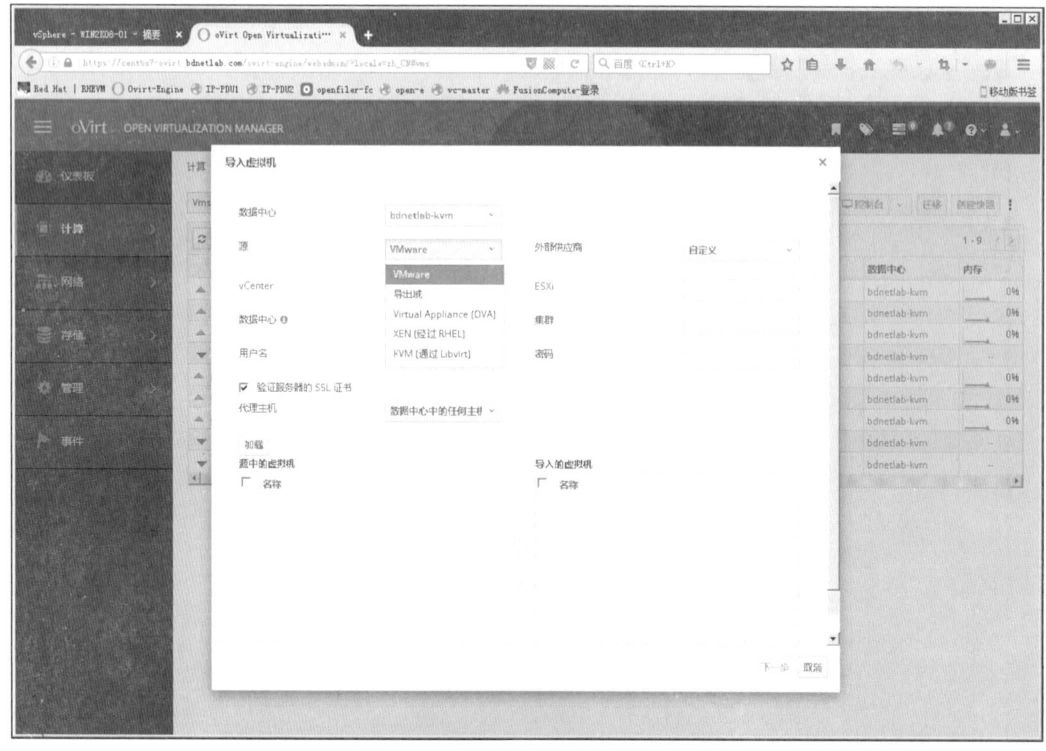

图 4-9-4

第 5 步，输入 VMware vCenter Server 相关信息，如图 4-9-5 所示，单击“加载”按钮。

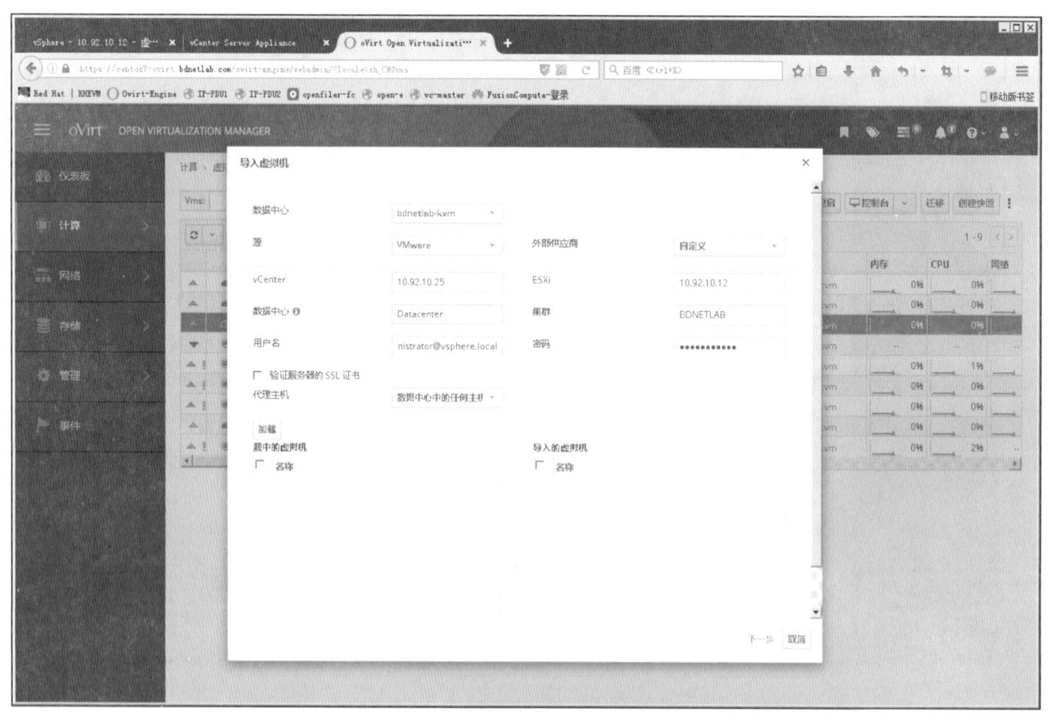

图 4-9-5

第 6 步，加载 VMware vCenter Server 主机虚拟机信息，如图 4-9-6 所示。注意：不建议勾选“验证服务器的 SSL 证书”，可能会导致无法加载。

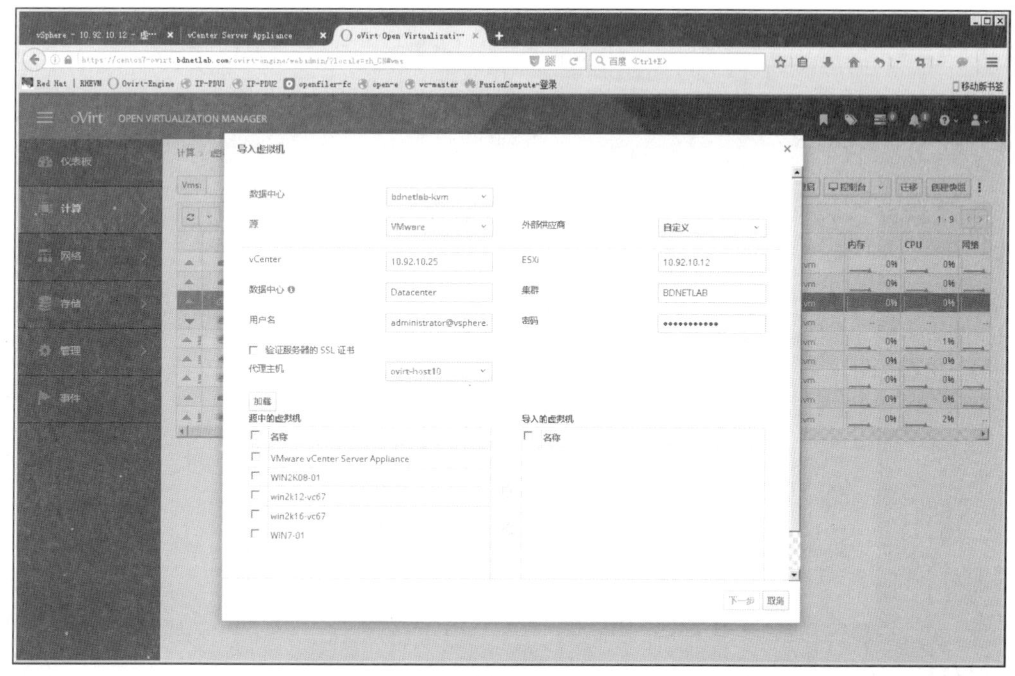

图 4-9-6

第 7 步，选择需要转换的虚拟机，如图 4-9-7 所示，单击“下一步”按钮。

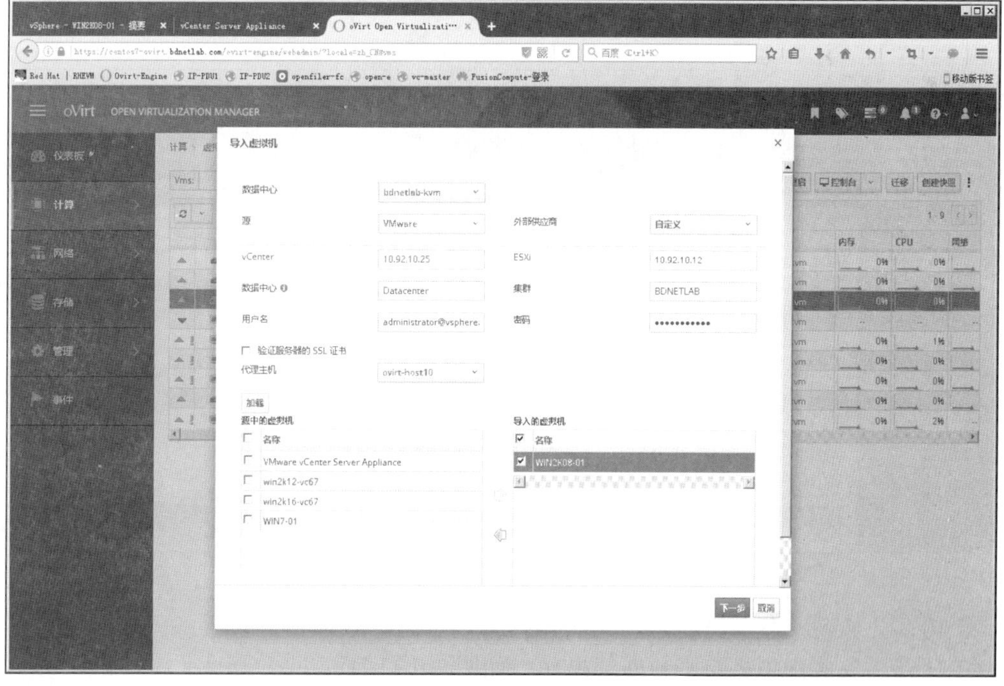

图 4-9-7

第 8 步，进入“导入虚拟机”参数配置界面，如图 4-9-8 所示，可以根据实际情况进行微调。

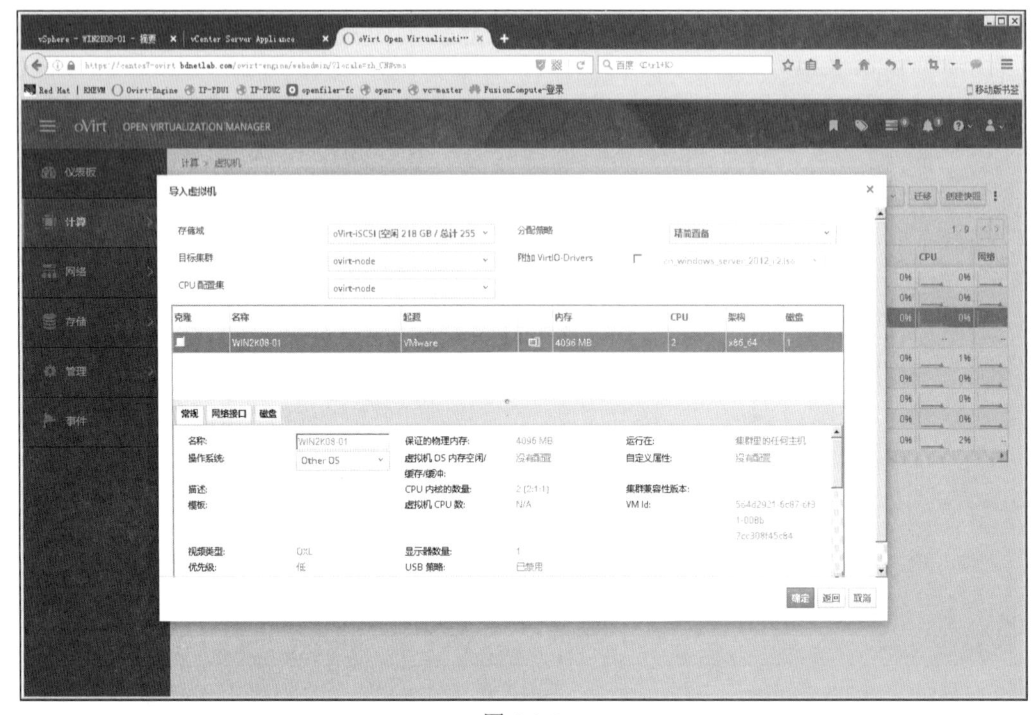

图 4-9-8

第 9 步，开始对虚拟机进行转换，如图 4-9-9 所示。注意：在 VMware vCenter Server 平台，该虚拟机需要处于关闭状态。

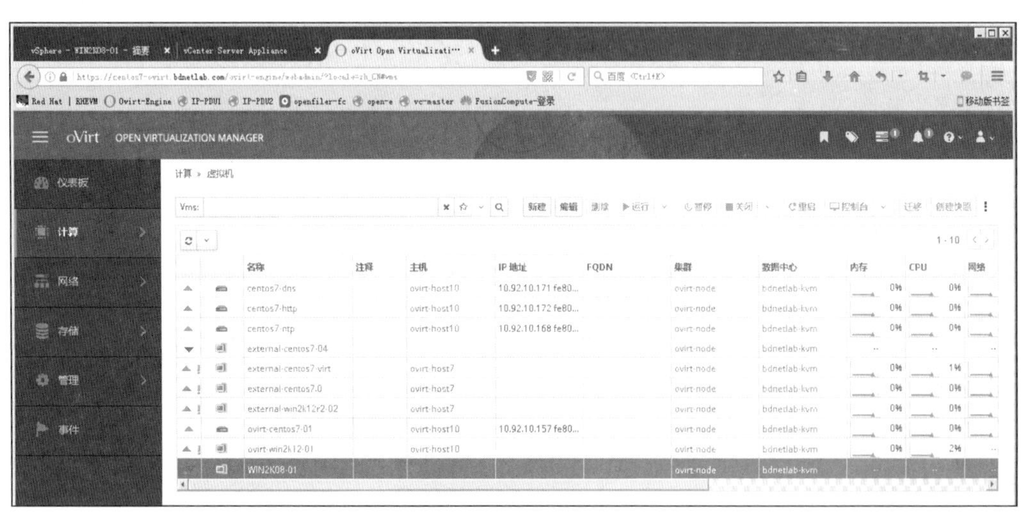

图 4-9-9

第 10 步，查看“任务”状态，如图 4-9-10 所示。

第 11 步，完成虚拟机跨平台转换操作，如图 4-9-11 所示。

第 12 步，在 oVirt Engine 管理端查看转换后的虚拟机信息，如图 4-9-12 所示。

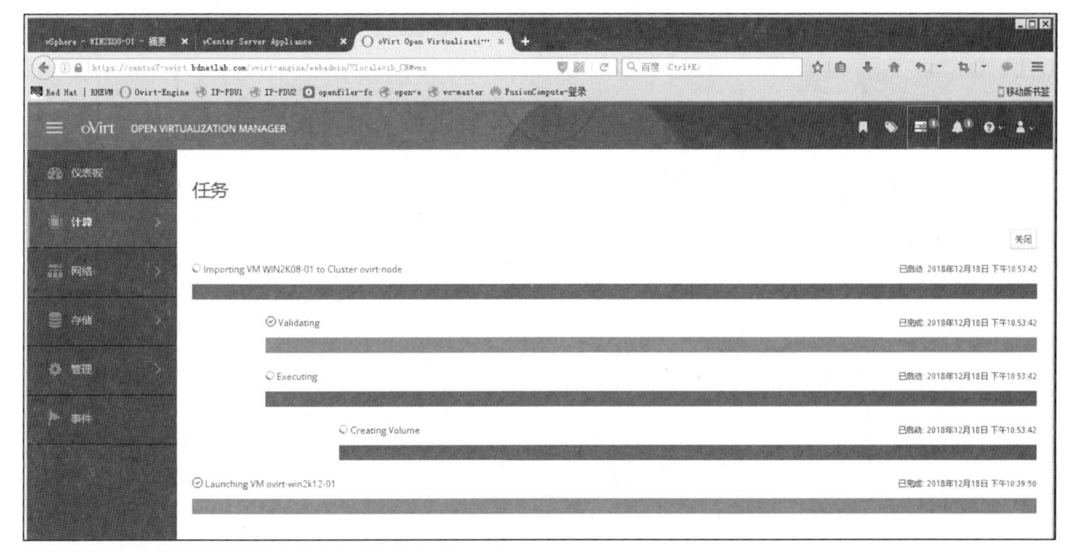

图 4-9-10

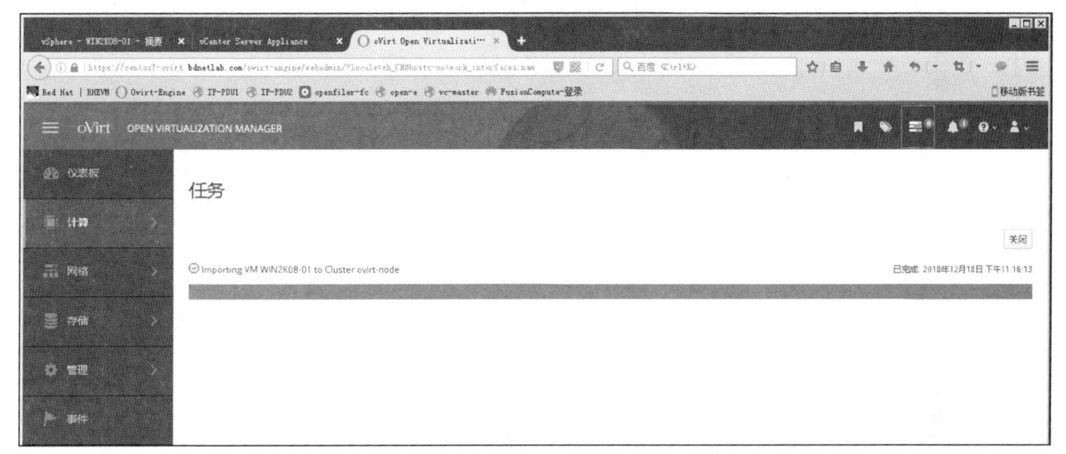

图 4-9-11

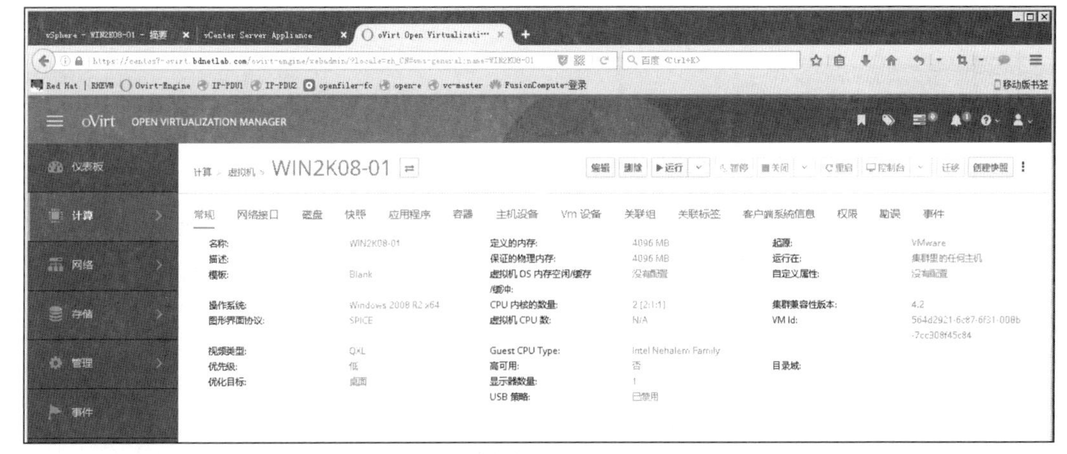

图 4-9-12

第 13 步，打开虚拟机电源，如图 4-9-13 所示。

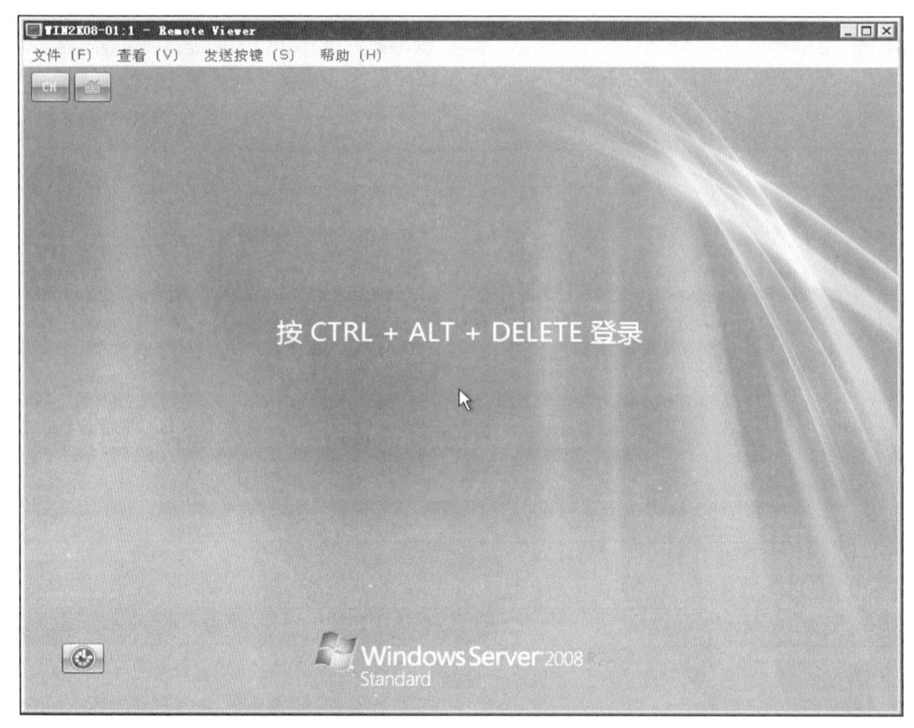

图 4-9-13

第 14 步，登录虚拟机查看相关信息，如图 4-9-14 所示，注意转换后的虚拟机 CPU 类型变为 Intel Core i7 系列。

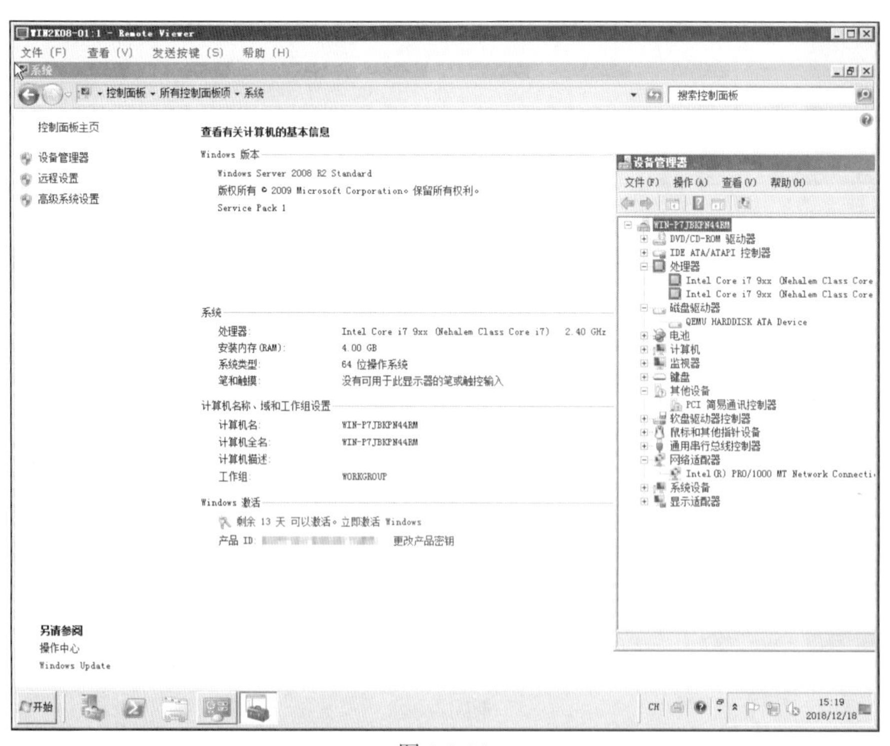

图 4-9-14

第 15 步，测试网络的连通性，网络连接正常，如图 4-9-15 所示。

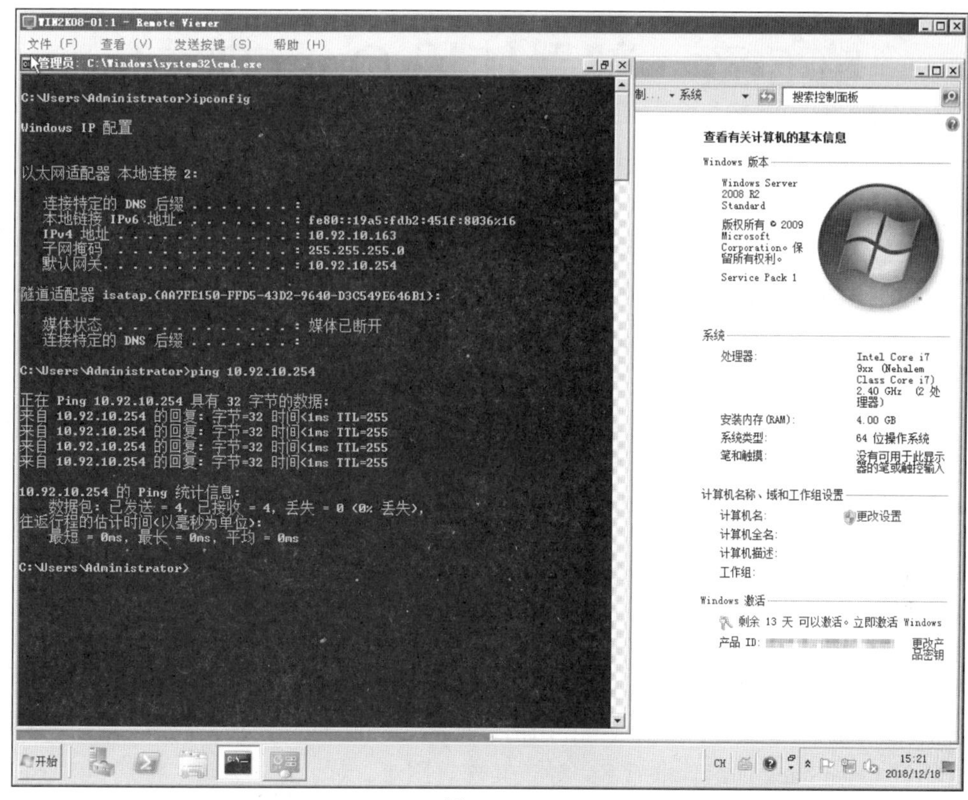

图 4-9-15

至此，将 VMware 虚拟机转换到 oVirt 平台操作完成。整体来说难度不算太大，跨平台转换虚拟机要注意时钟和 SSL 证书验证问题，可以在 oVirt Engine 管理端事件中查看问题的原因，解决后即可进行转换操作。另外，Windows 虚拟机建议安装相应的驱动程序以便更好地运行。

4.10 本章小结

本章介绍了企业级 oVirt 平台的使用，包括基本的部署、高可用以及跨平台迁移等。作为企业级的开源平台，oVirt 有大量的用户，如果单从成本上看，oVirt 是一个很好的选择，能够为企业节省授权的费用，但必须考虑后期的运维问题。由于 oVirt 是社区版本，不能提供官方的售后技术支持，所以完成 oVirt 构建后，后续对运维人员的能力要求非常高。

第 5 章　部署使用 OpenStack

前文已经介绍过让开源虚拟化架构运行起来。OpenStack 可以说是云计算的门槛，特别对于新手来说，其部署是相当困难的。部署使用 OpenStack 可以单独用一本书来进行详细介绍，本书的目的是尽量简化部署方式，让读者能够相对轻松地在企业环境中部署使用 OpenStack。本章介绍部署 OpenStack。

本章要点

- OpenStack 部署方式。
- 使用 DevStack 部署 OpenStack。
- 使用 RDO 部署 OpenStack。

5.1　OpenStack 部署方式简介

以 OpenStack 为代表的基础设施即服务（Infrastructure as a Service，IaaS）开源技术和以 Docker 为代表的平台即服务（Platform as a Service，PaaS）/通信即服务（Communication as a Service，CaaS）容器技术日益成熟。如何让二者强强联合，一直是业界颇为关心的焦点。前文介绍过 OpenStack 的组件非常多，导致其部署困难。对于企业来说，特别是对于刚接触 OpenStack 的运维人员人来说，OpenStack 的安装部署无疑是一大挑战，同时也直接提高了学习 OpenStack 云计算的技术门槛。

目前 OpenStack 常用的部署方式有以下 9 类。

5.1.1　DevStack 部署方式

开发使用 OpenStack，首推使用 DevStack，这是 OpenStack 项目为开发者运行 OpenStack 提供的单机安装程序。这种部署方式主要是通过配置参数，执行 Shell 脚本来安装一个 OpenStack 开发环境。在相当长一段时间内，DevStack 仍将是众多开发者常用的安装工具，支持 CentOS、Debian 等操作系统。

5.1.2　RDO 部署方式

RDO 是由 Red Hat 开源的一个自动化部署 OpenStack 的工具，支持单节点（all-in-one）和多节点（multi-node）部署。需要注意的是，RDO 只支持 RHEL、CentOS 系列操作系统，同时该项目并不属于 OpenStack 官方社区项目。

5.1.3　Puppet 部署方式

Puppet 是用 Ruby 语言编写的。Puppet 是早期进入 OpenStack 自动化部署的一个项目。目前，它的活跃开发群体有 Red Hat、Mirantis、UnitedStack 等。Mirantis 出品的 Fuel 部署工具，其大量的模块代码使用的便是 Puppet 的代码。

5.1.4 Ansible 部署方式

Ansible 是一个自动化部署配置管理工具，已被 Red Hat 收购。它基于 Python 开发，集合了众多运维工具（Puppet、Chef、SaltStack 等）的优点，实现了批量系统配置、批量程序部署以及批量运行命令等功能。

5.1.5 SaltStack 部署方式

SaltStack 也是一个开源的自动化部署工具，基于 Python 开发，实现了批量系统配置、批量程序部署以及批量运行命令等功能，和 Ansible 很相似。不同之处是，基于 SaltStack 的 Master 和 Minion 的认证机制及工作方式，需要在被控端安装 Minion 客户端。

5.1.6 TripleO 部署方式

TripleO 项目最早由 HP 于 2013 年 4 月在 Launchpad 上注册，用于完成 OpenStack 的安装与部署。TripleO 全称为“OpenStack On OpenStack”，意思为“云上云”，可以简单理解为利用 OpenStack 来部署 OpenStack，即首先基于 V2P（把虚拟机的镜像迁移到物理机上，和 P2V 相反）的理念事先准备好一些 OpenStack 节点（计算、存储以及控制节点）的镜像，然后利用已有 OpenStack 环境的 Ironic 裸机服务和软件安装部分的 diskimage-builder 部署裸机，最后通过 Heat 项目和镜像内的自动化部署工具（Puppet 或 Chef）在裸机上配置运行 OpenStack。和其他部署工具不同的是，TripleO 是利用 OpenStack 已有的基础设施来部署 OpenStack 的。

5.1.7 Fuel 部署方式

Fuel 是针对 OpenStack 的一个可以“界面部署”的工具，它大量采用了 Python、Ruby、Java 等语言。其功能涵盖了通过自动化 PXE 方式安装操作系统、DHCP 服务、Orchestration 编排服务和通过 Puppet 方式安装相关服务等，此外还有 OpenStack 的关键业务如健康检查和 log 实时查看等非常好用的功能。

5.1.8 Kolla 部署方式

Kolla 是具有广阔应用前景和市场的一个自动化部署工具。相比于其他部署工具，Kolla 完全革新地使用了 Docker 容器技术，将每一个 OpenStack 服务运行在不同的 Docker 容器中。

5.1.9 手动部署方式

按照 OpenStack 官方社区提供的文档，可以使用手动方式部署单节点、多节点以及 HA 节点环境。

综上所述，OpenStack 的安装部署方式多种多样。新手应该如何选择呢？这里，作者推荐使用 RDO 部署方式或手动部署方式（过程是艰难的，但能很好地加深新手对 OpenStack 的理解）；对于老手而言，可以尝试使用 Kolla 部署方式，体验 Docker 和 OpenStack 融合的新方式。

5.2 使用 RDO 部署 OpenStack

前文我们说过 OpenStack 的部署安装是一个难题，组件众多，如果手工部署 OpenStack，可能需要好几天甚至更长时间，最后的结果还可能是部署失败。本节介绍使用 RDO 部署单节点 OpenStack。

5.2.1 RDO 部署的前提条件

随着 OpenStack 的重要性的逐渐提高，OpenStack 越来越被大家认可。Red Hat 也推出了 OpenStack 的快捷安装部署项目，这个项目就是 RDO。RDO 项目的原理是整合上游的 OpenStack 版本，然后根据 Red Hat 操作系统做裁剪和定制，帮助用户进行选择，用户只需要简单几步就可以完成 OpenStack 的部署。使用 RDO，只需要几个命令，再加一两小时的等待即可。当然，RDO 的部署牺牲了灵活性，但是对最终用户来说，需要的是部署简单、使用稳定。RDO 部署虽然简单，但也有一些前提条件。

1. 服务器硬件

目前市场主流服务器都能够很好地支持虚拟化，更多的是内存和网络配置方面的需求，推荐使用 2 路或 4 路 CPU 的服务器，物理内存配置在 128GB 或以上。

2. 服务器安装 Linux 操作系统

由于 RDO 部署方式不是 OpenStack 项目发布的，目前只支持 RHEL、CentOS 系列操作系统，所以物理服务器需要安装 RHEL、CentOS 系列操作系统，不建议安装其他 Linux 操作系统。

3. 网络

采用 RDO 部署方式需要访问外部 YUM 源，所以需要确保物理服务器能够访问外网。同时，由于需要下载的组件较多，还需要确保网速的稳定，否则可能在部署过程中出现延时导致部署失败等问题。

5.2.2 部署单节点 OpenStack

准备好物理服务器后就可以开始部署 OpenStack，本节所述的物理服务器已安装好 CentOS 7 操作系统。

第 1 步，使用命令“systemctl stop firewalld”停止防火墙，以避免其与 Openstack 网络冲突。

```
[root@openstack ~]# systemctl stop firewalld
[root@openstack ~]# systemctl disable firewalld
Removed symlink /etc/systemd/system/multi-user.target.wants/firewalld.service.
Removed symlink /etc/systemd/system/dbus-org.fedoraproject.FirewallD1.service.
```

第 2 步，使用命令“cat /etc/selinux/config”查看 selinux 是否处于“disabled”状态，处于“enforcing”状态可能会对 Open vSwitch 的运行造成影响。

```
[root@openstack ~]# cat /etc/selinux/config
# This file controls the state of SELinux on the system.
# SELINUX= can take one of these three values:
#     enforcing - SELinux security policy is enforced.
#     permissive - SELinux prints warnings instead of enforcing.
#     disabled - No SELinux policy is loaded.
SELINUX= disabled
# SELINUXTYPE= can take one of three values:
#     targeted - Targeted processes are protected,
#     minimum - Modification of targeted policy. Only selected processes are protected.
#     mls - Multi Level Security protection.
SELINUXTYPE=targeted
```

第 3 步，使用命令“sudo yum update -y”更新 CentOS 操作系统。

```
[root@openstack ~]# sudo yum update -y
Loaded plugins: fastestmirror
Loading mirror speeds from cached hostfile
Resolving Dependencies
--> Running transaction check
---> Package NetworkManager.x86_64 1:1.18.0-5.el7 will be updated
……（省略）
  systemd-sysv.x86_64 0:219-67.el7_7.3
  tuned.noarch 0:2.11.0-5.el7_7.1
  tzdata.noarch 0:2019c-1.el7
  util-linux.x86_64 0:2.23.2-61.el7_7.1
Complete!
```

第 4 步，使用命令“sudo yum install -y https://rdoproject.org/repos/rdo-release.rpm”安装 OpenStack 的 YUM 源。

```
[root@openstack ~]# sudo yum install -y https://rdoproject.org/repos/rdo-release.rpm
Loaded plugins: fastestmirror
rdo-release.rpm
| 6.7 kB  00:00:00
Examining /var/tmp/yum-root-fLe2br/rdo-release.rpm: rdo-release-train-1.noarch
Marking /var/tmp/yum-root-fLe2br/rdo-release.rpm to be installed
Resolving Dependencies
……（省略）
Dependencies Resolved
================================================================================
 Package          Arch        Version      Repository              Size
================================================================================
Installing:
 rdo-release      noarch      train-1      /rdo-release            3.1 k
Transaction Summary
================================================================================
Install  1 Package
Total size: 3.1 k
Installed size: 3.1 k
Downloading packages:
Running transaction check
Running transaction test
Transaction test succeeded
Running transaction
  Installing : rdo-release-train-1.noarch                    1/1
  Verifying  : rdo-release-train-1.noarch                    1/1
Installed:
  rdo-release.noarch 0:train-1
Complete!
```

第 5 步，使用命令“sudo yum install -y openstack-packstack”安装 packstack 安装器。

```
[root@openstack ~]# sudo yum install -y openstack-packstack
Loaded plugins: fastestmirror
Loading mirror speeds from cached hostfile
 * openstack-train: mirrors.aliyun.com
 * rdo-qemu-ev: mirrors.aliyun.com
openstack-train                                    |  2.9 kB   00:00:00
rdo-qemu-ev                                        |  2.9 kB   00:00:00
(1/2): rdo-qemu-ev/x86_64/primary_db               |  73 kB    00:00:00
(2/2): openstack-train/x86_64/primary_db           |  962 kB   00:00:00
Resolving Dependencies
Complete!
```

第 6 步，使用命令“sudo packstack --allinone”部署 OpenStack。需要注意的是，RDO 部署的时间长短与访问外部网络的速度快慢有很大的关系，一般来说 1 小时左右能够完成部署。若访问外部网络速度较慢可能会出现错误提示，重复执行命令即可，一定要确保各个组件安装无错误提示，最后出现“Installation completed successfully”提示才能说明部署成功。

```
[root@openstack ~]# sudo packstack --allinone
Welcome to the Packstack setup utility
The installation log file is available at: /var/tmp/packstack/20200211-132300-Vilew7/openstack-setup.log
Packstack changed given value  to required value /root/.ssh/id_rsa.pub
Installing:
Clean Up                                             [ DONE ]
Discovering ip protocol version                      [ DONE ]
Setting up ssh keys                                  [ DONE ]
Preparing servers                                    [ DONE ]
Pre installing Puppet and discovering hosts' details [ DONE ]
Preparing pre-install entries                        [ DONE ]
Setting up CACERT                                    [ DONE ]
Preparing AMQP entries                               [ DONE ]
Preparing MariaDB entries                            [ DONE ]
Fixing Keystone LDAP config parameters to be undef if empty  [ DONE ]
Preparing Keystone entries                           [ DONE ]
Preparing Glance entries                             [ DONE ]
Checking if the Cinder server has a cinder-volumes vg [ DONE ]
Preparing Cinder entries                             [ DONE ]
Preparing Nova API entries                           [ DONE ]
Creating ssh keys for Nova migration                 [ DONE ]
Gathering ssh host keys for Nova migration           [ DONE ]
Preparing Nova Compute entries                       [ DONE ]
Preparing Nova Scheduler entries                     [ DONE ]
Preparing Nova VNC Proxy entries                     [ DONE ]
Preparing OpenStack Network-related Nova entries     [ DONE ]
Preparing Nova Common entries                        [ DONE ]
Preparing Neutron API entries                        [ DONE ]
Preparing Neutron L3 entries                         [ DONE ]
Preparing Neutron L2 Agent entries                   [ DONE ]
Preparing Neutron DHCP Agent entries                 [ DONE ]
Preparing Neutron Metering Agent entries             [ DONE ]
Checking if NetworkManager is enabled and running    [ DONE ]
Preparing OpenStack Client entries                   [ DONE ]
Preparing Horizon entries                            [ DONE ]
Preparing Swift builder entries                      [ DONE ]
Preparing Swift proxy entries                        [ DONE ]
Preparing Swift storage entries                      [ DONE ]
Preparing Gnocchi entries                            [ DONE ]
Preparing Redis entries                              [ DONE ]
Preparing Ceilometer entries                         [ DONE ]
Preparing Aodh entries                               [ DONE ]
Preparing Puppet manifests                           [ DONE ]
Copying Puppet modules and manifests                 [ DONE ]
Applying 10.92.10.124_controller.pp
10.92.10.124_controller.pp:                          [ DONE ]
Applying 10.92.10.124_network.pp
10.92.10.124_network.pp:                             [ DONE ]
Applying 10.92.10.124_compute.pp
10.92.10.124_compute.pp:                             [ DONE ]
Applying Puppet manifests                            [ DONE ]
Finalizing                                           [ DONE ]
 **** Installation completed successfully ******
```

```
    Additional information:
     * Parameter CONFIG_NEUTRON_L2_AGENT: You have chosen OVN Neutron backend. Note that this backend does not support the VPNaaS or FWaaS services. Geneve will be used as the encapsulation method for tenant networks
     * A new answerfile was created in: /root/packstack-answers-20200211-132301.txt
     * Time synchronization installation was skipped. Please note that unsynchronized time on server instances might be problem for some OpenStack components.
     * Warning: NetworkManager is active on 10.92.10.124. OpenStack networking currently does not work on systems that have the Network Manager service enabled.
     * File /root/keystonerc_admin has been created on OpenStack client host 10.92.10.124. To use the command line tools you need to source the file.
     * To access the OpenStack Dashboard browse to http://10.92.10.124/dashboard.
    Please, find your login credentials stored in the keystonerc_admin in your home directory.
     * Because of the kernel update the host 10.92.10.124 requires reboot.
     * The installation log file is available at: /var/tmp/packstack/20200211-132300-Vilew7/openstack-setup.log
     * The generated manifests are available at: /var/tmp/packstack/20200211-132300-Vilew7/manifests
```

第 7 步，使用命令“cat /root/keystonerc_admin”查看默认 admin 密码。

```
[root@openstack ~]# cat /root/keystonerc_admin
unset OS_SERVICE_TOKEN
    export OS_USERNAME=admin
    export OS_PASSWORD='9b0426c0e83c487e'
    export OS_REGION_NAME=RegionOne
    export OS_AUTH_URL=http://10.92.10.124:5000/v3
    export PS1='[\u@\h \W(keystone_admin)]\$ '

export OS_PROJECT_NAME=admin
export OS_USER_DOMAIN_NAME=Default
export OS_PROJECT_DOMAIN_NAME=Default
export OS_IDENTITY_API_VERSION=3
```

第 8 步，使用浏览器登录 OpenStack 仪表盘，如图 5-2-1 所示。

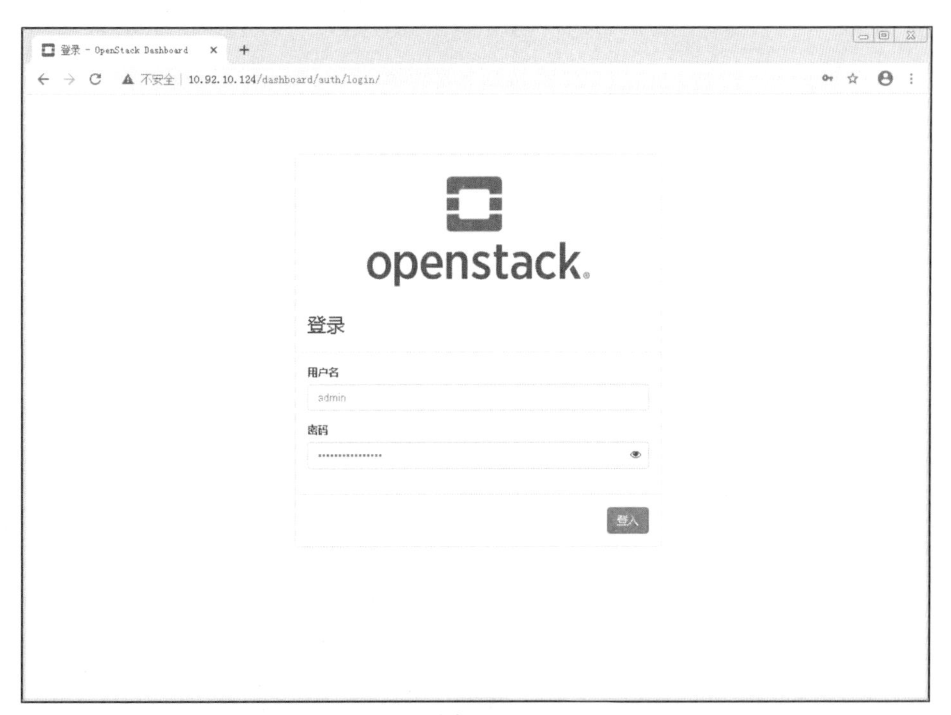

图 5-2-1

第 9 步，登录成功后的界面如图 5-2-2 所示。

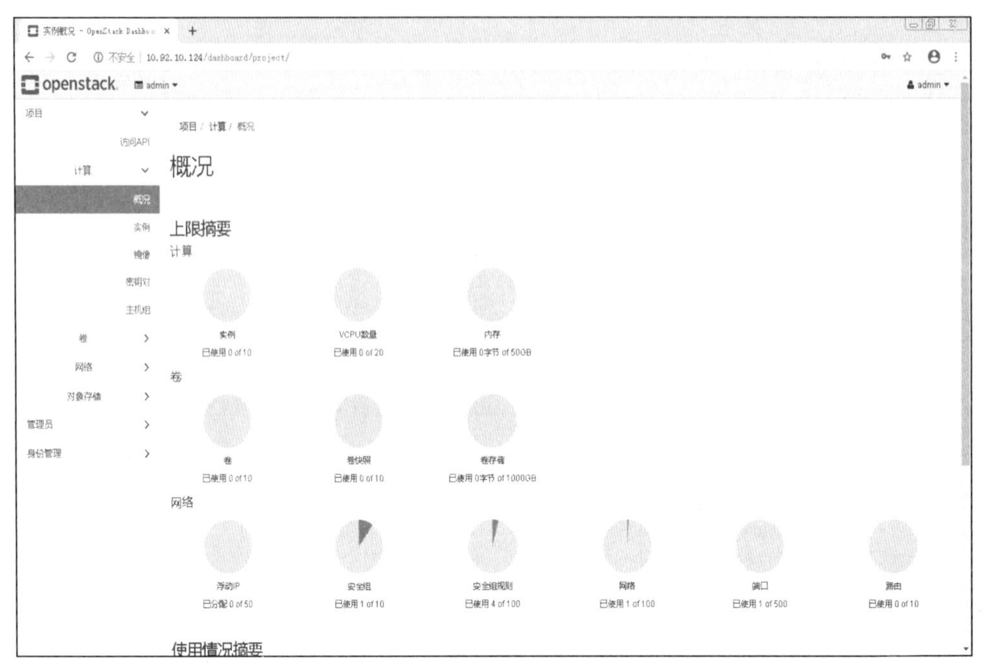

图 5-2-2

第 10 步，查看 API 服务相关信息，可以看到所有组件均部署在一台主机上，如图 5-2-3 所示。

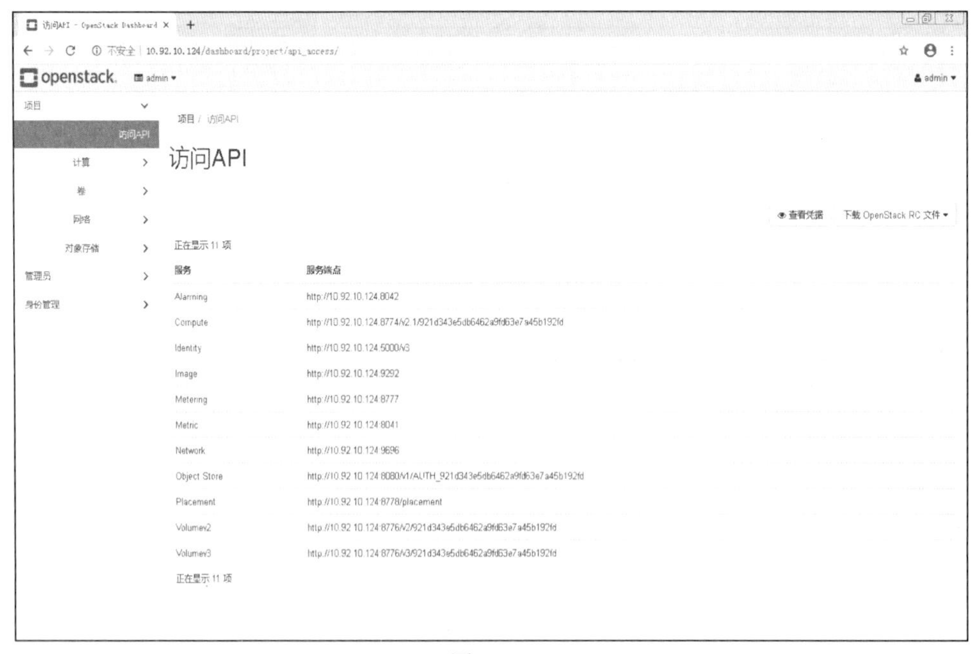

图 5-2-3

第 11 步，部署过程中生成的 admin 密码不易于记忆，推荐修改密码，如图 5-2-4 所示。

图 5-2-4

至此，使用 RDO 部署单节点 OpenStack 就完成了。自动化部署解决了很多问题，让企业可以快速部署使用 OpenStack。DevStack 不建议用于生产环境，RDO 部署的 OpenStack 是完全可以用于生产环境的，并且 RDO 也能够支持多节点部署。

虽然 RDO 部署非常简单，但在部署过程可能因为网速等因素导致部署失败。若部署失败可以结合日志分析失败的具体原因，重新进行部署。

5.2.3 部署多节点 OpenStack

前文介绍了使用 RDO 成功部署单节点 OpenStack，部署多节点 OpenStack 相对容易。部署多节点环境时，需要注意一些问题。

部署 OpenStack 使用的 packstack 命令理论上可以重复运行，而不需要清除之前的操作，但在生产环境建议全新部署。

在计算节点上可以提前使用命令“yum update”“yum makecache”来加速安装过程中对软件的部署。

使用命令“packstack --gen-answer-file”生成默认配置文件需要注意修改的地方，一般来说不需要配置的地方不要改动，目前 packstack 还没有提供一致性检查的保障，只会逐条进行解释执行。

第 1 步，使用命令“packstack --gen-answer-file”生成配置文件。

```
[root@RDO-Controller ~]#packstack --gen-answer-file=bdnetlab.txt
Additional information:
 * Parameter CONFIG_NEUTRON_L2_AGENT: You have chosen OVN Neutron backend. Note that this backend does
not support the VPNaaS or FWaaS services. Geneve will be used as the encapsulation method for tenant networks
```

第 2 步，编辑配置文件，调整 CONFIG_COMPUTE_HOSTS 参数，也可以根据实际情况调整其他节点参数。

```
[root@RDO-Controller ~]# vi bdnetlab.txt
[general]
```

```
# Path to a public key to install on servers. If a usable key has not
# been installed on the remote servers, the user is prompted for a
# password and this key is installed so the password will not be
# required again.
CONFIG_SSH_KEY=/root/.ssh/id_rsa.pub
……（省略）
# Default password to be used everywhere (overridden by passwords set
# for individual services or users).
CONFIG_DEFAULT_PASSWORD=
# Server on which to install OpenStack services specific to the
# controller role (for example, API servers or dashboard).
CONFIG_CONTROLLER_HOST=10.92.10.105
# List the servers on which to install the Compute service.
CONFIG_COMPUTE_HOSTS=10.92.10.128  #配置主机为 10.92.10.128
# List of servers on which to install the network service such as
# Compute networking (nova network) or OpenStack Networking (neutron).
CONFIG_NETWORK_HOSTS=10.92.10.105
……（省略）
```

第 3 步，使用命令“packstack --answer-file=bdnetlab.txt”部署 OpenStack。需要注意的是，部署的时间与网速有很大的关系，一般 1 小时左右能够完成部署。网速慢部署可能会出现错误提示，一般来说，重复执行命令即可，一定要确保各个组件安装无错误提示，最后出现“Installation completed successfully”提示才能说明部署成功。

```
[root@RDO-Controller ~]# packstack --answer-file=bdnetlab.txt
Welcome to the Packstack setup utility
The installation log file is available at: /var/tmp/packstack/20200302-163351-WOCrRA/openstack-setup.log
Installing:
Clean Up                                             [ DONE ]
Discovering ip protocol version                      [ DONE ]
root@10.92.10.105's password:  #输入主机口令
root@10.92.10.128's password:  #输入计算主机口令
Setting up ssh keys                                  [ DONE ]
Preparing servers                                    [ DONE ]
Preparing pre-install entries                        [ DONE ]
Setting up CACERT                                    [ DONE ]
Preparing AMQP entries                               [ DONE ]
Preparing MariaDB entries                            [ DONE ]
……（省略）
Preparing Gnocchi entries                            [ DONE ]
Preparing Redis entries                              [ DONE ]
Preparing Ceilometer entries                         [ DONE ]
Preparing Aodh entries                               [ DONE ]
Preparing Puppet manifests                           [ DONE ]
Copying Puppet modules and manifests                 [ DONE ]
Applying 10.92.10.105_controller.pp
10.92.10.105_controller.pp:                          [ DONE ]
Applying 10.92.10.105_network.pp
10.92.10.105_network.pp:                             [ DONE ]
Applying 10.92.10.128_compute.pp
10.92.10.128_compute.pp:                             [ DONE ]
Applying Puppet manifests                            [ DONE ]
Finalizing                                           [ DONE ]
 **** Installation completed successfully ******
Additional information:
  * Parameter CONFIG_NEUTRON_L2_AGENT: You have chosen OVN Neutron backend. Note that this backend does
not support the VPNaaS or FWaaS services. Geneve will be used as the encapsulation method for tenant networks
  * Time synchronization installation was skipped. Please note that unsynchronized time on server
instances might be problem for some OpenStack components.
```

```
    * Warning: NetworkManager is active on 10.92.10.105, 10.92.10.128. OpenStack networking currently
does not work on systems that have the Network Manager service enabled.
    * File /root/keystonerc_admin has been created on OpenStack client host 10.92.10.105. To use the command
line tools you need to source the file.
    * To access the OpenStack Dashboard browse to http://10.92.10.105/dashboard.
    Please, find your login credentials stored in the keystonerc_admin in your home directory.
    * The installation log file is available at: /var/tmp/packstack/20200302-163351-WOCrRA/openstack-
setup.log
    * The generated manifests are available at: /var/tmp/packstack/20200302-163351-WOCrRA/manifests
```

整体来说，使用 RDO 方式多节点部署 OpenStack 相对简单。一体化的部署方式解决了分离式部署的很多问题，也让运维人员能够快速部署使用 OpenStack。

5.3 OpenStack 的基础使用

完成 OpenStack 的部署后，后续的重点是如何使用。当然，OpenStack 的使用也是非常复杂的，本节我们介绍最基础的使用，需要复杂使用的读者可以参考其他有关的专业书籍。

5.3.1 OpenStack 基础配置

OpenStack 最基础的使用需要创建运行实例，其实质是在 OpenStack 上运行虚拟机，创建实例前还需要进行一些基础配置。

第 1 步，开始使用 OpenStack，推荐创建一个新的项目，或根据实际情况创建多个项目，如图 5-3-1 所示，输入项目相关信息，单击“创建项目”按钮。

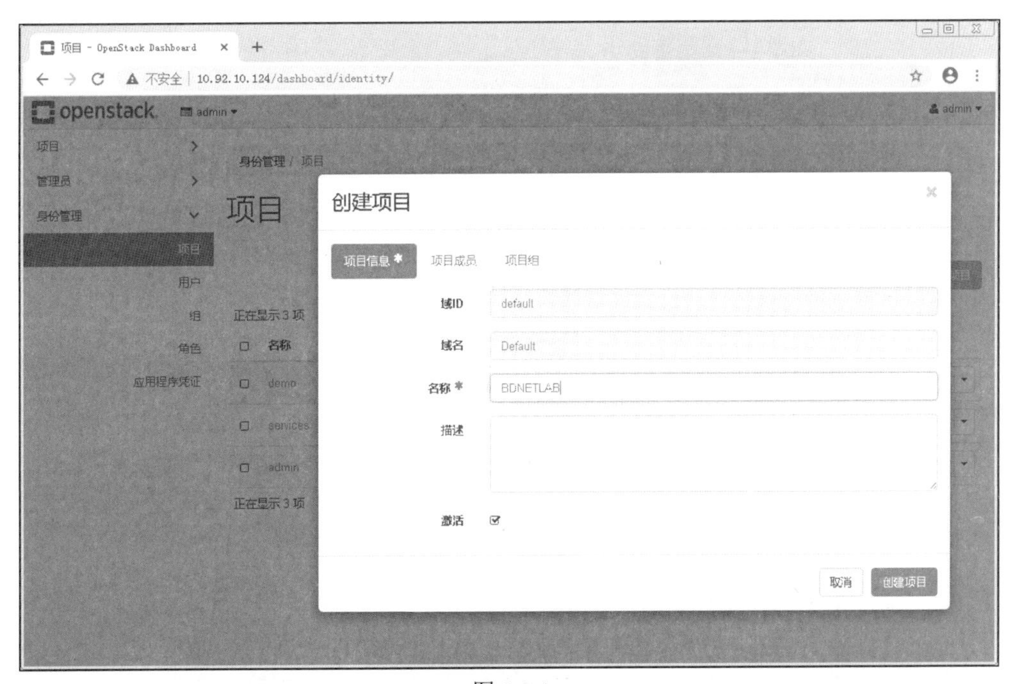

图 5-3-1

第 2 步，日常管理不能一直使用 admin 用户，根据不同的项目也需创建不同的用户。创建用户并输入相关信息，如图 5-3-2 所示，单击“创建用户”按钮。

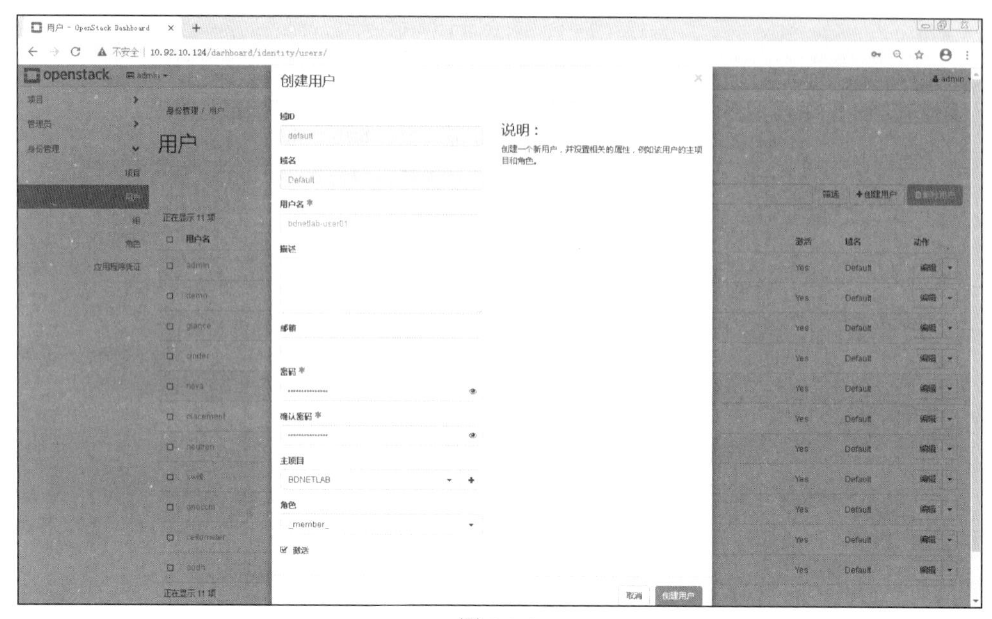

图 5-3-2

第 3 步，编辑刚创建的 BDNETLAB 项目，将新创建的用户添加为项目成员，如图 5-3-3 所示，单击“保存”按钮。

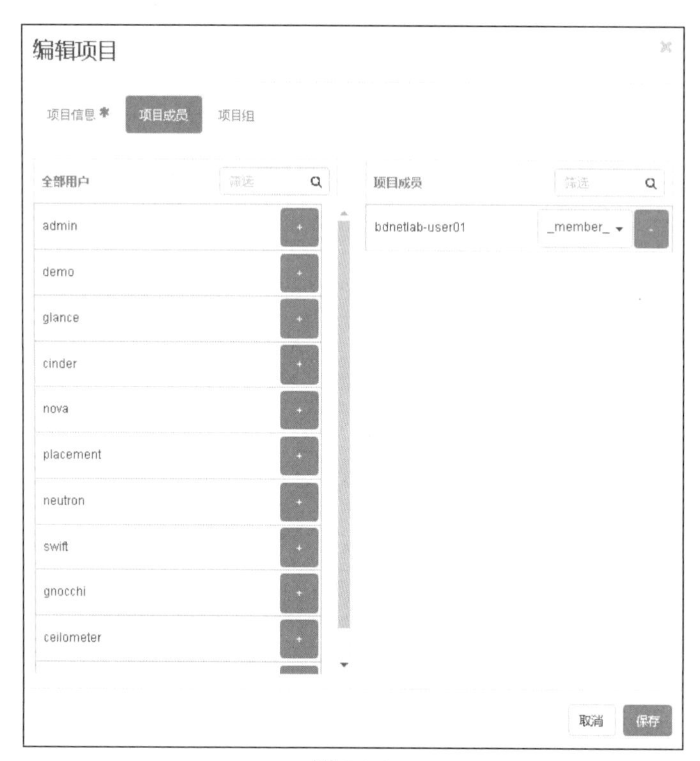

图 5-3-3

第 4 步，部署 OpenStack 的时候默认下载安装 cirros 镜像文件，如图 5-3-4 所示，该镜像非常小，可以用于基础的测试使用，单击“创建镜像”按钮创建新的镜像。

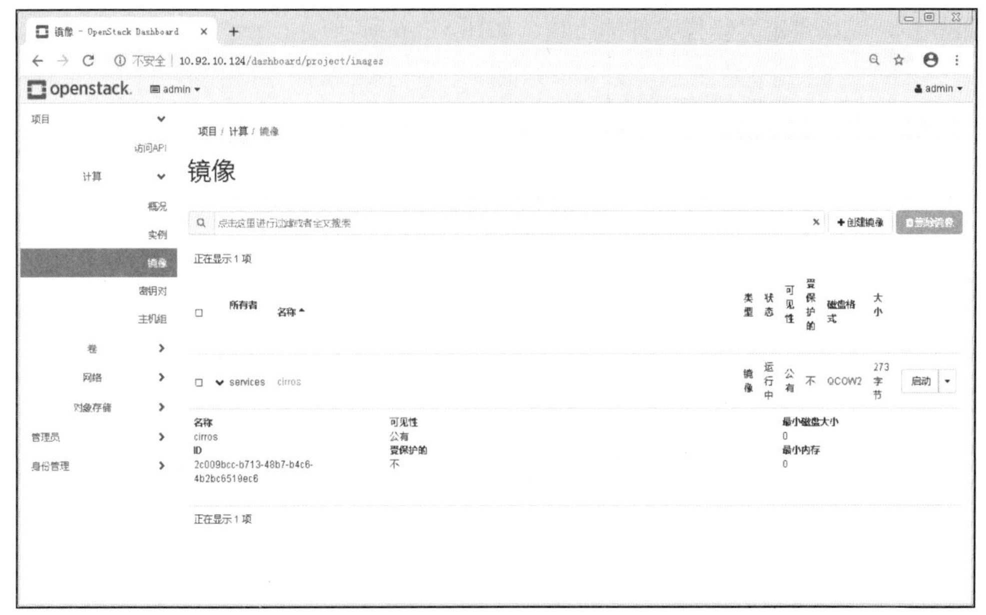

图 5-3-4

第 5 步，输入镜像名称，在“镜像源”的“文件”处选择下载好的文件，如图 5-3-5 所示，单击“创建镜像”按钮。

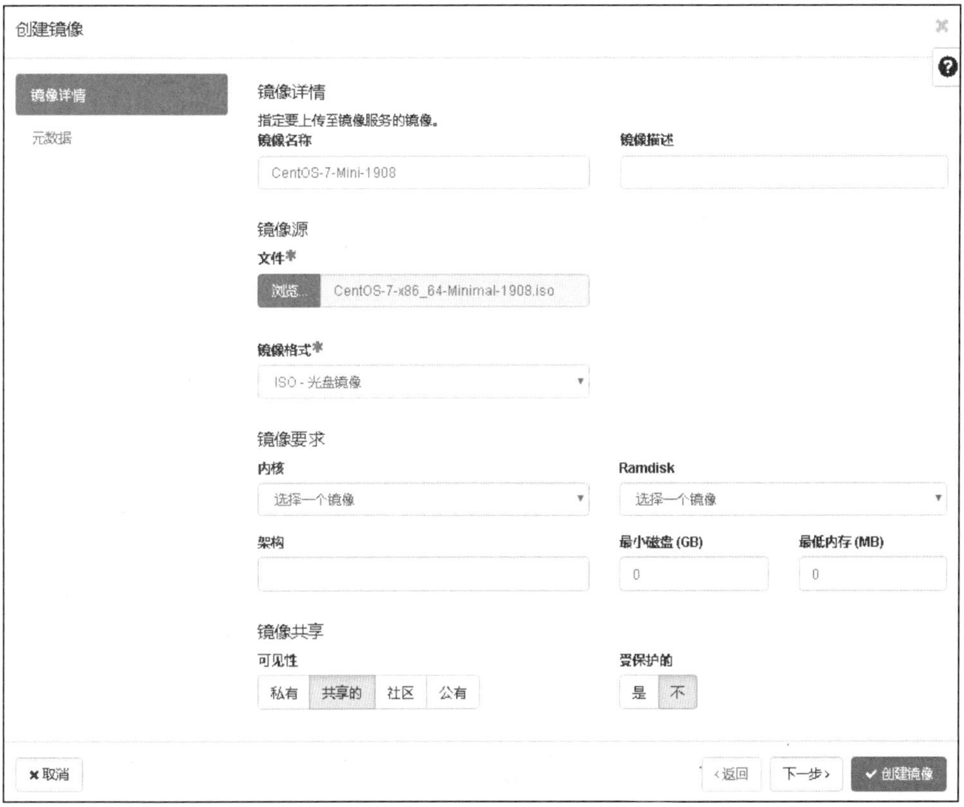

创建镜像
镜像详情
元数据
镜像详情
指定要上传至镜像服务的镜像。
镜像名称
CentOS-7-Mini-1908
镜像描述
镜像源
文件*
浏览... CentOS-7-x86_64-Minimal-1908.iso
镜像格式*
ISO - 光盘镜像
镜像要求
内核
选择一个镜像
Ramdisk
选择一个镜像
架构
最小磁盘 (GB)
0
最低内存 (MB)
0
镜像共享
可见性
私有 共享的 社区 公有
受保护的
是 不
× 取消
‹ 返回
下一步 ›
✔ 创建镜像

图 5-3-5

第 6 步，完成自定义镜像文件的上传，如图 5-3-6 所示。

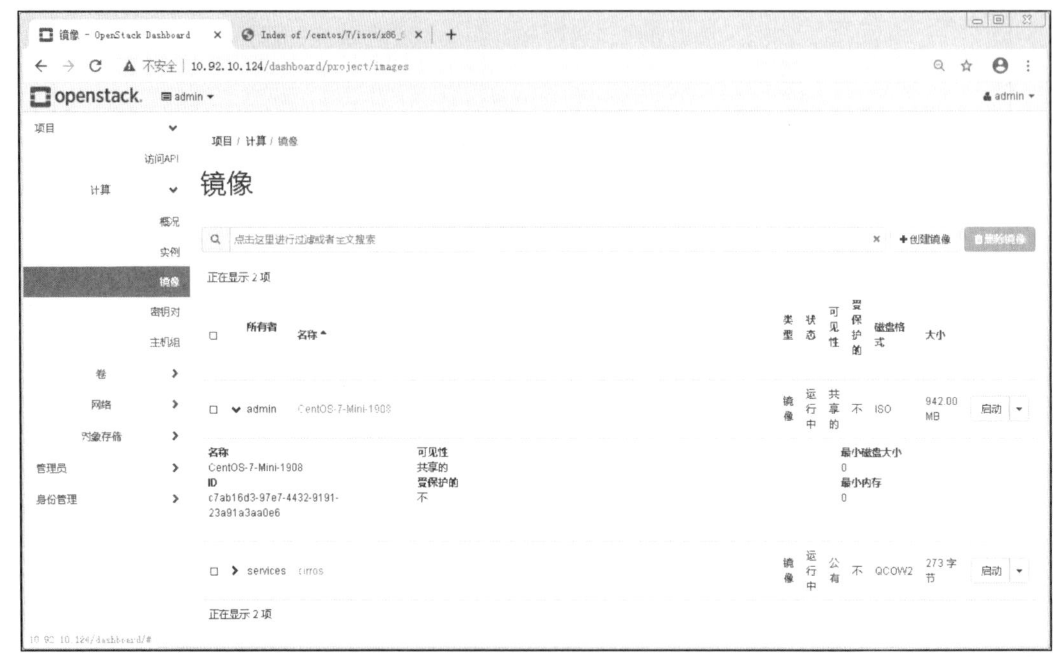

图 5-3-6

第 7 步，部署 OpenStack 的时候默认创建了网络，如图 5-3-7 所示，生产环境可以根据需要进行创建配置，单击“创建网络”按钮。

图 5-3-7

第 8 步，输入需要创建的网络信息，如图 5-3-8 所示，单击“下一步”按钮。

图 5-3-8

第 9 步，输入网络子网信息，如图 5-3-9 所示，单击“下一步”按钮。

创建网络

网络 * 子网 子网详情

子网名称

10.92.10.0

创建关联到这个网络的子网。您必须输入有效的"网络地址"和"网关IP"。如果您不输入"网关IP"，将默认使用该网络的第一个IP地址。如果您不想使用网关，请勾选"禁用网关"复选框。点击"子网详情"标签可进行高级配置。

网络地址

10.92.10.0/24

IP版本

IPv4

网关IP

10.92.10.254

禁用网关

取消 « 返回 下一步 »

图 5-3-9

第 10 步，配置子网其他信息，如图 5-3-10 所示，单击“创建”按钮。

第 11 步，部署 OpenStack 的时候默认创建了路由，如图 5-3-11 所示，单击“新建路由”按钮创建新的路由条目。

创建网络

网络 * 子网 子网详情

☑ 激活DHCP

为子网指定扩展属性

分配地址池

10.92.10.170,10.92.10.180

DNS服务器

10.92.10.51

主机路由

取消 « 返回 创建

图 5-3-10

图 5-3-11

第 12 步，输入新路由条目的名称等信息，如图 5-3-12 所示，单击“新建路由”按钮。

第 13 步，完成新路由条目配置，如图 5-3-13 所示，单击新路由名称进入接口配置界面。

第 14 步，进入新路由接口配置界面，如图 5-3-14 所示，单击“增加接口”按钮。

图 5-3-12

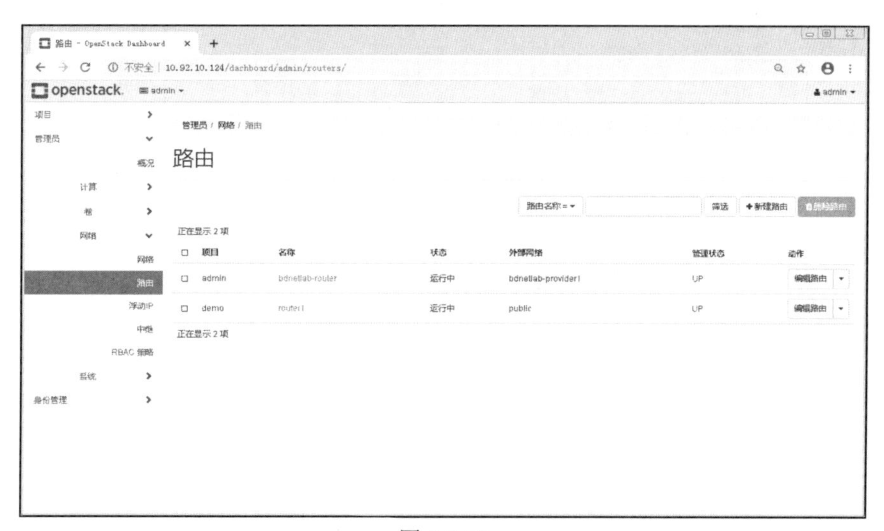

图 5-3-13

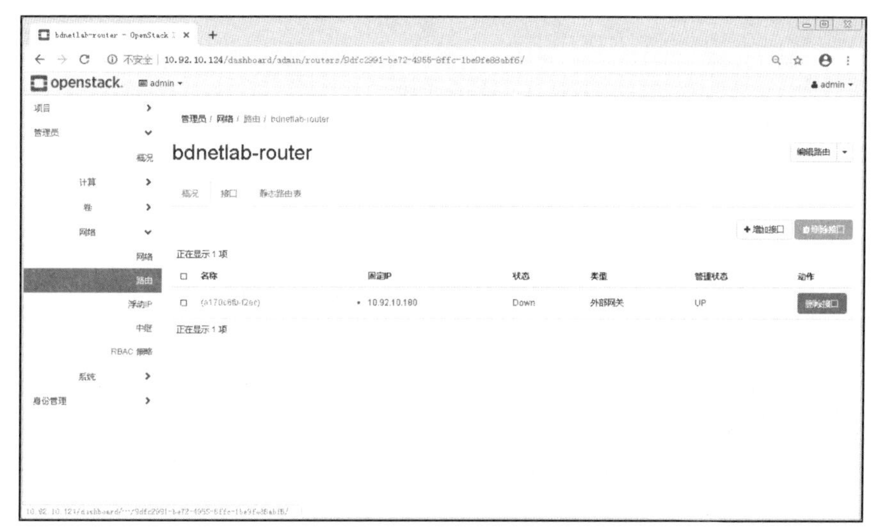

图 5-3-14

第 15 步，选择接口对应的子网，如图 5-3-15 所示，单击“提交”按钮。

图 5-3-15

第 16 步，完成路由接口配置，如图 5-3-16 所示。

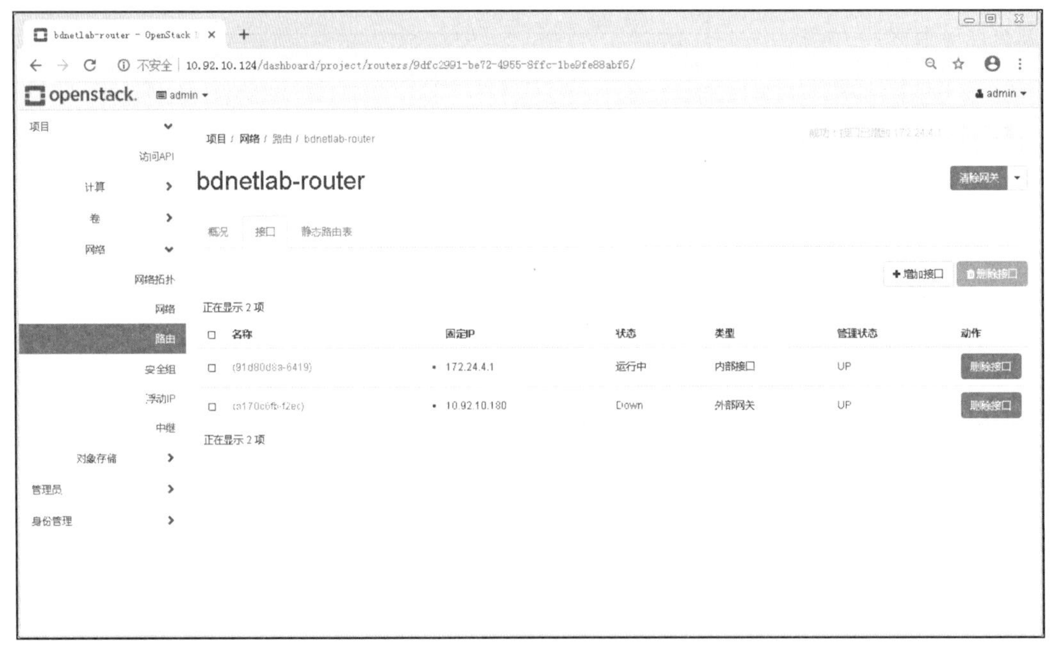

图 5-3-16

至此，基础的 OpenStack 配置完成，读者可以根据实际情况进行配置。复杂的镜像和网络配置不在本书的讨论范围，读者可以参考相关书籍。

5.3.2 创建基础 OpenStack 实例

完成 OpenStack 的基础配置后就可以创建运行实例，本节操作创建运行 CentOS 操作系统的实例。

第 1 步，打开 OpenStack 主界面中的实例界面，刚完成部署的 OpenStack 没有实例，如图 5-3-17 所示，单击“创建实例”按钮。

第 2 步，选择实例需要的源，选择自定义 CentOS-7-Mini-1908 镜像，如图 5-3-18 所示。

第 3 步，选择实例类型，由于是测试使用，选择 m1.small 类型即可，如图 5-3-19 所示。

图 5-3-17

图 5-3-18

第 4 步，选择实例需要的网络，如图 5-3-20 所示。

第 5 步，选择默认的安全组，如图 5-3-21 所示。

第 6 步，创建密钥对，如图 5-3-22 所示，其他非必须参数可以根据需求选择是否配置，单击“创建实例”按钮。

第 7 步，完成 CentOS-01 实例的创建，目前处于“运行中”状态，如图 5-3-23 所示。

第 8 步，为新创建的实例绑定浮动 IP，选择“绑定浮动 IP”，如图 5-3-24 所示。

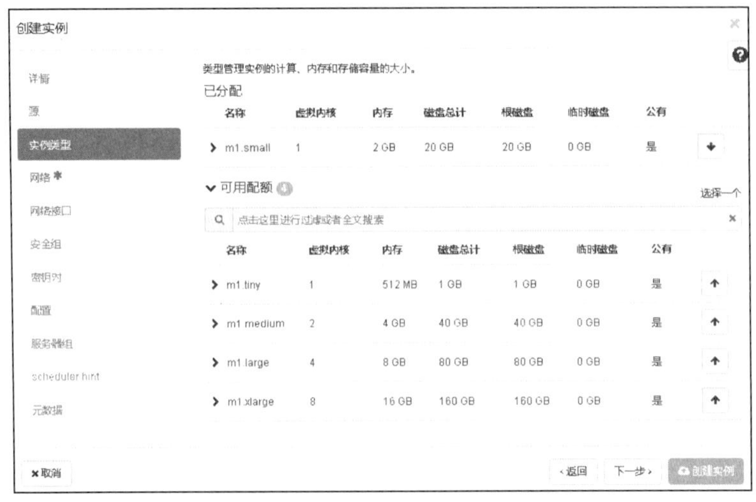

图 5-3-19

图 5-3-20

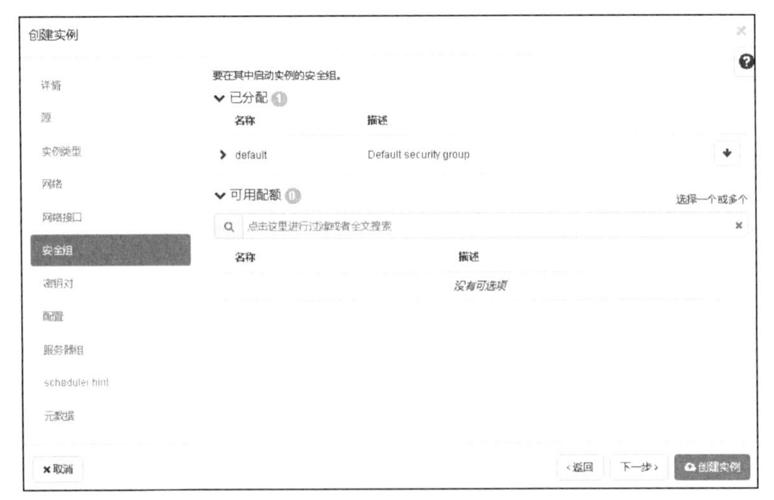

图 5-3-21

图 5-3-22

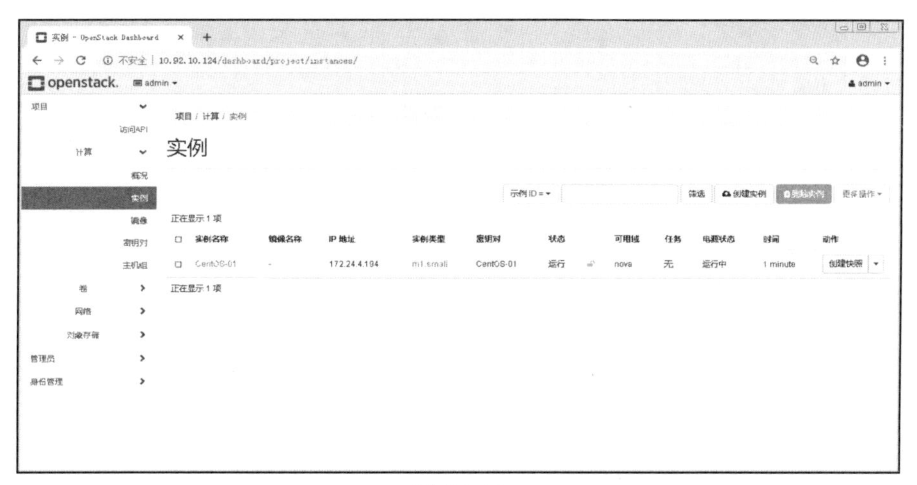

图 5-3-23

图 5-3-24

第 9 步，进入浮动 IP 配置界面，“待连接的端口”选择创建的实例，如图 5-3-25 所示，单击“关联”按钮。

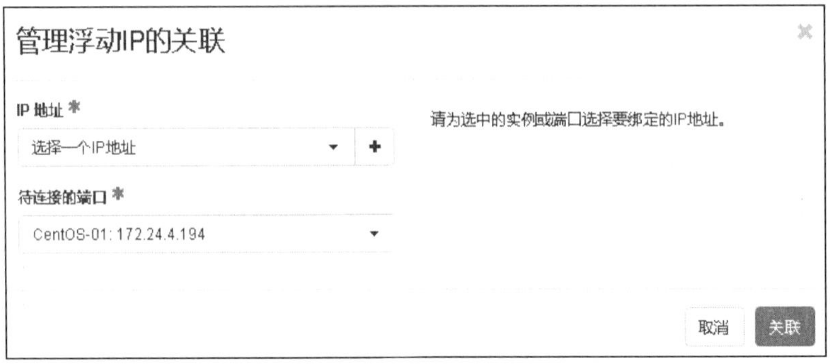

管理浮动IP的关联

IP 地址 *

选择一个IP地址

待连接的端口 *

CentOS-01: 172.24.4.194

请为选中的实例或端口选择要绑定的IP地址。

取消 关联

图 5-3-25

第 10 步，为实例分配浮动 IP，如图 5-3-26 所示，单击“分配 IP”按钮。

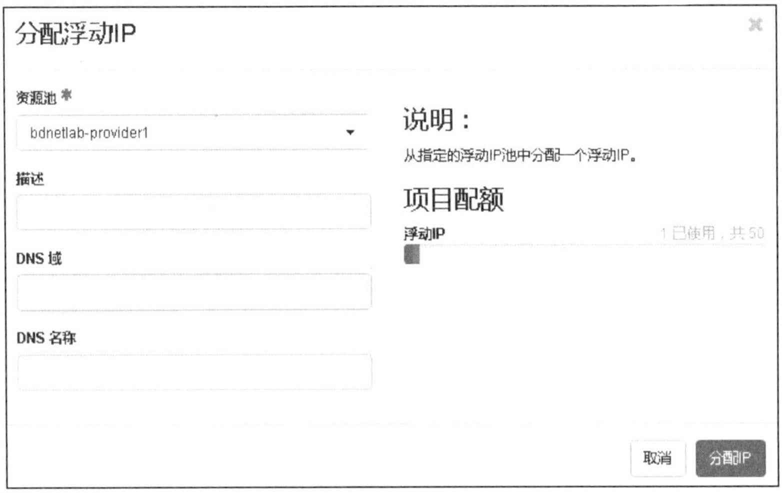

分配浮动IP

资源池 *

bdnetlab-provider1

描述

DNS 域

DNS 名称

说明：

从指定的浮动IP池中分配一个浮动IP。

项目配额

浮动IP 1 已使用，共 50

取消 分配IP

图 5-3-26

第 11 步，完成为实例关联浮动 IP，如图 5-3-27 所示，单击“关联”按钮。

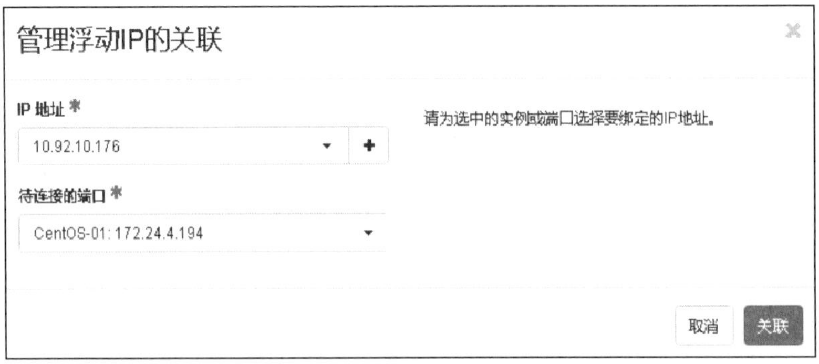

管理浮动IP的关联

IP 地址 *

10.92.10.176

待连接的端口 *

CentOS-01: 172.24.4.194

请为选中的实例或端口选择要绑定的IP地址。

取消 关联

图 5-3-27

第 12 步，查看实例相关信息，在“IP 地址”处可以看到分配的浮动 IP，如图 5-3-28 所示。

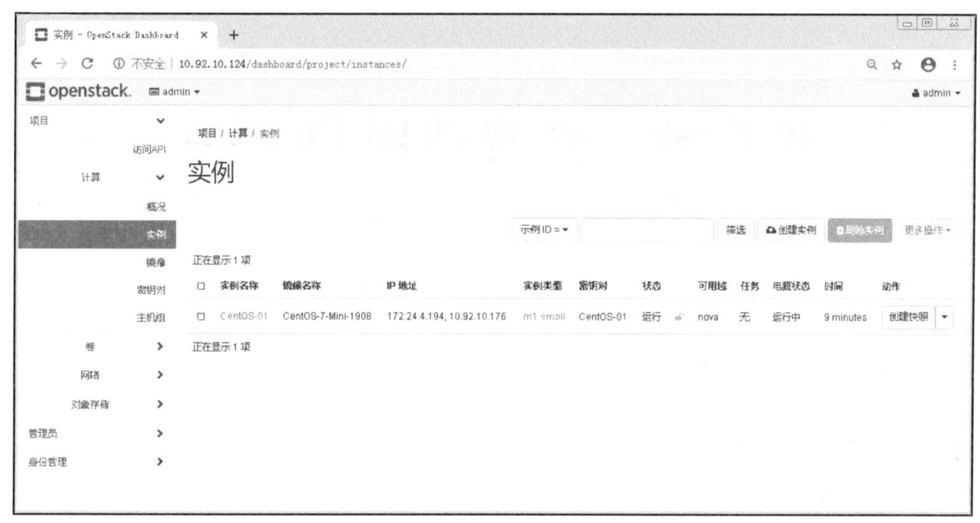

图 5-3-28

至此，使用 GUI 创建基本的实例完成。OpenStack 实例的创建本质是虚拟机的创建，生产环境可以根据实际情况进行细化配置。

5.4　本章小结

本章对 OpenStack 进行了基本的介绍。生产环境的多节点和大规模 OpenStack 应用是非常复杂的场景，从部署到运维，都需要一个团队，甚至会根据需求进行二次开发。作者遇到过非常多的运维人员想部署 OpenStack，但由于各种原因部署失败，失去了继续学习的兴趣。本章介绍了使用简易的 RDO 部署方式部署 OpenStack 环境，可以帮助读者轻松部署成功，感受 OpenStack 环境。同时，通过 RDO 部署方式还可以进行单节点和多节点的部署。在不少生产环境中，RDO 部署方式也常使用。

第 6 章　部署使用 Docker

最近几年，随着云计算的发展，以 Docker 为代表的容器技术也得到了广泛的应用。与传统虚拟机相比，容器小且易于运行，无论是开发人员还是生产环境中对容器的使用越来越多。本章介绍 Docker 容器的基本部署和使用。

本章要点

- Docker 与虚拟化。
- 部署 Docker。
- 使用 Docker。

6.1 Docker 与虚拟化

6.1.1 什么是 Docker

Docker 是 PaaS 提供商 dotCloud 开源的一个基于 Linux Container（LXC）的高级容器引擎，其源代码托管在 GitHub 上，基于 Go 语言并遵从 Apache2.0 协议开源。Docker 自 2013 年以来一直“火热”，无论是从 GitHub 上的代码活跃度来看，还是 Red Hat 在 RHEL 6.5 中集成对 Docker 的支持来看。就连 Google 的 Compute Engine 也支持 Docker 在其上运行。国内阿里云和腾讯云都提供对 Docker 的支持。

Docker 是一个开源的应用容器引擎，让开发者可以打包他们的应用和依赖包到一个可移植的容器中，然后发布到任何流行的 Linux 计算机或 Windows 计算机上，也可以实现虚拟化。容器完全使用沙箱机制，相互之间不会有任何接口。

Docker 的核心分别为主机、镜像、容器以及仓库，简单介绍如下。

1）Docker 主机（Docker Host）

Docker 主机，指安装了 Docker 程序的主机，用于运行 Docker 守护进程。

2）Docker 镜像（Docker Image）

Docker 镜像，指将软件环境打包好的模板，用来创建容器文件，一个镜像可以创建多个容器。如官方镜像包含了一套完整的 Ubuntu 最小系统。

3）Docker 容器（Docker Container）

Docker 容器，指运行镜像后生成的实例。运行一次镜像就会产生一个容器，容器可以启动、停止或删除。容器使用的是沙箱机制，相互隔离，是独立的、安全的。也可以把容器看作一个简易版的 Linux 环境，包括用户权限、文件系统以及运行的应用等。

4）Docker 仓库（Docker Repository）

Docker 仓库可以当作代码控制中心，用来保存镜像，仓库中包含许多镜像，每个镜像都有不同的标签。

6.1.2　Docker 与虚拟化

很多人会问，已经有了虚拟化技术，为什么还需要容器技术？实际上 Docker 容器技术与传统的虚拟技术存在区别，下面简单说明一下两者的特点。

1. 传统虚拟化技术

先虚拟出一套硬件，然后在其上安装操作系统，最后在操作系统上运行应用程序，相当于模拟一个完整的服务器操作系统。

2. Docker 容器技术

不虚拟硬件设备，也不模拟一个完整的操作系统，而是对进程进行隔离，将之封装成容器，容器内的应用程序直接使用宿主机的内核，且容器之间是相互隔离的，互不影响。

容器技术与传统虚拟化技术相比较，更轻便、效率更高、启动更快，特别适用于开发环境。

6.2　部署 Docker

Docker 可以让开发人员打包他们的应用和依赖包到一个轻量级、可移植的容器中，然后发布到任何流行的 Linux 计算机上，也可以实现虚拟化。Docker 的基本部署也相当简单，在部署使用 Docker 前，需要学习并掌握 Linux 的常用命令。

6.2.1　部署 Docker 前提条件

Docker 从 17.03 版本之后分为社区版（Community Edition，CE）和企业版（Enterprise Edition，EE），一般情况下使用社区版。

Docker 支持多种环境的部署，常见的 Ubuntu、CentOS、Debian 等 Linux 操作系统都可以部署使用 Docker，另外 MacOS 和 Windows 操作系统也能够部署使用 Docker，国内的阿里云和腾讯云等云平台也提供对 Docker 的支持。

整体来说，部署使用 Docker 环境要求很简单，生产环境部署 Docker 推荐使用 Linux 操作系统。

6.2.2　在 CentOS 上部署 Docker

CentOS 操作系统是常见的 Linux 发行版之一，在生产环境中大量被使用。本节实战操作使用 CentOS 操作系统部署 Docker。

第 1 步，使用命令“yum install docker”就可以部署 Docker，这是最简单的方式，后续步骤会介绍其他部署方式。

```
[root@docker ~]# yum install docker
Loaded plugins: fastestmirror
Loading mirror speeds from cached hostfile
Resolving Dependencies
--> Running transaction check
---> Package docker.x86_64 2:1.13.1-108.git4ef4b30.el7.centos will be installed
……（省略）
  subscription-manager-rhsm-certificates.x86_64 0:1.24.13-3.el7.centos
  yajl.x86_64 0:2.0.4-4.el7
```

```
Complete!
```

第 2 步，使用命令“systemctl start docker”启动，再使用命令“systemctl enable docker”设定自动运行。

```
[root@docker ~]# systemctl start docker
[root@docker ~]# systemctl enable docker
Created symlink from /etc/systemd/system/multi-user.target.wants/docker.service to /usr/lib/systemd/
system/docker.service.
```

第 3 步，使用命令“docker version”查看 Docker 版本。

```
[root@docker ~]# docker version
Client:
 Version:            1.13.1
 API version:        1.26
 Package version:    docker-1.13.1-108.git4ef4b30.el7.centos.x86_64
 Go version:         go1.10.3
 Git commit:         4ef4b30/1.13.1
 Built:              Tue Jan 21 17:16:25 2020
 OS/Arch:            linux/amd64
Server:
 Version:            1.13.1
 API version:        1.26 (minimum version 1.12)
 Package version:    docker-1.13.1-108.git4ef4b30.el7.centos.x86_64
 Go version:         go1.10.3
 Git commit:         4ef4b30/1.13.1
 Built:              Tue Jan 21 17:16:25 2020
 OS/Arch:            linux/amd64
 Experimental:       false
```

第 4 步，前面的操作是使用 CentOS 自带的 YUM 源部署的 Docker，也可以手动使用 Docker 仓库进行部署，使用命令“sudo yum install -y yum-utils device-mapper-persistent-data lvm2”安装所需的软件包。

```
[root@docker ~]# sudo yum install -y yum-utils device-mapper-persistent-data lvm2
Loaded plugins: fastestmirror
Loading mirror speeds from cached hostfile
Resolving Dependencies
--> Running transaction check
---> Package yum-utils.noarch 0:1.1.31-52.el7 will be installed
……（省略）
Dependencies Resolved
============================================================================
 Package        Arch          Version                Repository          Size
============================================================================
Installing:
 yum-utils      noarch        1.1.31-52.el7          base                121 k
……（省略）
Transaction Summary
============================================================================
Install  1 Package (+3 Dependent packages)

Total download size: 862 k
Installed size: 4.3 M
Downloading packages:
Downloading packages:
(1/4): libxml2-python-2.9.1-6.el7_2.3.x86_64.rpm      | 247 kB  00:00:00
(2/4): python-chardet-2.2.1-3.el7.noarch.rpm          | 227 kB  00:00:00
(3/4): python-kitchen-1.1.1-5.el7.noarch.rpm          | 267 kB  00:00:00
(4/4): yum-utils-1.1.31-52.el7.noarch.rpm             | 121 kB  00:00:00
```

```
--------------------------------------------------------------------------------------------------
Total                              1.1 MB/s | 862 kB  00:00:00
Running transaction check
Running transaction test
Transaction test succeeded
Running transaction
  Installing : python-chardet-2.2.1-3.el7.noarch          1/4
……(省略)
  Verifying  : python-chardet-2.2.1-3.el7.noarch          4/4
Installed:
  yum-utils.noarch 0:1.1.31-52.el7
Dependency Installed:
  libxml2-python.x86_64 0:2.9.1-6.el7_2.3    python-chardet.noarch 0:2.2.1-3.el7
  python-kitchen.noarch 0:1.1.1-5.el7
Complete!
```

第 5 步，使用命令“sudo yum-config-manager --add-repo https://download.docker.com/linux/centos/docker-ce.repo”设置仓库。

```
[root@docker ~]# sudo yum-config-manager --add-repo https://download.docker.com/linux/centos/docker-ce.repo
Loaded plugins: fastestmirror
adding repo from: https://download.docker.com/linux/centos/docker-ce.repo
grabbing file https://download.docker.com/linux/centos/docker-ce.repo to /etc/yum.repos.d/docker-ce.repo
repo saved to /etc/yum.repos.d/docker-ce.repo
```

第 6 步，使用命令“sudo yum install docker-ce docker-ce-cli containerd.io”安装 Docker 的 CE 版本。

```
[root@docker ~]# sudo yum install docker-ce docker-ce-cli containerd.io
Loaded plugins: fastestmirror
Loading mirror speeds from cached hostfile
 * openstack-train: mirrors.aliyun.com
 * rdo-qemu-ev: mirrors.cn99.com
docker-ce-stable                                    |   3.5 kB    00:00:00
(1/2): docker-ce-stable/x86_64/updateinfo           |   55 B      00:00:00
(2/2): docker-ce-stable/x86_64/primary_db           |   38 kB     00:00:03
Resolving Dependencies
--> Running transaction check
……(省略)
Complete!
```

第 7 步，使用命令“yum list docker-ce --showduplicates | sort -r”查看存储库中可用的版本。

```
[root@docker ~]# yum list docker-ce --showduplicates | sort -r
Loading mirror speeds from cached hostfile
Loaded plugins: fastestmirror
docker-ce.x86_64          3:19.03.6-3.el7                docker-ce-stable
docker-ce.x86_64          3:19.03.5-3.el7                docker-ce-stable
docker-ce.x86_64          3:19.03.4-3.el7                docker-ce-stable
docker-ce.x86_64          3:19.03.3-3.el7                docker-ce-stable
……(省略)
docker-ce.x86_64          17.03.2.ce-1.el7.centos        docker-ce-stable
docker-ce.x86_64          17.03.1.ce-1.el7.centos        docker-ce-stable
docker-ce.x86_64          17.03.0.ce-1.el7.centos        docker-ce-stable
Available Packages
```

第 8 步，使用命令“systemctl start docker”查看当前安装的 Docker 版本信息。

```
[root@docker ~]# systemctl start docker
[root@docker ~]# systemctl enable docker
Created symlink from /etc/systemd/system/multi-user.target.wants/docker.service to /usr/lib/systemd/
system/docker.service.
[root@docker ~]# docker version
Client: Docker Engine - Community
 Version:           19.03.8
 API version:       1.40
 Go version:        go1.12.17
 Git commit:        afacb8b
 Built:             Wed Mar 11 01:27:04 2020
 OS/Arch:           linux/amd64
 Experimental:      false
Server: Docker Engine - Community
 Engine:
  Version:          19.03.8
  API version:      1.40 (minimum version 1.12)
  Go version:       go1.12.17
  Git commit:       afacb8b
  Built:            Wed Mar 11 01:25:42 2020
  OS/Arch:          linux/amd64
  Experimental:     false
 containerd:
  Version:          1.2.13
  GitCommit:        7ad184331fa3e55e52b890ea95e65ba581ae3429
 runc:
  Version:          1.0.0-rc10
  GitCommit:        dc9208a3303feef5b3839f4323d9beb36df0a9dd
 docker-init:
  Version:          0.18.0
  GitCommit:        fec3683
```

第 9 步，使用命令“sudo docker run hello-world”验证是否正确安装 Docker。

```
[root@docker ~]# sudo docker run hello-world
Unable to find image 'hello-world:latest' locally
Trying to pull repository docker.io/library/hello-world ...

latest: Pulling from docker.io/library/hello-world
1b930d010525: Pull complete
Digest: sha256:fc6a51919cfeb2e6763f62b6d9e8815acbf7cd2e476ea353743570610737b752
Status: Downloaded newer image for docker.io/hello-world:latest

Hello from Docker!
This message shows that your installation appears to be working correctly.

To generate this message, Docker took the following steps:
 1. The Docker client contacted the Docker daemon.
 2. The Docker daemon pulled the "hello-world" image from the Docker Hub.
    (amd64)
 3. The Docker daemon created a new container from that image which runs the
    executable that produces the output you are currently reading.
 4. The Docker daemon streamed that output to the Docker client, which sent it
    to your terminal.

To try something more ambitious, you can run an Ubuntu container with:
 $ docker run -it ubuntu bash

Share images, automate workflows, and more with a free Docker ID:
 https://hub.docker.com/
```

```
For more examples and ideas, visit:
 https://docs.docker.com/get-started/
```

至此，在 CentOS 操作系统上部署 Docker 完成，非常简单。下文会继续介绍 Docker 的基本操作和如何在 Docker 上创建实例。

6.3 使用 Docker

部署完 Docker 后即可使用 Docker 创建实例，但在创建实例前还需要进行一些基本的操作。本节介绍 Docker 的基本使用和使用 Docker 创建实例，复杂环境下的 Dockr 应用如集群不在本书的讨论范围，请参考其他相关书籍。

6.3.1 Docker 基本使用

部署完 Docker 后，其基本的使用比较简单，可以直接输入命令执行 Docker 的操作。本节介绍 Docker 的基本使用。

1. Docker 获取镜像

第 1 步，刚部署完 Docker，本地主机是没有镜像的，使用命令“docker images ls”查看本地主机镜像情况，hello-world 镜像是部署测试时创建的。

```
[root@docker ~]# docker images ls
REPOSITORY             TAG          IMAGE ID        CREATED          SIZE
docker.io/hello-world  latest       fce289e99eb9    14 months ago    1.84 kB
```

参数解释。

1）REPOSITORY：表示镜像的仓库源。

2）TAG：镜像的标签。

3）IMAGE ID：镜像 ID。

4）CREATED：镜像在多久前创建。

5）SIZE：镜像大小。

第 2 步，使用命令“docker search mysql”查找镜像。

```
[root@docker ~]# docker search mysql
INDEX      NAME                DESCRIPTION        STARS      OFFICIAL     AUTOMATED
docker.io  docker.io/mysql     MySQL is a widely used, open-source relati...   9182 [OK]
docker.io  docker.io/mariadb   MariaDB is a community-developed fork of M...  3269 [OK]
……（省略）
```

第 3 步，使用命令“docker pull centos:centos7”下载需要的镜像。

```
[root@docker ~]# docker pull centos:centos7
Trying to pull repository docker.io/library/centos ...
centos7: Pulling from docker.io/library/centos
……（省略）
[root@docker ~]# docker images
REPOSITORY             TAG        IMAGE ID        CREATED          SIZE
docker.io/centos       centos7    5e35e350aded    3 months ago     203 MB
docker.io/hello-world  latest     fce289e99eb9   14 months ago     1.84 kB
```

第 4 步，由于 Docker 默认镜像仓库位于国外，下载速度比较慢，所以建议将镜像服务器修改为国内服务器，使用命令“vi /etc/docker/daemon.json”增加国内镜像仓库。

```
[root@docker ~]# vi /etc/docker/daemon.json
{
  "registry-mirrors": [ "https://registry.docker-cn.com"]
}
[root@docker ~]# systemctl restart docker  #重启服务
```

第 5 步，由于使用的需要，可能会下载多个镜像，运维人员应该定期清理不使用的镜像文件，使用命令“docker rmi”和“docker image rm”删除下载的镜像。

```
[root@docker ~]# docker images
REPOSITORY              TAG         IMAGE ID         CREATED          SIZE
docker.io/mysql         latest      7a3923452254     6 days ago       465 MB
docker.io/nginx         latest      a1523e859360     11 days ago      127 MB
docker.io/python        3.5         0320ef7199ca     11 days ago      909 MB
docker.io/centos        latest      470671670cac     7 weeks ago      237 MB
docker.io/busybox       latest      6d5fcfe5ff17     2 months ago     1.22 MB
docker.io/centos        centos7     5e35e350aded     3 months ago     203 MB
docker.io/hello-world   latest      fce289e99eb9     14 months ago    1.84 kB
[root@docker ~]# docker rmi busybox   #使用 docker rmi 命令删除
Untagged: busybox:latest
Untagged:
docker.io/busybox@sha256:6915be4043561d64e0ab0f8f098dc2ac48e077fe23f488ac24b665166898115a
Deleted: sha256:6d5fcfe5ff170471fcc3c8b47631d6d71202a1fd44cf3c147e50c8de21cf0648
Deleted: sha256:195be5f8be1df6709dafbba7ce48f2eee785ab7775b88e0c115d8205407265c5
[root@docker ~]# docker images
REPOSITORY              TAG         IMAGE ID           CREATED           SIZE
docker.io/mysql         latest      7a3923452254       6 days ago        465 MB
docker.io/nginx         latest      a1523e859360       11 days ago       127 MB
docker.io/python        3.5         0320ef7199ca       11 days ago       909 MB
docker.io/centos        latest      470671670cac       7 weeks ago       237 MB
docker.io/centos        centos7     5e35e350aded       3 months ago      203 MB
docker.io/hello-world   latest      fce289e99eb9       14 months ago     1.84 kB
[root@docker ~]# docker image rm centos:latest  #使用 docker image rm 命令删除
Untagged: centos:latest
Untagged:
docker.io/centos@sha256:fe8d824220415eed5477b63addf40fb06c3b049404242b31982106ac204f6700
Deleted: sha256:470671670cac686c7cf0081e0b37da2e9f4f768ddc5f6a26102ccd1c6954c1ee
Deleted: sha256:0683de2821778aa9546bf3d3e6944df779daba1582631b7ea3517bb36f9e4007
[root@docker ~]# docker images
REPOSITORY              TAG         IMAGE ID         CREATED          SIZE
docker.io/mysql         latest      7a3923452254     6 days ago       465 MB
docker.io/nginx         latest      a1523e859360     11 days ago      127 MB
docker.io/python        3.5         0320ef7199ca     11 days ago      909 MB
docker.io/centos        centos7     5e35e350aded     3 months ago     203 MB
docker.io/hello-world   latest      fce289e99eb9     14 months ago    1.84 kB
```

镜像是使用 Docker 容器的基础，下载需要的镜像非常重要，请根据生产环境的需求下载相应的镜像，也可以根据实际需求自定义镜像。

2. 使用 Docker 容器

可以根据需要下载各种镜像，镜像下载后就可以通过命令使用镜像启动容器。

第 1 步，使用命令“docker run -itd --name centos-test centos:centos7”运行容器。

```
[root@docker ~]# docker run -itd --name centos-test centos:centos7
c0fca1cc594581724387b5a634700a2541c38640caf24254d59dc3667cfc7b0c
[root@docker ~]# docker ps  #查看容器运行的信息
CONTAINER ID   IMAGE            COMMAND       CREATED          STATUS   PORTS      NAMES
c0fca1cc5945   centos:centos7   "/bin/bash"   9 seconds ago    Up 8 seconds        centos-test
```

第 2 步，使用命令“docker exec -it centos-test /bin/bash”进入容器执行常用的 Linux

命令。

```
[root@docker ~]# docker exec -it centos-test /bin/bash
[root@c0fca1cc5945 /]# ls  #在容器中执行 ls 命令
anaconda-post.log  bin  dev  etc  home  lib  lib64  media  mnt  opt  proc  root  run  sbin  srv  sys
tmp  usr  var
[root@c0fca1cc5945 /]# pwd  #在容器中执行 pwd 命令
/
[root@c0fca1cc5945 /]# uname -a  #在容器中执行 uname -a 命令
Linux c0fca1cc5945 3.10.0-1062.el7.x86_64 #1 SMP Wed Aug 7 18:08:02 UTC 2019 x86_64 x86_64 x86_64
GNU/Linux
[root@c0fca1cc5945 /]# ping www.baidu.com #网络连接测试
PING www.a.shifen.com (14.215.177.39) 56(84) bytes of data.
64 bytes from 14.215.177.39 (14.215.177.39): icmp_seq=1 ttl=54 time=34.8 ms
64 bytes from 14.215.177.39 (14.215.177.39): icmp_seq=2 ttl=54 time=34.0 ms
64 bytes from 14.215.177.39 (14.215.177.39): icmp_seq=3 ttl=54 time=34.7 ms
```

第 3 步，使用命令“docker stop c0fca1cc5945”停止容器。

```
[root@docker ~]# docker stop c0fca1cc5945
c0fca1cc5945
[root@docker ~]# docker ps
CONTAINER ID  IMAGE  COMMAND    CREATED    STATUS  PORTS   NAMES
[root@docker ~]#
```

第 4 步，使用命令“docker start c0fca1cc5945”启动已停止的容器。

```
[root@docker ~]# docker start  c0fca1cc5945
c0fca1cc5945
[root@docker ~]# docker ps
CONTAINER ID  IMAGE           COMMAND      CREATED         STATUS      PORTS     NAMES
c0fca1cc5945  centos:centos7  "/bin/bash"  11 minutes ago  Up 2 seconds          centos-test
```

第 5 步，使用命令“docker rm -f c0fca1cc5945”删除创建的容器。

```
[root@docker ~]# docker rm -f c0fca1cc5945
c0fca1cc5945
[root@docker ~]# docker ps
CONTAINER ID    IMAGE    COMMAND  CREATED    STATUS     PORTS     NAMES
```

通过以上几条命令可以对容器进行最基本的操作，当然不是只有这几条命令，读者可以通过命令“docker command –help”更深入地了解 Docker 命令使用方法。推荐使用“docker exec”命令进入容器操作，因为使用此命令进入容器操作后，退出容器终端时不会导致容器的停止。如果使用“docker attach”命令进入容器操作，输入“exit”退出时会导致容器停止。

6.3.2 使用 Docker 安装 Nginx

Nginx 是一款轻量级的 Web 服务器、反向代理服务器及电子邮件（IMAP/POP3）代理服务器工具，在 BSD-like 协议下发行。其特点是占用内存少和并发能力强。Nginx 能够在 Linux 环境下部署，也可以通过容器方式运行。本节介绍如何使用 Docker 安装 Nginx。

第 1 步，使用命令“docker search nginx”搜索 Nginx 的版本。

```
[root@docker ~]# docker search nginx
INDEX      NAME                    DESCRIPTION                    STARS   OFFICIAL  AUTOMATED
docker.io  docker.io/nginx         Official build of Nginx.       12739     [OK]
docker.io  docker.io/bitnami/nginx Bitnami nginx Docker Image     77        [OK]
……（省略）
```

第 2 步，使用命令“docker pull nginx:latest”下载 Nginx 镜像。

```
[root@docker ~]# docker pull nginx:latest
Trying to pull repository docker.io/library/nginx ...
latest: Pulling from docker.io/library/nginx
Digest: sha256:380eb808e2a3b0dd954f92c1cae2f845e6558a15037efefcabc5b4e03d666d03
Status: Downloaded newer image for docker.io/nginx:latest
[root@docker ~]# docker images
REPOSITORY          TAG         IMAGE ID        CREATED         SIZE
docker.io/nginx     latest      a1523e859360    5 days ago      127 MB
docker.io/centos    latest      470671670cac    6 weeks ago     237 MB
docker.io/centos    centos7     5e35e350aded    3 months ago    203 MB
……（省略）
```

第3步，使用命令“docker run --name nginx-test -p 5555:80 -d nginx”运行容器。

```
[root@docker ~]# docker run --name nginx-test -p 5555:80 -d nginx
ed76c2fd1bbba2934215a5fabd7c7b8b4680ce33c9cc0881a33bff652e0005eb
```

参数解释。

1）--name nginx-test：容器名称。

2）-p 5555:80：通过端口进行映射，将本地5555端口映射到容器内部的80端口。

3）-d nginx：设置容器在后台一直运行。

第4步，查看容器运行情况。

```
[root@docker ~]# docker ps
CONTAINER ID  IMAGE  COMMAND   CREATED     STATUS      PORTS      NAMES
ed76c2fd1bbb nginx "nginx -g 'daemon ..."   12 seconds ago Up 10 seconds   0.0.0.0:5555->80/tcp
nginx-test
[root@docker ~]# ip addr  #Docker主机IP地址
1: lo: <LOOPBACK,UP,LOWER_UP> mtu 65536 qdisc noqueue state UNKNOWN group default qlen 1000
    link/loopback 00:00:00:00:00:00 brd 00:00:00:00:00:00
    inet 127.0.0.1/8 scope host lo
      valid_lft forever preferred_lft forever
    inet6 ::1/128 scope host
      valid_lft forever preferred_lft forever
2: ens192: <BROADCAST,MULTICAST,UP,LOWER_UP> mtu 1500 qdisc mq state UP group default qlen 1000
    link/ether 00:50:56:93:4c:63 brd ff:ff:ff:ff:ff:ff
    inet 10.92.10.105/24 brd 10.92.10.255 scope global noprefixroute dynamic ens192
      valid_lft 52980sec preferred_lft 52980sec
    inet6 fe80::6d60:70a:54a:109/64 scope link noprefixroute
      valid_lft forever preferred_lft forever
3: docker0: <BROADCAST,MULTICAST,UP,LOWER_UP> mtu 1500 qdisc noqueue state UP group default
    link/ether 02:42:a7:f7:53:d8 brd ff:ff:ff:ff:ff:ff
    inet 172.17.0.1/16 scope global docker0
      valid_lft forever preferred_lft forever
    inet6 fe80::42:a7ff:fef7:53d8/64 scope link
      valid_lft forever preferred_lft forever
13: vethb97e7bf@if12: <BROADCAST,MULTICAST,UP,LOWER_UP> mtu 1500 qdisc noqueue master docker0 state
UP group default
    link/ether 22:9f:d5:96:50:ad brd ff:ff:ff:ff:ff:ff link-netnsid 0
    inet6 fe80::209f:d5ff:fe96:50ad/64 scope link
      valid_lft forever preferred_lft forever
```

第5步，使用浏览器访问Nginx服务，运行正常，如图6-3-1所示。

第6步，进入容器查看Nginx目录，默认目录下有两个.html文件。

```
[root@docker ~]# docker exec -it nginx-test bash
root@ed76c2fd1bbb:/# cd /usr/share/nginx/html/
root@ed76c2fd1bbb:/usr/share/nginx/html# ls
50x.html  index.html
```

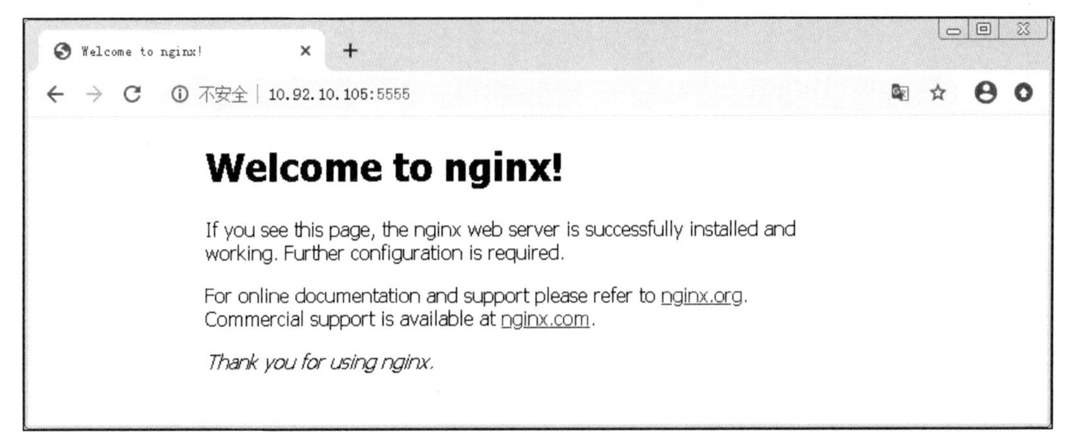

图 6-3-1

第 7 步，生产环境下经常会在宿主机上创建卷，关联到容器用于数据的保存，使用命令“docker volume create nginx-test02”为容器创建卷。

```
[root@docker ~]# docker volume ls  #查看卷
DRIVER              VOLUME NAME
[root@docker ~]# docker volume create nginx-test02  #创建卷
nginx-test02
[root@docker ~]# docker volume ls
DRIVER              VOLUME NAME
local               nginx-test02
```

第 8 步，使用命令“docker volume inspect nginx-test02”查看卷的详细信息。

```
[root@docker ~]# docker volume inspect nginx-test02
[
    {
        "CreatedAt": "2020-02-13T14:52:45+08:00",
        "Driver": "local",
        "Labels": {},
        "Mountpoint": "/var/lib/docker/volumes/nginx-test02/_data",
        "Name": "nginx-test02",
        "Options": {},
        "Scope": "local"
    }
]
```

第 9 步，重新创建容器关联到新创建的卷。

```
[root@docker ~]# docker run -itd --name nginx-test02 -p 5555:80 --mount src=nginx-test,
dst=/usr/share/nginx/html nginx
5c4fbc34a10fdf1a5a91524cd2828c06c128975ac45a9c3407cdd1ca60a77b33
```

第 10 步，使用命令“docker exec -it nginx-test02 bash”进入容器查看目录。

```
[root@docker ~]# docker exec -it nginx-test02 bash
root@5c4fbc34a10f:/# ls /usr/share/nginx/html/
50x.html  index.html
[root@docker _data]# cd  /var/lib/docker/volumes/nginx-test02/_data/
[root@docker _data]# ll
total 12
-rw-r--r--. 1 root root 494 Mar  3 22:32 50x.html
-rw-r--r--. 1 root root 612 Mar  3 22:32 index.html
```

第 11 步，使用命令“vi test.html”在宿主机上创建 test.html 文件，并查看该文件是否同步到容器。

```
[root@docker _data]# vi test.html
<h1>Welcome to Docker-Nginx!!!<h1>
[root@docker _data]# ll
total 12
-rw-r--r--. 1 root root 494 Mar  3 22:32 50x.html
-rw-r--r--. 1 root root 612 Mar  3 22:32 index.html
-rw-r--r--. 1 root root  36 Mar 13 15:01 test.html
[root@docker _data]# docker exec -it nginx-test02 bash
root@5c4fbc34a10f:/# ls /usr/share/nginx/html/
50x.html  index.html  test.html   #test.html 文件同步到容器
```

第 12 步，使用浏览器访问 test.html，运行正常，如图 6-3-2 所示。

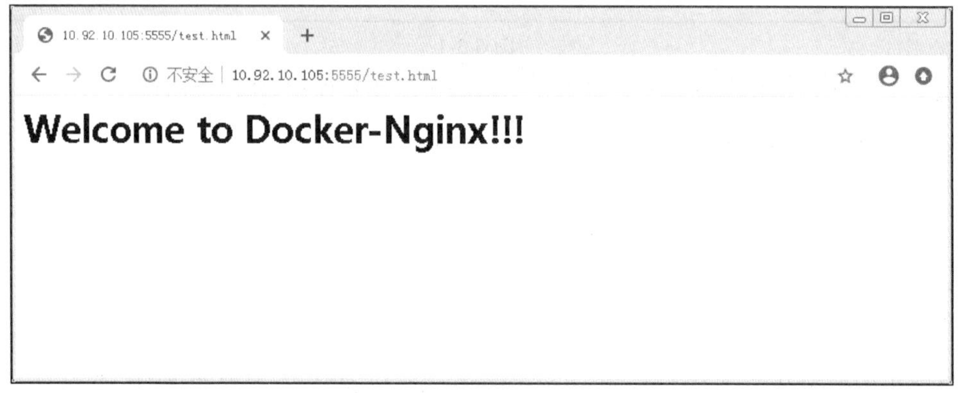

图 6-3-2

至此，使用 Docker 安装 Nginx 完成，读者可以将文件放置到相关的目录即可运行服务。通过容器方式进行安装比直接部署要节省很多时间，因为仓库中已经部署好相应的模板，基本上直接下载对应的镜像运行即可。这也是开发人员大量使用容器的原因之一。

6.3.3 使用 Docker 安装 MySQL

MySQL 是流行的关系型数据库管理系统之一，在 Web 应用方面，MySQL 是较好的应用软件之一。MySQL 是一种关系型数据库管理系统，关系数据库将数据保存在不同的表中，而不是将所有数据放在一个大仓库内，这样就提高了数据库的访问速度并增强了灵活性。

MySQL 所使用的结构化查询语言（Structured Query Language，SQL）是用于访问数据库的较常用标准化语言。MySQL 软件采用了双授权政策，分为社区版和商业版。由于其体积小、速度快、总体拥有成本低，尤其是开放源码等特点，一般中小型网站的开发都选择 MySQL 作为网站数据库。MySQL 能够在 Linux 环境下部署，也可以通过容器方式运行，如果结合其他容器可以形成一个整体架构。本节介绍如何使用 Docker 安装 MySQL。

第 1 步，使用命令“docker pull mysql:latest”下载最新 MySQL 镜像。

```
[root@docker ~]# docker pull mysql:latest
latest: Pulling from library/mysql
68ced04f60ab: Already exists
f9748e016a5c: Pull complete
……（省略）
Digest: sha256:4a30434ce03d2fa396d0414f075ad9ca9b0b578f14ea5685e24dcbf789450a2c
Status: Downloaded newer image for mysql:latest
```

```
docker.io/library/mysql:latest
[root@docker ~]# docker images
REPOSITORY          TAG         IMAGE ID          CREATED           SIZE
docker.io/nginx     latest      a1523e859360      10 days ago       127 MB
mysql               latest      9b51d9275906      10 days ago       547MB
centos              centos7     5e35e350aded      4 months ago      203MB
hello-world         latest      fce289e99eb9      14 months ago     1.84kB
```

第 2 步，使用命令“docker run -itd --name mysql-test -p 3306:3306 -e MYSQL_ROOT_PASSWORD=123456 mysql”运行 MySQL。

```
[root@Docker ~]# docker run -itd --name mysql-test -p 3306:3306 -e MYSQL_ROOT_PASSWORD=123456 mysql
b5674c1acab31a6558e19895babd415cdffb3d90ad94df6e3b0fac797076dd8a
[root@Docker ~]# docker ps
CONTAINER ID        IMAGE                      COMMAND                      CREATED                 STATUS
PORTS                                   NAMES
b5674c1acab3        mysql                  "docker-entrypoint.s…"   6 seconds ago       Up 5 seconds
0.0.0.0:3306->3306/tcp, 33060/tcp   mysql-test
5c4fbc34a10f        nginx                  "nginx -g 'daemon of…"   46 hours ago        Up 46 hours
0.0.0.0:5555->80/tcp                    nginx-test02
```

第 3 步，在容器中使用命令“mysql -h localhost -u root -p”访问 MySQL 数据库。

```
[root@Docker ~]# docker exec -it mysql-test bash  #进入容器命令行模式
root@b5674c1acab3:/#
root@b5674c1acab3:/# mysql -h localhost -u root -p  #访问 MySQL 数据库
Enter password:  #输入创建容器使用的密码
Welcome to the MySQL monitor.  Commands end with ; or \g.
Your MySQL connection id is 8
Server version: 8.0.19 MySQL Community Server - GPL
Copyright (c) 2000, 2020, Oracle and/or its affiliates. All rights reserved.
Oracle is a registered trademark of Oracle Corporation and/or its
affiliates. Other names may be trademarks of their respective
owners.
Type 'help;' or '\h' for help. Type '\c' to clear the current input statement.
mysql>
mysql> show databases;  #查看 MySQL 数据库
+--------------------------------+
| Database               |
+--------------------------------+
| information_schema     |
| mysql                  |
| performance_schema     |
| sys                    |
+--------------------------------+
4 rows in set (0.01 sec)
```

第 4 步，在宿主机上使用命令“mysql -h10.92.10.105 -uroot -p”登录 MySQL，注意宿主机需要安装 MySQL 连接组件。

```
[root@Docker ~]# mysql -h10.92.10.105 -uroot -p
Enter password:
Welcome to the MySQL monitor.  Commands end with ; or \g.
Your MySQL connection id is 8
Server version: 8.0.19 MySQL Community Server - GPL
Copyright (c) 2000, 2020, Oracle and/or its affiliates. All rights reserved.
Oracle is a registered trademark of Oracle Corporation and/or its
affiliates. Other names may be trademarks of their respective
owners.
Type 'help;' or '\h' for help. Type '\c' to clear the current input statement.
mysql>
```

至此，使用 Docker 创建 MySQL 完成，完成后可以运行 MySQL 相关命令创建数据库提供给相关应用使用。

6.4 本章小结

本章介绍了 Docker 容器的基本部署和使用。与传统的虚拟化相比较，容器更小、更精简，常用的应用通过下载镜像进行简单的配置即可使用，非常适合开发人员和中小生产环境使用。与介绍 OpenStack 类似，本章介绍的都是一些基础的部署、使用方法，让读者能够轻松部署、使用 Docker。

第 7 章　部署使用 Hadoop

Hadoop 是一个由 Apache 开发的分布式系统基础架构。用户可以使用 Hadoop 在不了解分布式底层细节的情况下，开发分布式程序。Hadoop 充分利用集群的威力进行高速运算和存储，实现了 Hadoop 分布式文件系统（Hadoop Distributed File System，HDFS）。HDFS 有高容错性的特点，被设计用来部署在低廉（low-cost）的硬件上；它提供高吞吐量（high throughput）来访问应用程序的数据，适合那些有着超大数据集（large data set）的应用程序。HDFS 放宽了可移植操作系统接口（Portable Operating System Interface，POSIX）的要求，可以以流的形式访问（streaming access）文件系统中的数据。Hadoop 的框架最核心的设计就是 HDFS 和 MapReduce。HDFS 为海量的数据提供了存储，而 MapReduce 则为海量的数据提供了计算能力。本章介绍 Hadoop 的概念和基本的操作，不涉及大数据方面的内容。

本章要点

- Hadoop 介绍。
- 部署使用 Hadoop。

7.1　Hadoop 简介

Hadoop 起源于 Apache Nutch 项目，始于 2002 年，是 Apache Lucene 的子项目之一。2004 年，Google 在操作系统设计与实现会议上公开发表了题为“简化大规模集群上的数据处理”的论文之后，受到启发的 Doug Cutting 等人开始尝试实现大规模集群上的数据计算框架（MapReduce），并将它与 Nutch 分布式文件系统（Nutch Distributed File System，NDFS）结合，用以支持 Nutch 引擎的主要算法。NDFS 和 MapReduce 在 Nutch 引擎中有着良好的应用，它们于 2006 年 2 月被分离出来，成为一套完整而独立的软件，并被命名为 Hadoop。到了 2008 年初，Hadoop 已成为 Apache 的顶级项目，包含众多子项目，被应用到很多互联网公司。

7.1.1　什么是 Hadoop

Hadoop 由许多元素构成，其最底部是 HDFS，用于存储 Hadoop 集群中所有存储节点上的文件；HDFS 的上一层是 MapReduce 引擎，该引擎由 JobTrackers 和 TaskTrackers 组成。分布式文件系统 HDFS、MapReduce 处理过程，以及数据仓库工具 Hive 和分布式数据库 Hbase，共同构成了 Hadoop 分布式平台的所有技术核心内容。

Hadoop 框架可在单一的 Linux 平台上使用（开发和调试时），官方提供 MiniCluster 作为单元测试使用，不过使用存放在机架上的商业服务器才能发挥它的力量。这些机架组成一个 Hadoop 集群，它通过集群拓扑知识决定如何在整个集群中分配作业和文件。Hadoop 假定节点可能失败，因此采用本机方法处理单个计算机甚至所有机架的失败。

7.1.2 Hadoop 和虚拟化的关系

Hadoop 最初的设计是运行在传统的 x86 物理服务器上的，使企业能够运行分析任务。但是使用物理服务器的问题在于比较僵化，不能根据实际情况进行调整和改变，并且许多大数据分析组件并没有内置高可用方案。而虚拟化的高可用特征很容易实现，Hadoop 虚拟化后可以显著提高 Hadoop 集群的管理效率。

7.2 部署使用 Hadoop

Hadoop 兼容性非常强，可以部署在多个平台中，常见的物理服务器、虚拟化平台以及云平台都能够部署使用 Hadoop。本节介绍在 Linux 环境下 Hadoop 的部署使用。

7.2.1 部署 Hadoop 的前提条件

Hadoop 有很多发行版本，3 大发行版本为 Apache、Hortonworks、Cloudera。用户可以根据实际需求选择使用。

1. Apache 版本

Apache 软件基金会开源版本，也是较原始较基础的版本，这个版本的 Hadoop 适用于入门学习。

2. Hortonworks 版本

属于完全开源版本，每一行代码都贡献回 Apache Hadoop，参考文档非常多。

3. Cloudera 版本

该版本包括 Apache Hadoop 核心功能，同时可以购买其他框架，如 Cloudera Manager 等，在大型互联网企业中用得较多。

同时，Hadoop 分为本地部署、伪分布式部署以及完全分布式部署等多种安装模式，用户可以根据需求进行选择。

1. 本地部署

适用于本地开发调试，或者快速安装体验 Hadoop。

2. 伪分布式部署

适用于学习 Hadoop。伪分布式部署是指在一台计算机的各个进程上运行 Hadoop 各个模块的模式。伪分布式的意思是：虽然各个模块是在各个进程上分开运行的，但是只是运行在一个操作系统上的，并不是真正的分布式。

3. 完全分布式部署

完全分布式部署一般是生产环境采用的部署模式，Hadoop 运行在服务器集群上。为解决 Hadoop 单点故障问题，生产环境一般采用分布式高可用部署。

7.2.2 本地部署使用 Hadoop

基于 CentOS 操作系统本地部署 Hadoop 在开发环境中大量使用，本节实战操作使用

CentOS 操作系统本地部署 Hadoop。

第 1 步，为避免网络问题，关闭防火墙。

```
[root@CentOS7-Hadoop01 ~]# systemctl stop firewalld
[root@CentOS7-Hadoop01 ~]# systemctl status firewalld
  firewalld.service - firewalld - dynamic firewall daemon
  Loaded: loaded (/usr/lib/systemd/system/firewalld.service; enabled; vendor preset: enabled)
  Active: inactive (dead) since Sun 2020-02-15 16:38:25 CST; 11s ago
    Docs: man:firewalld(1)
 Process: 917 ExecStart=/usr/sbin/firewalld --nofork --nopid $FIREWALLD_ARGS (code=exited, status=0/SUCCESS)
Main PID: 917 (code=exited, status=0/SUCCESS)
Mar 15 16:30:01 CentOS7-Hadoop01 systemd[1]: Starting firewalld - dynamic firewall daemon...
Mar 15 16:30:01 CentOS7-Hadoop01 systemd[1]: Started firewalld - dynamic firewall daemon.
Mar 15 16:38:24 CentOS7-Hadoop01 systemd[1]: Stopping firewalld - dynamic firewall daemon...
Mar 15 16:38:25 CentOS7-Hadoop01 systemd[1]: Stopped firewalld - dynamic firewall daemon.
```

第 2 步，访问 Hadoop 网站下载 Hadoop，同时需要下载 Java 安装文件。

```
[root@CentOS7-Hadoop01 Downloads]# ll
total 407500
-rw-r--r--. 1 root root 197657687 Mar 15 16:57 hadoop-2.7.2.tar.gz
-rw-r--r--. 1 root root 189815615 Mar 15 17:09 jdk-8u162-linux-x64.tar.gz
```

第 3 步，使用命令“tar -zxvf /Downloads/jdk-8u162-linux-x64.tar.gz -C /usr/lib/jvm/”解压 Java 安装文件。

```
[root@CentOS7-Hadoop01 lib]# mkdir jvm
[root@CentOS7-Hadoop01 lib]# tar -zxvf /Downloads/jdk-8u162-linux-x64.tar.gz -C /usr/lib/jvm/
jdk1.8.0_162/
jdk1.8.0_162/javafx-src.zip
jdk1.8.0_162/bin/
jdk1.8.0_162/bin/jmc
……（省略）
```

第 4 步，使用命令“vi ~/.bashrc”修改 Java 环境变量。

```
[root@CentOS7-Hadoop01 jvm]# cd ~
[root@CentOS7-Hadoop01 ~]# vi ~/.bashrc
export JAVA_HOME=/usr/lib/jvm/jdk1.8.0_162
export JRE_HOME=${JAVA_HOME}/jre
export CLASSPATH=.:${JAVA_HOME}/lib:${JRE_HOME}/lib
export PATH=${JAVA_HOME}/bin:$PATH
[root@CentOS7-Hadoop01 jvm]# source ~/.bashrc  #使变量设置生效
```

第 5 步，使用命令检查 Java 环境变量安装是否正确。

```
[root@CentOS7-Hadoop01 jvm]# echo $JAVA_HOME
/usr/lib/jvm/jdk1.8.0_162
[root@CentOS7-Hadoop01 jvm]# $JAVA_HOME/bin/java -version
java version "1.8.0_162"
Java(TM) SE Runtime Environment (build 1.8.0_162-b12)
Java HotSpot(TM) 64-Bit Server VM (build 25.162-b12, mixed mode)
```

第 6 步，选择将 Hadoop 解压到/usr/local/目录，Hadoop 解压后即可使用，使用命令“tar -zvxf hadoop-2.7.2.tar.gz -C /usr/local/”来检查 Hadoop 是否可用，可用则会显示 Hadoop 的版本信息。

```
[root@CentOS7-Hadoop01 Downloads]# tar -zvxf hadoop-2.7.2.tar.gz -C /usr/local/
hadoop-2.7.2/
hadoop-2.7.2/NOTICE.txt
```

```
hadoop-2.7.2/etc/
……（省略）
[root@CentOS7-Hadoop01 hadoop-2.7.2]# ./bin/hadoop version
Hadoop 2.7.2
Subversion Unknown -r Unknown
Compiled by root on 2017-05-22T10:49Z
Compiled with protoc 2.5.0
From source with checksum d0fda26633fa762bff87ec759ebe689c
This command was run using /usr/local/hadoop-2.7.2/share/hadoop/common/hadoop-common-2.7.2.jar
```

第 7 步，现在可以执行例子来感受 Hadoop 的运行。选择运行 grep 例子，将 input 文件夹中的所有文件作为输入，筛选当中符合正则表达式 dfs 的单词并统计出现的次数，最后输出结果到 output 文件夹中。若执行成功的话会输出很多作业的相关信息，最后的输出信息如下所示。作业的结果会输出在指定的 output 文件夹中，通过命令“cat ./output/*”查看结果，符合正则表达式的单词 dfsadmin 出现了 1 次。

```
[root@CentOS7-Hadoop01 hadoop]# mkdir ./input
[root@CentOS7-Hadoop01 hadoop]# cp ./etc/hadoop/*.xml ./input
[root@CentOS7-Hadoop01 hadoop]# ./bin/hadoop jar ./share/hadoop/mapreduce/hadoop-mapreduce-examples-*.jar grep ./input ./output 'dfs[a-z.]+'
20/02/15 17:43:56 INFO Configuration.deprecation: session.id is deprecated. Instead, use dfs.metrics.session-id
        File System Counters
                FILE: Number of bytes read=1158674
                FILE: Number of bytes written=2216432
                FILE: Number of read operations=0
                FILE: Number of large read operations=0
                FILE: Number of write operations=0
        Map-Reduce Framework
                Map input records=1
                Map output records=1
                Map output bytes=17
                Map output materialized bytes=25
                Input split bytes=121
                Combine input records=0
                Combine output records=0
                Reduce input groups=1
                Reduce shuffle bytes=25
                Reduce input records=1
                Reduce output records=1
                Spilled Records=2
                Shuffled Maps =1
                Failed Shuffles=0
                Merged Map outputs=1
                GC time elapsed (ms)=0
                Total committed heap usage (bytes)=655360000
        Shuffle Errors
                BAD_ID=0
                CONNECTION=0
                IO_ERROR=0
                WRONG_LENGTH=0
                WRONG_MAP=0
                WRONG_REDUCE=0
        File Input Format Counters
                Bytes Read=123
        File Output Format Counters
                Bytes Written=23
[root@CentOS7-Hadoop01 hadoop]# cat ./output/*
1       dfsadmin
```

至此，本地部署使用 Hadoop 完成。本地部署不需要其他安装，解压即可使用 Hadoop。本地部署常见错误是 Java 版本与环境变量配置错误。

7.2.3 伪分布式部署使用 Hadoop

Hadoop 可以在单节点上以伪分布式的方式运行，Hadoop 进程以分离的 Java 进程来运行，节点既作为 NameNode 也作为 DataNode，同时读取的是 HDFS 中的文件。本节实战操作使用 CentOS 操作系统进行伪分布式部署 Hadoop。

第 1 步，进行 Hadoop 伪分布式配置前，需要设置 Hadoop 环境变量，使用命令“vi ~/.bashrc”进行配置。

```
[root@CentOS7-Hadoop01 jvm]# vi ~/.bashrc
Hadoop Environment Variables
export HADOOP_HOME=/usr/local/hadoop
export HADOOP_INSTALL=$HADOOP_HOME
export HADOOP_MAPRED_HOME=$HADOOP_HOME
export HADOOP_COMMON_HOME=$HADOOP_HOME
export HADOOP_HDFS_HOME=$HADOOP_HOME
export YARN_HOME=$HADOOP_HOME
export HADOOP_COMMON_LIB_NATIVE_DIR=$HADOOP_HOME/lib/native
export PATH=$PATH:$HADOOP_HOME/sbin:$HADOOP_HOME/bin
[root@CentOS7-Hadoop01 jvm]# source ~/.bashrc
```

第 2 步，Hadoop 的配置文件位于/usr/local/hadoop/etc/hadoop/，需要修改 2 个配置文件 core-site.xml 和 hdfs-site.xml。Hadoop 的配置文件是.xml 文件，以声明 property 的 name 和 value 的方式来实现配置。

```
[root@CentOS7-Hadoop01 jvm]# vi /usr/local/hadoop/etc/hadoop/core-site.xml
Hadoop Environment Variables
<configuration>
    <property>
        <name>hadoop.tmp.dir</name>
        <value>file:/usr/local/hadoop/tmp</value>
        <description>Abase for other temporary directories.</description>
    </property>
    <property>
        <name>fs.defaultFS</name>
        <value>hdfs://localhost:9000</value>
    </property>
</configuration>
[root@CentOS7-Hadoop01 jvm]# vi /usr/local/hadoop/etc/hadoop/hdfs-site.xml
<configuration>
    <property>
        <name>dfs.replication</name>
        <value>1</value>
    </property>
    <property>
        <name>dfs.namenode.name.dir</name>
        <value>file:/usr/local/hadoop/tmp/dfs/name</value>
    </property>
    <property>
        <name>dfs.datanode.data.dir</name>
        <value>file:/usr/local/hadoop/tmp/dfs/data</value>
    </property>
</configuration>
```

第 3 步，使用命令“./bin/hdfs namenode -format”执行 NameNode 格式化。

```
[root@CentOS7-Hadoop01 /]# cd /usr/local/hadoop
[root@CentOS7-Hadoop01 hadoop]# ./bin/hdfs namenode -format
20/03/15 18:24:16 INFO namenode.NameNode: STARTUP_MSG:
/************************************************************
STARTUP_MSG: Starting NameNode
STARTUP_MSG:   host = CentOS7-Hadoop01.bdnetlab.com/127.0.1.1
STARTUP_MSG:   args = [-format]
STARTUP_MSG:   version = 2.7.2
……（省略）
20/03/15 18:24:18 INFO namenode.NameNode: SHUTDOWN_MSG:
/************************************************************
SHUTDOWN_MSG: Shutting down NameNode at CentOS7-Hadoop01.bdnetlab.com/127.0.1.1
************************************************************/
```

第 4 步，使用命令“./sbin/start-dfs.sh”开启 NaneNode 和 DataNode 守护进程。若出现提示“WARN util.NativeCodeLoader: Unable to load native-hadoop library for your platform… using builtin-java classes where applicable”可以忽略，不会影响 Hadoop 的正常运行。

```
[root@CentOS7-Hadoop01 hadoop]# ./sbin/start-dfs.sh
20/03/15 18:24:52 WARN util.NativeCodeLoader: Unable to load native-hadoop library for your platform... using builtin-java classes where applicable
Starting namenodes on [localhost]
The authenticity of host 'localhost (::1)' can't be established.
ECDSA key fingerprint is SHA256:XIPMp8tSp3i/cdMCUpef+HctsVHeg81XqAcqfGBrBm0.
ECDSA key fingerprint is MD5:77:3d:a5:a2:98:96:b8:0b:73:0c:20:ac:d0:a6:8a:31.
Are you sure you want to continue connecting (yes/no)? yes
localhost: Warning: Permanently added 'localhost' (ECDSA) to the list of known hosts.
root@localhost's password:
localhost: starting namenode, logging to /usr/local/hadoop/logs/hadoop-root-namenode-CentOS7-Hadoop01.out
root@localhost's password:
localhost: starting datanode, logging to /usr/local/hadoop/logs/hadoop-root-datanode-CentOS7-Hadoop01.out
Starting secondary namenodes [0.0.0.0]
The authenticity of host '0.0.0.0 (0.0.0.0)' can't be established.
ECDSA key fingerprint is SHA256:XIPMp8tSp3i/cdMCUpef+HctsVHeg81XqAcqfGBrBm0.
ECDSA key fingerprint is MD5:77:3d:a5:a2:98:96:b8:0b:73:0c:20:ac:d0:a6:8a:31.
Are you sure you want to continue connecting (yes/no)? yes
0.0.0.0: Warning: Permanently added '0.0.0.0' (ECDSA) to the list of known hosts.
root@0.0.0.0's password:
0.0.0.0: starting secondarynamenode, logging to /usr/local/hadoop/logs/hadoop-root-secondarynamenode-CentOS7-Hadoop01.out
20/03/15 18:25:50 WARN util.NativeCodeLoader: Unable to load native-hadoop library for your platform... using builtin-java classes where applicable
```

第 5 步，启动完成后，使用命令“jps”检查是否成功启动，若成功启动则会列出如下进程。

```
[root@CentOS7-Hadoop01 hadoop]# jps
19842 DataNode
20004 SecondaryNameNode
20120 Jps
19723 NameNode
```

第 6 步，成功启动后，使用浏览器登录查看 NameNode 和 Datanode 信息，如图 7-2-1 所示。

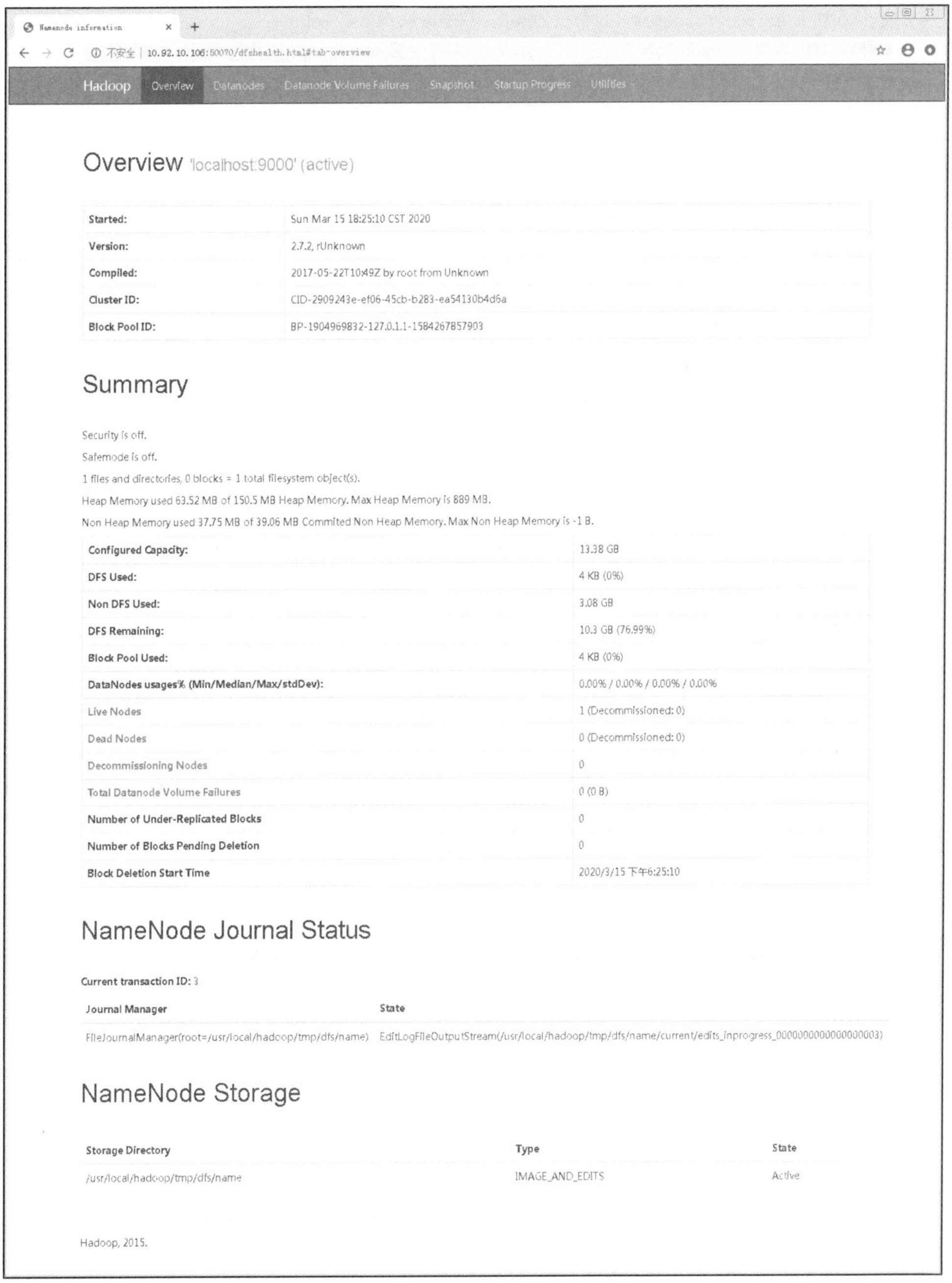

Namenode information

① 不安全 | 10.92.10.106:50070/dfshealth.html#tab-overview

Hadoop　Overview　Datanodes　Datanode Volume Failures　Snapshot　Startup Progress　Utilities

Overview 'localhost:9000' (active)

Started:	Sun Mar 15 18:25:10 CST 2020
Version:	2.7.2, rUnknown
Compiled:	2017-05-22T10:49Z by root from Unknown
Cluster ID:	CID-2909243e-ef06-45cb-b283-ea54130b4d6a
Block Pool ID:	BP-1904969832-127.0.1.1-1584267857903

Summary

Security is off.

Safemode is off.

1 files and directories, 0 blocks = 1 total filesystem object(s).

Heap Memory used 63.52 MB of 150.5 MB Heap Memory. Max Heap Memory is 889 MB.

Non Heap Memory used 37.75 MB of 39.06 MB Commited Non Heap Memory. Max Non Heap Memory is -1 B.

Configured Capacity:	13.38 GB
DFS Used:	4 KB (0%)
Non DFS Used:	3.08 GB
DFS Remaining:	10.3 GB (76.99%)
Block Pool Used:	4 KB (0%)
DataNodes usages% (Min/Median/Max/stdDev):	0.00% / 0.00% / 0.00% / 0.00%
Live Nodes	1 (Decommissioned: 0)
Dead Nodes	0 (Decommissioned: 0)
Decommissioning Nodes	0
Total Datanode Volume Failures	0 (0 B)
Number of Under-Replicated Blocks	0
Number of Blocks Pending Deletion	0
Block Deletion Start Time	2020/3/15 下午6:25:10

NameNode Journal Status

Current transaction ID: 3

Journal Manager	State
FileJournalManager(root=/usr/local/hadoop/tmp/dfs/name)	EditLogFileOutputStream(/usr/local/hadoop/tmp/dfs/name/current/edits_inprogress_0000000000000000003)

NameNode Storage

Storage Directory	Type	State
/usr/local/hadoop/tmp/dfs/name	IMAGE_AND_EDITS	Active

Hadoop, 2015.

图 7-2-1

第 7 步，ResourceManager 是 Hadoop 集群管理的主守护进程，新版的 Hadoop 使用了新的 MapReduce 框架 Yet Another Resource Negotiator（YARN）。YARN 是从 MapReduce 中分离出来的，负责资源管理与任务调度。YARN 运行于 MapReduce 之上，提供了高可用性和高扩展性。使用命令“./sbin/start-dfs.sh”启动 Hadoop，仅仅是启动了 MapReduce 环境，我们可以启动 YARN，让 YARN 来负责资源管理与任务调度。需要修改配置文件 mapredsite.xml 和 yarn-site.xml。

```
[root@CentOS7-Hadoop01 hadoop]# vi ./etc/hadoop/mapred-site.xml
configuration>
   <property>
      <name>mapreduce.framework.name</name>
      <value>yarn</value>
   </property>
</configuration>
[root@CentOS7-Hadoop01 hadoop]# vi ./etc/hadoop/yarn-site.xml
<configuration>
   <property>
      <name>yarn.nodemanager.aux-services</name>
      <value>mapreduce_shuffle</value>
      </property>
</configuration>
```

第 8 步，使用命令“./sbin/start-yarn.sh”启动 YARN，可以看到增加了 NodeManager 和 ResourceManager 两个后台进程。

```
[root@CentOS7-Hadoop01 hadoop]# ./sbin/start-yarn.sh
starting yarn daemons
starting resourcemanager, logging to /usr/local/hadoop/logs/yarn-root-resourcemanager-CentOS7-Hadoop01.out
root@localhost's password:
localhost: starting nodemanager, logging to /usr/local/hadoop/logs/yarn-root-nodemanager-CentOS7-Hadoop01.out
[root@CentOS7-Hadoop01 hadoop]# ./sbin/mr-jobhistory-daemon.sh start historyserver
starting historyserver, logging to /usr/local/hadoop/logs/mapred-root-historyserver-CentOS7-Hadoop01.out
[root@CentOS7-Hadoop01 hadoop]# jps
21728 Jps
19842 DataNode
20004 SecondaryNameNode
21526 NodeManager
19723 NameNode
21244 ResourceManager
21660 JobHistoryServer
```

第 9 步，使用浏览器查看任务的运行情况，如图 7-2-2 所示。

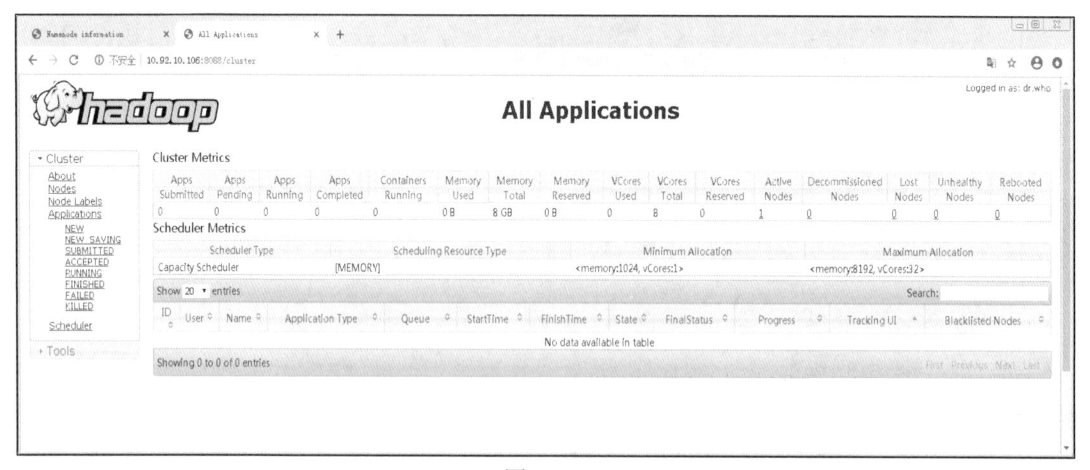

图 7-2-2

至此，Hadoop 的部署完成，生产环境可以根据实际情况选择分布式部署以满足更多的需求。大数据分析开发相关内容不在本书讨论范围内，请参考其他相关书籍。

7.3　本章小结

本章介绍了 Hadoop 的基本部署和使用。Hadoop 的应用场景基于大数据分析，本地部署和伪分布式部署非常适合开发人员和中小生产环境。与介绍 OpenStack 类似，本章介绍的都是一些基础的部署使用，让读者能够轻松部署使用 Hadoop。

第 8 章　认识 SDN 架构

软件定义网络（Software Defined Network，SDN）是美国斯坦福大学 Clean State 研究组提出的一种新型网络创新架构，可通过软件编程的形式定义和控制网络。其控制平面和转发平面分离及开放性可编程的特点，被认为是网络领域的一场革命，为新型互联网体系结构研究提供了新的实验途径，也极大地推动了互联网的发展。本章介绍 SDN 架构及其理论。

本章要点

■　SDN 基本概念。

■　主流 SDN 技术介绍。

8.1　SDN 的基本概念

传统网络世界是水平标准和开放的，每个网元可以和周边网元进行互联。而在计算机的世界里，不仅水平是标准和开放的，同时垂直也是标准和开放的，从下到上有硬件、驱动程序、操作系统、编程平台以及应用软件等，编程者可以很容易地创造各种应用。从某个角度和计算机对比，在垂直方向上，网络是“相对封闭”和“没有框架”的，在垂直方向创造应用和部署业务是相对困难的。但 SDN 将整个网络在垂直方向变得开放、标准化、可编程，从而让人们更容易、更有效地使用网络资源。因此，SDN 技术能够有效降低设备负载，协助网络运营商更好地控制基础设施，降低整体运营成本，成为极具前途的网络技术之一。

SDN 的思想是通过控制与转发分离，将网络中交换设备的控制逻辑集中到一个计算设备上，为提升网络管理配置能力带来新的思路。SDN 的本质特点是控制平面和数据平面的分离以及开放可编程性。通过分离控制平面和数据平面以及开放的通信协议，SDN 打破了传统网络设备的封闭性。此外，南北向和东西向的开放接口和可编程性，也使得网络管理变得更加简单、动态以及灵活。

利用分层的思想，SDN 将数据与控制相分离：在控制层面，包括具有逻辑中心化和可编程的控制器，可掌握全局网络信息，方便运营商和科研人员管理配置网络和部署新协议等；在数据层面，与传统的二层交换机不同，它仅提供简单的数据转发功能，可以快速处理匹配的数据报，适应流量日益增长的需求。两层之间采用开放的统一接口（如 OpenFlow 等）进行交互，控制器通过标准接口向交换机下发统一标准规则，交换机仅需按照这些规则执行相应的动作。

8.2 主流 SDN 技术

8.2.1 Open vSwitch 简介

Open vSwitch（OVS）是开源 Apache 2.0 许可下的产品级质量的多层虚拟交换标准。它旨在通过编程扩展，使庞大的网络自动化（配置、管理、维护等），同时还支持标准的管理接口和协议（如 NetFlow、sFlow、SPAN、RSPAN、CLI、LACP、802.1ag）。总的来说，它被设计为支持分布在多个物理服务器，如 VMware 分布式交换机或 Cisco 的 Nexus 1000V 交换机。

OVS 是产品级的虚拟交换机，大量应用在生产环境中，支撑整个数据中心虚拟网络的运转。OVS 基于 SDN 思想，将整个核心架构分为控制面和数据面，数据面负责数据的交换工作，控制面实现交换策略，指导数据面工作。从整体上看，OVS 可以分为三大块：管理面、数据面以及控制面。

OVS 架构如图 8-2-1 所示，整体架构由内核模块 datapath、用户空间 vswitchd 以及 ovsdb 组成。其中 datapath 模块负责数据转发，从网卡中读取数据并在流表项中进行匹配，成功则转发，失败则交给 vswitchd 进行处理；vswitchd 模块是 OVS 的主程序，负责与 OpenFlow 控制器和第三方软件进行通信；ovsdb 模块是用于存储配置的数据库模块。

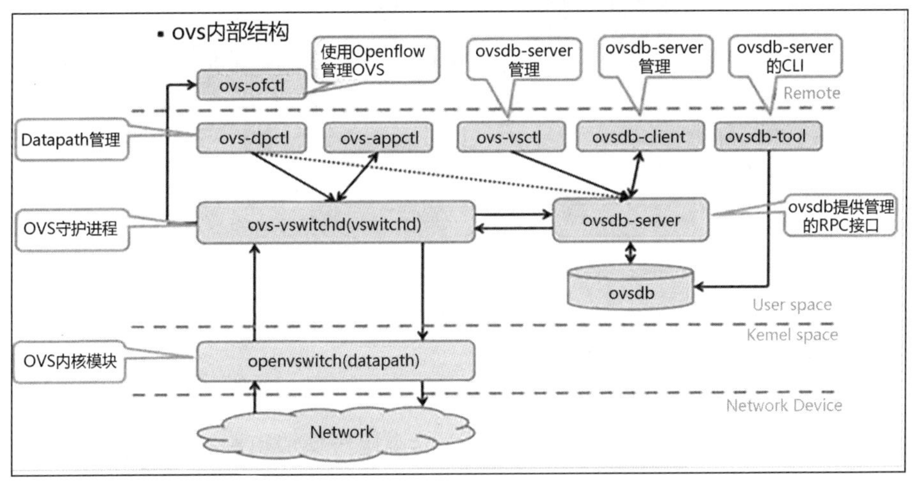

图 8-2-1

8.2.2 Cisco ACI 简介

Cisco ACI 是 Cisco 公司提出的 SDN 和网络虚拟化解决方案，旨在为新一代数据中心和云应用提供转变性的运营模式。在 Cisco ACI 框架中，应该是应用指导网络行为，而不是网络行为指导应用。预定义的应用要求和说明（策略配置文件）让网络、应用服务、安全策略、租户子网以及工作负载分配的调配实现自动化。通过将整个应用网络的调配实现自动化，Cisco ACI 可以帮助降低 IT 成本、减少错误、加快部署以及使业务更加敏捷。

新 Cisco ACI 模式使用基于交换矩阵的方法。此方法经过全新设计，用于支持新出现的行业要求，同时为现有的架构维护迁移路径。使用这一关键技术，传统的企业应用和内部开发的应用可以通过动态和灵活的方式，在为其提供支持的网络基础设施上一同运行。传统上，网络策略和逻辑拓扑决定应用设计。而现在，我们需要根据应用需求来确定如何应用网络策略和逻辑拓扑。交换矩阵旨在支持向管理自动化、以编程方式定义策略以及在任意位置、任意设备上实现动态工作负载的转变。Cisco ACI 利用紧密结合的软件和硬件组合实现此目标，从而具有其他模式无法提供的优势，如图 8-2-2 所示。

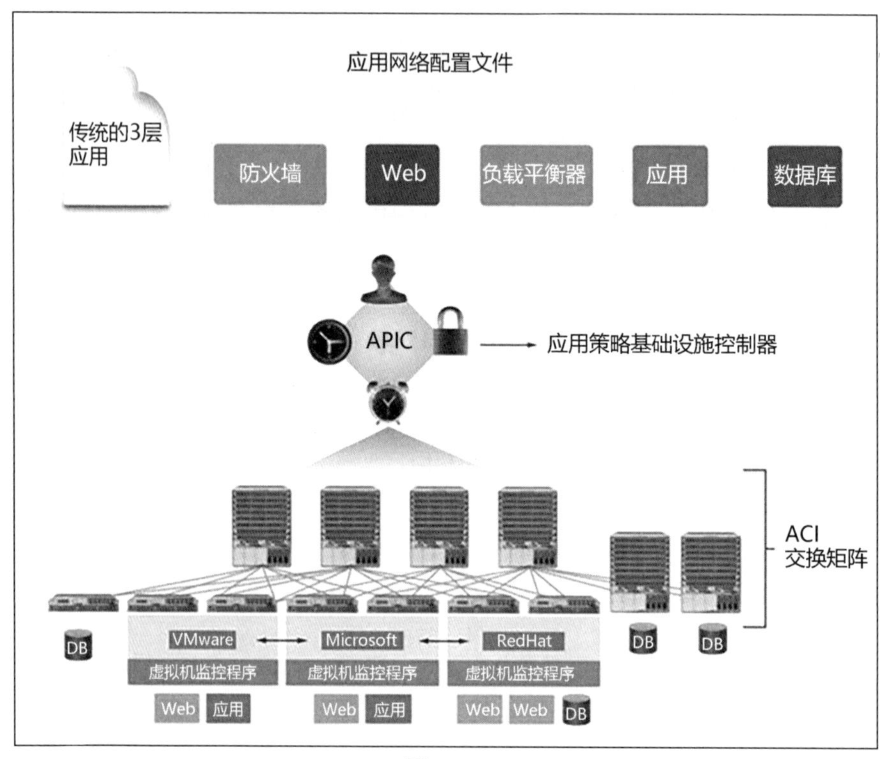

图 8-2-2

Cisco ACI 让所有应用部署、安全、网络服务和网络配置人员都能够通过通用平台相互协作，有助于消除 IT 孤岛。主要有以下优势。

1）应用速度——任何位置，任何应用。

2）可完整查看应用的系统架构，具有集中的应用级集成可视性，以及跨物理和虚拟环境的实时的应用运行状况监控功能。

3）可管理物理、虚拟以及基于云环境的通用平台。

4）可对应用和租户进行严格控制。

5）整合了软件灵活性和硬件性能的可扩展性能。

6）卓越的应用性能，和传统模式相比，可节约多达 80% 的应用流完成时间。

7）运营简单，具有跨应用、网络以及安全资源的通用策略、管理以及运营模式。

8）开放式 API、开放式标准以及开源元素，可带来非常好的软件灵活性，让开发和运营（DevOps）团队能够与生态系统合作伙伴轻松整合。

Cisco ACI 组件包括 Cisco 应用策略基础设施控制器、应用网络配置文件以及 ACI 交换矩阵。

1. Cisco 应用策略基础设施控制器

Cisco 应用策略基础设施控制器（Cisco Application Policy Infrastructure ControUer，Cisco APIC）是 Cisco ACI 解决方案的主要架构组件。它是 Cisco ACI 交换矩阵、策略实施以及健康状态监控实现自动化和管理的统一点。Cisco APIC 是一个集中的群集式控制器。它优化了性能，能够在任何位置支持任何应用，并统一了物理环境和虚拟环境的操作。Cisco APIC 负责的任务包括交换矩阵激活、交换机固件维护以及网络策略配置和实例化。Cisco APIC 已完全从数据路径中移除，这意味着即使与 Cisco APIC 的通信中断，交换矩阵也仍然可以转发流量。

2. 应用网络配置文件

交换矩阵内的应用网络配置文件是由端点组（代表一个应用层的相似端点的逻辑分组，或需要相似策略的服务集）、端点组连接以及定义这些连接的策略组成的集合。应用网络配置文件是应用的所有组件和该应用在应用交换矩阵上的依存关系的逻辑表示。应用网络配置文件通过与应用的设计和部署方法相匹配的逻辑方法进行模式化。随后操作系统（而不是管理员）通过 Cisco APIC，来执行策略的配置和实施，以及连接工作。

3. Cisco ACI 交换矩阵

Cisco 推出可同时用于传统部署和 Cisco ACI 数据中心部署的 Cisco Nexus 9000 系列交换机，以此扩大 Cisco Nexus 交换产品组合。Cisco Nexus 9000 系列提供模块化和非模块化的 1/10/40 千兆以太网交换机配置。它既可在 Cisco NX-OS 模式下运营，与当前的 Cisco Nexus 交换机兼容并保持一致；又可在 Cisco ACI 模式下运行，从而充分利用 Cisco ACI 应用基于策略的服务和基础设施自动化功能的优势。这种双功能能力可为客户提供投资保护，并让客户能够通过软件升级轻松迁移到 Cisco ACI。

8.2.3 VMware NSX 简介

NSX 是 VMware 公司提供的 SDN 解决方案，目前分为 NSX-V 和 NSX-T 两个版本，其中 NSX-V 运行在 VMware vSphere 架构上，NSX-T 可以运行在开源 Linux 平台上。NSX-V 和 NSX-T 基本架构是相同的，差别在于部署的方式不同。

NSX 由数据平面、控制平面以及管理平面组成，如图 8-2-3 所示。

1. 数据平面

NSX 数据平面包含 NSX 虚拟交换机，可以提供丰富的服务。NSX 附加组件包括在 Hypervisor 内核中运行的内核模块，可提供分布式路由和分布式防火墙等服务，并支持 VXLAN 桥接功能。

使用 VXLAN 协议和集中式网络配置为叠加网络连接提供支持。叠加的网络连接方式可实现下列功能。

1）在现有物理基础架构的 IP 网络上创建一个灵活的第 2 层逻辑叠加网络，而无须重新构建任何数据中心网络。

2）敏捷地调配通信流量（东西向和南北向），而租户之间仍保持相互隔离。

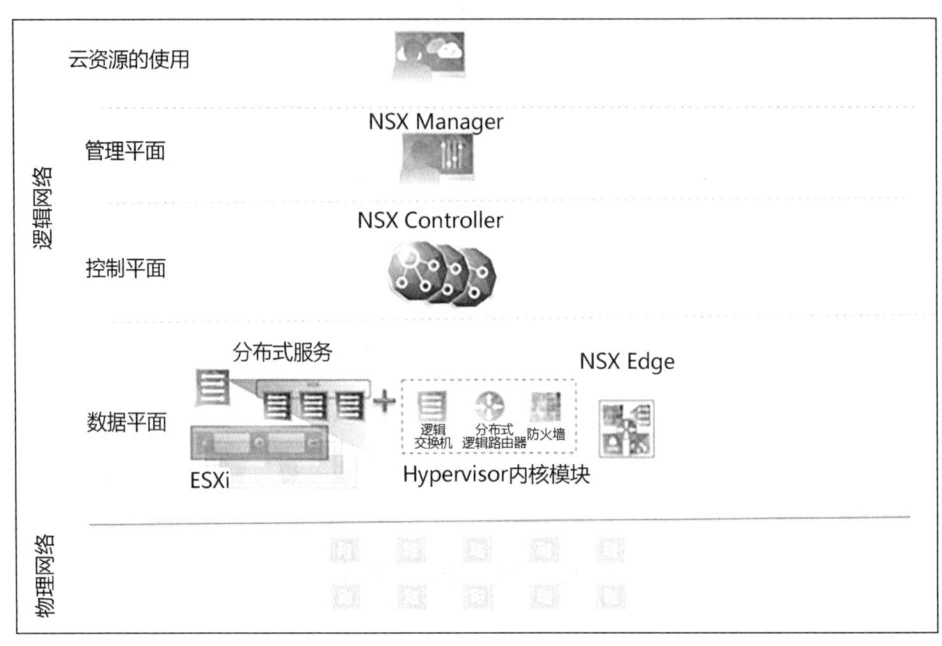

图 8-2-3

3）应用工作负载和虚拟机根本察觉不到叠加网络的存在，就像连接到第 2 层物理网络一样运行。

4）NSX 虚拟交换机可推动 Hypervisor 的大规模扩展。

5）诸如端口镜像、NetFlow、配置备份和还原、网络运行状况检查、服务质量以及链路汇聚控制协议（Link Aggregation Control Protocol，LACP）之类的多项功能可提供用于在虚拟网络中管理、监控流量，以及排查故障的综合性工具包。

另外，数据平面还包括可提供从逻辑网络连接空间到物理网络的通信的网关设备。可在第 2 层 NSX 桥接或第 3 层 NSX 路由使用此功能。

2. 控制平面

NSX 控制平面在 NSX Controller 中运行，Controller 支持无须多播虚拟扩展局域网（Virtual Extensible LAN，VXLAN）并可在控制平面对分布式逻辑路由等元素进行编程。在所有情况下，Controller 仅是控制平面的一部分，没有任何数据平面流量通过它。还会在具有奇数个成员的集群中部署 Controller 节点以实现高可用性和扩展。

3. 管理平面

NSX 管理平面由 NSX Manager 构建，可通过 NSX Manager UI 直接使用 NSX。终端用户通常会将网络虚拟化与其云计算管理平台结合使用，以便部署应用。NSX 可以通过 REST API 在几乎任意连接管理处理器（Connectivity Management Processor，CMP）中提供丰富的集成功能。

NSX 在虚拟网络上可以提供几乎所有的网络服务，可以将虚拟机和物理网络相隔离，做到了网络服务与具体的物理网络设备无关，使得用户在网络设备的选择和采购上有着更大的灵活性。NSX 能够提供常用的路由、负载均衡以及防火墙保护等，如图 8-2-4 所示。

1）交换：支持在结构中的任意位置扩展第 2 层网段，不受物理网络设计的影响。

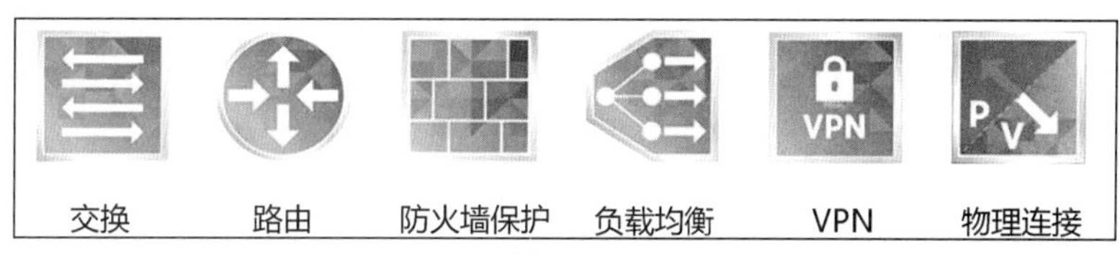

图 8-2-4

2）路由：IP 子网之间的路由可以在逻辑空间内实现，数据不会向外流向物理路由器。此路由在 Hypervisor 内核中执行，可最大限度减少 CPU/内存开销。此功能可以为虚拟基础架构中的路由流量（东西向通信）提供最佳数据路径。类似地，NSX Edge 可以提供一个与物理网络基础架构无缝集成的理想的集中位置，用于处理与外部网络之间的通信（南北向通信）。

3）防火墙保护：安全性在内核和虚拟网卡级别实施。这样，系统便能够以高度可扩展的方式强制实施防火墙规则，且不会导致物理设备出现瓶颈。防火墙分布在内核中，因此可最大限度减少 CPU 开销，并且能够以线速运行。

4）负载均衡：借助安全套接层（Secure Socket Layer，SSL）端接功能，为第 4 层至第 7 层负载均衡提供支持。

5）虚拟专用网络（Virtual Private Network，VPN）：通过 SSL VPN 服务启用第 2 层和第 3 层 VPN 服务。

6）物理连接：NSX 中支持第 2 层和第 3 层网关功能，可提供部署在逻辑和物理空间中的工作负载之间的通信。

整体来说，NSX 在 SDN 市场的占有率相当高，特别是支持开源平台的 NSX-T 版本。目前 NSX 主要有三大应用场景。

1. 数据中心网络安全

分布式软件防火墙和微分段大大简化了数据中心的网络安全管理，相比物理网络环境，能够实现更高等级的安全防护。

2. IT 自动化

虚拟网络的功能都是由软件来实现的，所以能够使用命令来动态地创建网络设备，调整网络配置和安全策略参数，实现数据中心的 IT 自动化。

3. 业务持续性

虚拟机网络环境都是由虚拟网络提供的，当发生故障转移时，虚拟机不用改变包括 IP 地址在内的任何网络参数，NSX 会负责把虚拟机所依赖的整个虚拟网络环境和对应的网络安全策略迁移到新的服务器上运行，从而保证业务的持续性。

8.3 本章小结

本章介绍了 SDN 技术的基本概念和主流 SDN 技术，不涉及实战操作。因为生产环境的实现一般会结合硬件厂商相关解决方案，如华为 SDN 解决方案、新华三 SDN 解决方案等，部署过程中会使用到厂商的硬件设备来实现最终的支持。了解 SDN 的架构对于后续的学习是有帮助的，SDN 数据中心是今后数据中心发展的趋势。